JS

DIGITAL FILTERS AND THEIR APPLICATIONS

Techniques of Physics

Editors

N. H. MARCH

Department of Theoretical Chemistry, University of Oxford, Oxford, England

H. N. DAGLISH

Post Offices Research Centre, Martlesham Heath, Ipswich, England

Techniques of physics find wide application in biology, medicine, engineering and technology generally. This series is devoted to techniques which have found and are finding application. The aim is to clarify the principles of each technique, to emphasize and illustrate the applications and to draw attention to new fields of possible employment.

1. D. C. Champeney: Fourier Transforms and their Physical Applications.
2. J. B. Pendry: Low Energy Electron Diffraction.
3. K. G. Beauchamp: Walsh Functions and their Applications.
4. V. Cappellini, A. G. Constantinides and P. Emiliani: Digital Filters and their Applications.

DIGITAL FILTERS AND THEIR APPLICATIONS

V. CAPPELLINI, A. G. CONSTANTINIDES
and P. EMILIANI

Istituto di Elettronica, Facoltà di Ingegneria, Università degli Studi, Florence, Italy
Department of Electrical Engineering, Imperial College of Science and Technology, London, England
Istituto di Ricerca sulle Onde Elettromagnetiche, Consiglio Nazionale delle Ricerche, Florence, Italy

ACADEMIC PRESS, INC.

(Harcourt Brace Jovanovich, Publishers)

London Orlando San Diego New York
Toronto Montreal Sydney Tokyo

ADADEMIC PRESS INC. (LONDON) LTD.
24/28 Oval Road
London NW1

United States Edition published by
ACADEMIC PRESS, INC.
Orlando, Florida 32887

Second Printing 1981

Library of Congress Catalog Card Number: 77-93214
ISBN: 0-12-159250-2

PRINTED IN THE UNITED STATES OF AMERICA

84 85 86 87 9 8 7 6 5 4 3 2

Preface

Signal processing methods and techniques now form the basis of very important developments in physics, electrical and electronic engineering, particularly in communications and radar–sonar systems, instrumentation and industrial process control. During the last decade digital methods for signal processing have become more significant in that now they not only replace the more classical analogue techniques in many relevant areas, but they are also being applied in many new areas.

There are several reasons for this development: the consistently high efficiency of digital techniques permits better signal processing and analysis; there is greater flexibility within applications; and there is an increasing availability of general purpose computers and minicomputers, or indeed special type digital processors, at a decreasing cost. Digital techniques have also become more important in two-dimensional (2-D) signal or image processing.

Of all the methods and techniques used for digital signal processing, *digital filtering* is the most important. In the past it has been limited to theoretical research, but recently has been used in many important practical applications for processing 1-D and 2-D signals. This fact may be attributed to:

(i) the availability of efficient and relatively simple design methods;

(ii) extremely fast and impressive technological advances in large and very large scale integration circuits for multipliers, accumulators, memories, with an increase in maximum working frequency and new devices, such as *charge coupled devices* (CCD) and *surface acoustic wave* (SAW) devices;

(iii) advances in computer *hardware* and *software*, particularly with the introduction of *microprocessors* and *microcomputers*, and implementation of fast *array processors*, which are useful as peripheral parts of a computing system or as the main processing system.

As a result of the above developments, digital filtering techniques have been introduced during the last few years in different and extremely important areas such as communications, radar–sonar, the processing of results of physical experiments, aerospace systems, biomedicine, earth resource satellites, etc.

In this book the theory and design of digital filters (1-D and 2-D) are presented together with a description of their practical utility and application in many areas of signal and image processing.

The book is divided into three main parts. The first part (Chapters 1 to 4) is essentially tutorial in character and summarizes the basic relationships for the different types of digital filters, presenting design methods and analysing the error and stability problems. From this mathematical basis, in the second part (Chapters 5 to 7, Appendices 1 to 4), criteria for the practical design of 1-D and 2-D digital filters are derived, and coefficient values, frequency responses and efficiency comparisons are presented. The implementation problems are also considered: useful computer programs are listed and hardware realizations are described. In Appendices 2 to 4, some relatively simple but efficient computer programs (FORTRAN IV) are presented, both for performing the design and the actual filtering. The third part (Chapter 8, Appendices 5 and 6) describes the impact of digital filters in research, operative and industrial areas, presenting interesting applications to signal and image processing in communications, radar, biomedicine, power systems protection and remote sensing.

We have included not only 1-D but also 2-D digital filtering methods and techniques. Thus we are able to compare 2-D problems with 1-D problems and we can clarify which of the methods and techniques used in 1-D can also be applied in 2-D.

It is hoped that the approach adopted in this book, with the inclusion of the theoretical aspects of the subject as well as the practical implementation procedures, will be of interest to researchers and also to practising engineers.

The co-operation of the research groups involved in signal processing at Imperial College, London, at Florence University and at IROE-CNR Institute has made the writing of this book possible. We have drawn both from our own individual experiences and from joint research efforts in bringing together the theoretical developments and the practical applications included in the book. We should also like to thank the British Council for their encouragement and support during the early stages of the above research and also the Consiglio Nazionale delle Ricerche for their support.

The understanding and support of Professor E. C. Cherry, Head Communication Section, Department of Electrical Engineering, Imperial College of Science and Technology, and also of Professor G. Francini, Dean of Facoltà di Ingegneria, Florence University, are gratefully acknowledged. Professor G. Toraldo di Francia, Director of IROE-CNR Institute and Professor N. Carrara, President of the Scientific Council of the Institute, have been constant and continuous supporters of the entire enterprise. Our colleagues at the Communication Section, Department of Electrical Engineering, Imperial College and at the Istituto di Elettronica, Facoltà di Ingegneria have been most helpful with criticisms and comments for which we are very grateful. The co-operation of Professor M. Fondelli at Facoltà di Ingegneria, the help of Dr D. Benelli, Dr M. Bernabo' and Dr E. Del Re of the Istituto di Elettronica is appreciated and in particular the contribution of Yusif Fakhro (Appendices 1 and 5) and Majid Ahmadi at Communication Section, Department of Electrical Engineering. The help of Manos Tzanettis and his comments on the manuscript were invaluable.

Other institutes were helpful for specific topics. In particular we acknowledge the contribution by Professor T. W. Parks, Rice University, for Remez exchange algorithm (Appendix 3), Dr K. G. Beauchamp, University of Lancaster (Section VIG), and Dr L. Fusco and co-workers at CSATA in Bari, Dr G. Mottola at Università di Trieste, and Dr G. Garibotto at CSELT in Turin for interesting digital image processing results.

June 1978

V. CAPPELLINI
A. G. CONSTANTINIDES
P. EMILIANI

Contents

To Grazia, Pamela and Silvana

Chapter 1

Digital Filters

IA Introduction

In this chapter the fundamental properties of linear digital filters are presented as linear transformations on discrete time functions. Firstly discrete functions are examined in general and subsequently the particular case of discrete functions derived by sampling continuous functions is considered in some detail for its direct relevance to practical applications. The fundamental properties of z-transform, the transform for discrete functions, are summarized. Then the definition of discrete linear systems is given, including descriptions in the time domain, frequency domain and z-transform domain.

Causality, recursivity and stability are discussed, and algorithms for stability tests and stabilization of unstable filters are included; these algorithms are of particular interest, as it will be observed, in the two-dimensional (2-D) case.

Realization structures from given transfer functions are discussed.

Finally the main properties of Discrete Fourier Transform (DFT) and of Fast Fourier Transform (FFT) are summarized, pointing out the aspects more useful for the design and implementation of digital filters.

IB Discrete functions

In the following, the properties of linear digital filters, defined as transformations of discrete functions, are described.[13,15,16]

Discrete functions are defined only for *discrete values* of their variables, that is they are *sequences of numbers.* These numbers can be obtained as quantized samples of a continuous function (representing an analogue signal or image) or they can be the values of a discrete variable such as the readings indicated by, or the output data of, a digital counter. The sequences thus obtained are then processed to obtain other sequences.

The notation which is used for one-dimensional (1-D) sequences is the following

$$\{x(n)\} \qquad N_1 \leq n \leq N_2 \tag{1.1}$$

where $N_1 = -\infty$ and $N_2 = \infty$ for infinite sequences.

For two-dimensional (2-D) sequences, the notation is

$$\{x(n_1, n_2)\} \qquad N_1 \leq n_1 \leq N_2, \qquad M_1 \leq n_2 \leq M_2 \tag{1.2}$$

where $N_1 = M_1 = -\infty$ and $N_2 = M_2 = \infty$ for double infinite sequences.

We can observe that within the terms *sequences of numbers* and *numerical transformations of sequences* the quantization aspect is conceptually included due to the implicit finite precision needed for any numerical representation. However in the development of the theory of discrete linear systems we consider the more general case of sequences of numbers with infinite precision. Thus we use functions whose variable is defined on a discrete set of values, but whose amplitude can assume any value in a continuous way within a specified range.

In Chapter 4 we discuss in some detail the effects and consequences of the amplitude quantization of the input and output sequences of a system, of the numbers which define a system and the arithmetic operations involved in a system.

In many applications, sequences of the type (1.1) and (1.2) are obtained by taking samples of continuous or analogue signals and images. Therefore we consider more precisely the way an analogue signal is related to its samples leading to the sampling theorem.

IC The 1-D sampling theorem

One of the most important applications of digital filtering techniques is in the processing of sequences of samples derived from continuous or analogue signals. This is made possible due to the results and implications of the *sampling theorem.*[8]

This theorem can be stated as follows: "A continuous analogue function $x(t)$ which has a limited Fourier spectrum, that is a spectrum $X(j\omega)$ such that $X(j\omega) = 0$ for $\omega \geq \omega_m$, is uniquely described from a knowledge of its values at uniformly spaced time instants, T units apart, where $T = 2\pi/\omega_s$ and $\omega_s \geq 2\omega_m$".

To prove this theorem, consider a function $x(t)$ with a Fourier transform $X(j\omega)$ given by

$$X(j\omega)=0 \qquad \text{for} \qquad \omega \geq \omega_m \tag{1.3}$$

as shown in Fig. 1.1(a).

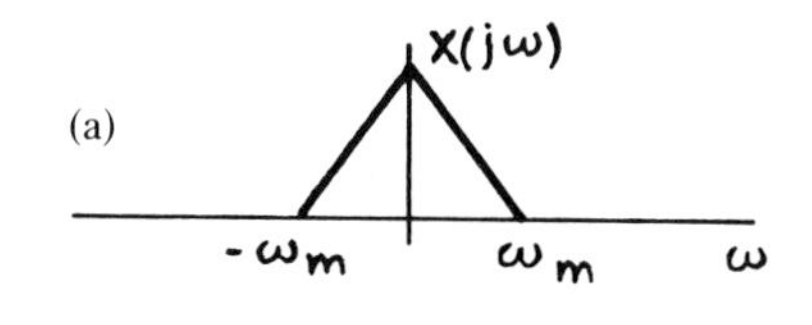

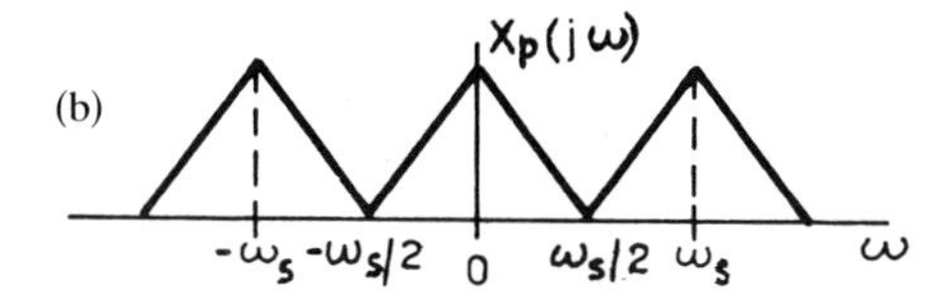

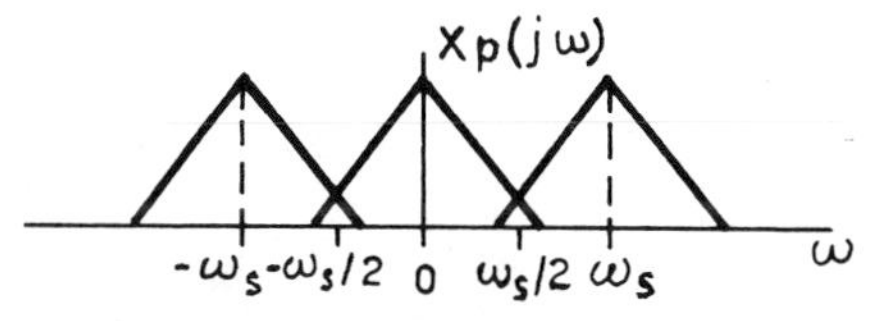

FIG. 1.1. (a) Fourier transform of a function $x(t)$; (b) sampled signal spectrum with $\omega_s/2=\omega_m$; (c) sampled signal spectrum with $\omega_s/2<\omega_m$, showing the aliasing phenomenon.

Consider now a periodic function $X_p(j\omega)$ with period ω_s, which is identical to $X(j\omega)$ in the interval $-\omega_s/2 \leq \omega \leq \omega_s/2$. This function can be expanded in a Fourier series in the form

$$X_p(j\omega)=\sum_{k=-\infty}^{\infty} x_k \, e^{-j\omega kT} \tag{1.4}$$

with coefficients

$$x_k=\frac{1}{\omega_s}\int_{-\omega_s/2}^{\omega_s/2} X(j\omega)\, e^{+j\omega kT}\, d\omega \tag{1.5}$$

Now the expression of $x(t)$ as a Fourier transform of $X(j\omega)$ is given by

$$x(t)=\frac{1}{2\pi}\int_{-\omega_s/2}^{\omega_s/2} X(j\omega)\, e^{j\omega t}\, d\omega \tag{1.6}$$

Setting $t=kT$, we obtain for the samples of $x(t)$, T units apart,

$$x(kT)=\frac{1}{2\pi}\int_{-\omega_s/2}^{\omega_s/2} X(j\omega)\, e^{j\omega kT}\, d\omega \tag{1.7}$$

and by comparing the two expressions (1.5) (1.7) it is easy to obtain

$$x_k = Tx(+kT) \tag{1.8}$$

Therefore, if we know the values of $x(kT)$ for $k=-\infty, \infty$, the Fourier series of the periodically repeated spectrum is uniquely determined by these samples. Further we can show that when $\omega_s \geq 2\omega_m$, an expression can be found for reconstructing the continuous analogue signal from its samples. Over the interval $-\omega_s/2$, $\omega_s/2$, the signal $x(t)$ can be expressed in the form

$$x(t)=\frac{1}{2\pi}\int_{-\omega_s/2}^{\omega_s/2}\left(\sum_{k=-\infty}^{\infty} x_k\, \mathrm{e}^{-j\omega kT}\right)\mathrm{e}^{j\omega t}\,\mathrm{d}\omega=\frac{1}{2\pi}\sum_{k=-\infty}^{\infty} x_k\int_{-\omega_s/2}^{\omega_s/2}\mathrm{e}^{-j\omega kT}\,\mathrm{e}^{j\omega t}\,\mathrm{d}\omega$$

$$=\frac{1}{2\pi}\sum_{k=-\infty}^{\infty} x_k\int_{-\omega_s/2}^{\omega_s/2}\mathrm{e}^{j(t-kT)\omega}\,\mathrm{d}\omega=\frac{1}{T}\sum_{k=-\infty}^{\infty} x_k\frac{\sin\left[\omega_s(t-kT)/2\right]}{\omega_s(t-kT)/2} \tag{1.9}$$

and by using (1.8) we obtain

$$x(t)=\sum_{k=-\infty}^{\infty} x(kT)\left(\frac{\sin\left[\omega_s(t-kT)/2\right]}{\omega_s(t-kT)/2}\right) \tag{1.10}$$

The relation (1.10) can be interpreted as the *convolution summation* of a sequence of pulses, with amplitude $x(kT)$ with the analogue filter impulse response

$$h(t)=\frac{\sin \omega_s t/2}{\omega_s t/2} \tag{1.11}$$

This is the impulse response of an ideal and therefore not physically realizable low-pass filter, having a constant amplitude transfer function equal to T in the interval $-\omega_s/2$, $\omega_s/2$ and zero elsewhere. It can be observed that, whilst this filter is indeed not physically realizable (its impulse response is anticipatory), it can be approximated nevertheless by using a sufficiently long time delay.

It is now important to point out that the samples of the signal determine only the periodic spectrum $X_p(j\omega)$, which is called the *sampled signal spectrum*. This spectrum has been obtained through the periodic repetition of the original band-limited spectrum, using a period equal to ω_s. The sampled signal spectrum can also be obtained in general by shifting the spectrum of the continuous signal to all the multiples of ω_s and by summing all these shifted spectra, that is

$$X_p(j\omega)=\frac{1}{T}\sum_{n=-\infty}^{\infty} X(j\omega-jn\omega_s) \tag{1.12}$$

If $\omega_s/2$ is greater than the maximum frequency present in the analogue signal, the sampled signal spectrum can reproduce with no distortion the spectrum of the analogue signal around the region $\omega = 0$. If $\omega_s/2 < \omega_m$, then the sampled signal spectrum is not equal to the continuous signal spectrum in the $-\omega_s/2$, $\omega_s/2$ interval: the spectrum represented by the sample series cannot determine the analogue signal spectrum. This phenomenon is commonly called *aliasing* and in general the aliased spectrum can behave as in the form shown in Fig. 1.1(c).

Finally we offer a comment on the notation we chose for the data sequences. If a sequence is obtained by sampling a continuous signal, the sampling interval T should be considered explicitly to define the frequency scale in an absolute way in the notation $x(nT)$. However, to simplify the notation, it is convenient to use a normalized frequency scale f/f_s. In this case $T=1$ and the multiples of ω_s in the absolute scale are substituted by the multiples of 2π. When an absolute scale is required, it is easy to transform the normalized frequency scale to the actual frequency scale.

ID The 2-D sampling theorem

A sampling theorem can be proved also for the 2-D case.[8]

This theorem can be stated as follows: "A function of two variables, $x(x_1, x_2)$, whose 2-D Fourier transform is equal to zero for $\omega_1 \geq \omega_{1m}$ and $\omega_2 \geq \omega_{2m}$, is uniquely determined by the values taken at uniformly spaced points in the x_1 and x_2 plane, if the spacings χ_1 and χ_2 satisfy the conditions $\chi_1 \leq \pi/\omega_{1m}$ and $\chi_2 \leq \pi/\omega_{2m}$".

The spectrum of the sampled function or signal $x(n_1\chi_1, n_2\chi_2)$ can be expressed in the form

$$X_p(j\omega_1, j\omega_2) = \frac{1}{\chi_1}\frac{1}{\chi_2}\sum_{n_1=-\infty}^{\infty}\sum_{n_2=-\infty}^{\infty} X(j\omega_1 - jn_1\omega_{s1}, j\omega_2 - jn_2\omega_{s2}) \tag{1.13}$$

where $X(j\omega_1, j\omega_2)$ is the continuous 2-D signal spectrum and $\omega_{s1} = 2\pi/\chi_1$, $\omega_{s2} = 2\pi/\chi_2$.

As in the 1-D case, the sampled signal spectrum is a periodically repeated spectrum, obtained now by superposing an infinite number of replicas of the continuous signal spectrum each of which is centred on a point from the set $(n_1\omega_{s1}, n_2\omega_{s2})$. Therefore it is clear that if the function $x(x_1, x_2)$ is sampled according to the sampling theorem, all the spectrum replicas are disjoint and it is possible to recover the spectrum of the analogue 2-D signal, using a 2-D ideal low-pass filter.

The reconstruction low-pass filter could be for instance a filter with a transfer function equal to 1 in the rectangular region with dimensions ω_{1m}, ω_{2m}, that is of a transfer function for $\omega_{1m}=\omega_{s1/2}$, $\omega_{2m}=\omega_{s2/2}$

$$H(j\omega_1, j\omega_2)=\text{rect}\left(j\frac{\omega_1}{\omega_{s1}}\right)\text{rect}\left(j\frac{\omega_2}{\omega_{s2}}\right) \tag{1.14}$$

where the function rect (x) is defined as

$$\text{rect}(x)=\begin{cases}1 & \text{for } -\frac{1}{2}\leq x\leq\frac{1}{2}\\ 0 & \text{elsewhere}\end{cases} \tag{1.15}$$

The spectrum of $x(x_1, x_2)$ can be obtained starting from the spectrum $X_p(j\omega_1, j\omega_2)$ by means of the relation

$$X(j\omega_1, j\omega_2)=X_p(j\omega_1, j\omega_2)\,\text{rect}\left(j\frac{\omega_1}{\omega_{s1}}\right)\text{rect}\left(j\frac{\omega_2}{\omega_{s2}}\right) \tag{1.16}$$

which in the space domain corresponds to

$$x(x_1, x_2)=\chi_1\chi_2 x(n_1\chi_1, n_2\chi_2) * h(x_1, x_2) \tag{1.17}$$

where $*$ means convolution and

$$\begin{aligned}h(x_1, x_2)&=\frac{1}{4\pi^2}\int_{-\infty}^{\infty}\int_{-\infty}^{\infty}\text{rect}\left(j\frac{\omega_1}{\omega_{s1}}\right)\text{rect}\left(j\frac{\omega_2}{\omega_{s2}}\right)e^{j(\omega_1x_1+\omega_2x_2)}\,d\omega_1\,d\omega_2\\ &=\frac{1}{\chi_1}\cdot\frac{1}{\chi_2}\cdot\frac{\sin(\omega_{s1}x_1/2)}{\omega_{s1}x_1/2}\cdot\frac{\sin(\omega_{s2}x_2/2)}{\omega_{s2}x_2/2}\end{aligned} \tag{1.18}$$

Substituting (1.18) in (1.17), we obtain the following relation

$$\begin{aligned}x(x_1, x_2)=&\sum_{n_1=-\infty}^{\infty}\sum_{n_2=-\infty}^{\infty}x(n_1\chi_1, n_2\chi_2)\frac{\sin[\omega_{s1}(x_1-n_1\chi_1)/2]}{\omega_{s1}(x_1-n_1\chi_1)/2}\\ &\cdot\sin\frac{[\omega_{s2}(x_2-n_2\chi_2)/2]}{\omega_{s2}(x_2-n_2\chi_2)/2}\end{aligned} \tag{1.19}$$

which is the *interpolation relation* for the 2-D case.

IE The 1-D *z*-transform

Given a sequence $\{x(n)\}$ with $-\infty\leq n\leq\infty$, its *z-transform* is defined as

$$X(z)=\sum_{n=-\infty}^{\infty}x(n)z^{-n} \tag{1.20}$$

where z is a complex variable.

Only if the sequence $x(n)$ is of finite duration, the z-transform converges everywhere in the z plane, except possibly in the points $z=0$ and $z=\infty$.

If the sequence is causal, $x(n)\neq 0$ for $0\leq N_1\leq n\leq\infty$, $X(z)$ converges everywhere outside a circle of some radius R_1.

If the sequence is noncausal, that is $x(n)$ is different from zero only for $-\infty\leq n\leq N_1<0$, then $X(z)$ converges inside a circle of some radius R_2.

Finally, if $x(n)$ is defined over $-\infty\leq N_1\leq n\leq N_2\leq\infty$, then $X(z)$ converges in an annular region between the circles of radii R_1 and R_2, respectively, with $R_1<R_2$.

The z-transform can be inverted and $x(n)$ can be obtained as

$$x(n)=\frac{1}{2\pi j}\oint_C X(z)z^{n-1}\,dz \tag{1.21}$$

where C is a counter-clockwise closed contour in the region of convergence of $X(z)$ and encircling the origin of the z plane.

Integrals of the preceding type can be evaluated by using the residue theorem

$$x(n)=\frac{1}{2\pi j}\oint_C X(z)z^{n-1}\,dz=\sum\,[\text{residues of } X(z)z^{n-1} \text{ at the poles inside } C] \tag{1.22}$$

In the following the main properties of the z-transform are summarized, and reference is given to appropriate texts for complete proofs.[13,17]

(a) Linearity

If

$$Z[x(n)]=X(z), \qquad R_{x-}<|z|<R_{x+}$$
$$Z[y(n)]=Y(z), \qquad R_{y-}<|z|<R_{y+}$$

then

$$Z[ax(n)+by(n)]=aX(z)+bY(z) \qquad R_-<|z|<R_+ \tag{1.23}$$

where the region of convergence is at least the overlap of the two convergence regions. If $X(z)$ and $Y(z)$ are rational, R_- will be the maximum of R_{x-} and R_{y-} and R_+ will be the minimum of R_{x+} and R_{y+}.

If the linear combination is such that some zeros are introduced which cancel poles, then the region of convergence may of course be larger.

(b) Shift of sequences

If

$$Z[x(n)]=X(z) \qquad R_{x-}<|z|<R_{x+}$$

then

$$Z[x(n+n_0)]=z^{n_0}X(z) \qquad R_{x-}<|z|<R_{x+} \tag{1.24}$$

The regions of convergence are identical with the possible exception of $z=0$ and $z=\infty$. In fact for n_0 positive, zeros are introduced at $z=0$ and poles at $z=\infty$; for n_0 negative, poles are introduced at the origin and zeros at infinity.[13]

(c) Multiplication by an exponential sequence

If

$$Z[x(n)]=X(z) \qquad R_{x-}<|z|<R_{x+}$$

then

$$Z[a^n x(n)]=X(a^{-1}z) \qquad |a|R_{x-}<|z|<R_{x+}|a| \tag{1.25}$$

If $X(z)$ has a pole at $z=z_1$, then $X(a^{-1}z)$ will have a pole at $z=az_1$. In general, all the pole-zero locations are scaled by a factor a. If a is a positive real number, this can be interpreted as a shrinking or expanding of the z-plane, i.e. the pole and zero locations change along radial lines in the z-plane.[13]

If a is complex with unity magnitude, the scaling corresponds to a rotation in the z-plane, i.e. the pole and zero locations change along circles centred at the origin.[13]

(d) Differentiation

If

$$Z[x(n)]=X(z) \qquad R_{x-}<|z|<R_{x+}$$

then

$$Z[nx(n)]=-z\frac{\mathrm{d}X(z)}{\mathrm{d}z} \qquad R_{x-}<|z|<R_{x+} \tag{1.26}$$

(e) Conjugation of a complex sequence

If

$$Z[x(n)]=X(z) \qquad R_{x-}<|z|<R_{x+}$$

then

$$Z[x^*(n)]=X^*(z^*) \qquad R_{x-}<|z|<R_{x+} \tag{1.27}$$

(f) Initial value theorem

If $\{x(n)\}$ is zero for $n<0$, then

$$x(0)=\lim_{z\to\infty} X(z) \tag{1.28}$$

(g) Convolution of sequences

If $\{w(n)\}$ is the convolution of the two sequences $\{x(n)\}$ and $\{y(n)\}$, then the z-transform of $\{w(n)\}$ is the product of the z-transforms of $\{x(n)\}$ and $\{y(n)\}$, that is

$$Z[w(n)] = Z\left[\sum_{k=-\infty}^{\infty} x(k)y(n-k)\right] = W(z) = X(z)Y(z)$$

$$R_{y_-} < |z| < R_{y_+}, \qquad R_{x_-} < |z| < R_{x_+} \tag{1.29}$$

where the region of convergence includes the intersection of the regions of convergence of $Y(z)$ and $X(z)$.

If a pole at the border of the region of convergence of one of the z-transforms is cancelled by a zero of the other, then the region of convergence of $W(z)$ will be larger.[13]

(h) Complex convolution theorem

In the continuous case a convolution of time functions leads to a product of transforms and similarly a convolution of transforms leads to a product of time functions. In the discrete case we would not expect an exact duality, because of the fact that sequences are discrete while their transforms are continuous.[13]

However a relationship can be derived in which the z-transform of a product of sequences has a form similar to a convolution. This relation is known as the complex convolution theorem and can be expressed as

$$Z[w(n)] = Z[x(n)y(n)] = W(z) = \frac{1}{2\pi j} \oint_{C_2} X(v) Y\left(\frac{z}{v}\right) v^{-1}\, dv \tag{1.30}$$

with the region of convergence $R_{x_-}R_{y_-} < |z| < R_{x_+}R_{y_+}$, if $X(z)$ is the z-transform of $\{x(n)\}$ with region of convergence $R_{x_-} < |z| < R_{x_+}$ and $Y(z)$ is the z-transform of $\{y(n)\}$ with region of convergence $R_{y_-} < |z| < R_{y_+}$. Again, in certain cases, the region of convergence may be larger than this, but will always include the region defined above and then extend inwards and outwards to the nearest poles.

(i) Parseval relation

The Parseval relation states that

$$\sum_{n=-\infty}^{\infty} x(n)y^*(n) = \frac{1}{2\pi j} \oint_C X(v) Y^*(1/v^*) v^{-1}\, dv \tag{1.31}$$

where the contours of integration is taken in the overlap of the regions of convergence of $X(v)$ and $Y^*(1/v^*)$.

IF The 2-D z-transform

The *2-D z-transform* of a sequence $\{x(n_1, n_2)\}$ with $-\infty \leq n_1, n_2 \leq \infty$ is defined as

$$X(z_1, z_2) = \sum_{n_1=-\infty}^{\infty} \sum_{n_2=-\infty}^{\infty} x(n_1, n_2) z_1^{-n_1} z_2^{-n_2} \tag{1.32}$$

where z_1 and z_2 are complex variables.

The inverse relation can be written in the form

$$x(n_1, n_2) = \left(\frac{1}{2\pi j}\right)^2 \oint_{C_1} \oint_{C_2} X(z_1, z_2) z_1^{n_1-1} z_2^{n_2-1}\, dz_1\, dz_2 \tag{1.33}$$

where C_1 and C_2 are closed contours encircling the origin and within the region of convergence.

A very important class of 2-D z-transforms is that from which we have separable z-transform, that is

$$X(z_1, z_2) = X_1(z_1) X_2(z_2) \tag{1.34}$$

$X(z_1, z_2)$ will be separable only if $x(n_1, n_2)$ is separable in the form $x_1(n_1)x_2(n_2)$. In this case we have: $Z[x_1(n_1)] = X_1(z_1)$ and $Z[x_2(n_2)] = X_2(z_2)$.

It is very difficult to study in detail the convergence of the 2-D z-transforms, save the particular case of finite length sequences, which are bounded. In this case the z-transforms converge everywhere in the z_1 and z_2 planes, except perhaps at $z_1 = 0$, $z_2 = 0$ or $z_1 = \infty$, $z_2 = \infty$.

We do not go in any detail into the theory of 2-D z-transforms, but we summarize in Table 1.1 the main properties.

IG Fundamental properties of 1-D digital systems

A digital system can be defined as an operator which transforms an input sequence $\{x(n)\}$ to an output sequence $\{y(n)\}$[13,15,16]

$$\{y(n)\} = T[\{x(n)\}] \tag{1.35}$$

A system of this type is said to be *linear*, indicated by the symbol L, if and only if the principle of superposition holds. This is equivalent to saying that if $\{y_1(n)\}$ and $\{y_2(n)\}$ are the outputs of the system with inputs $\{x_1(n)\}$ and $\{x_2(n)\}$, then

$$\begin{aligned} L[\{ax_1(n) + bx_2(n)\}] &= aL[\{x_1(n)\}] + bL[\{x_2(n)\}] \\ &= a\{y_1(n)\} + b\{y_2(n)\} \end{aligned} \tag{1.36}$$

where a and b are arbitrary constants.

Sequence	z-Transform
$x(n_1, n_2)$	$X(z_1, z_2)$
$y(n_1, n_2)$	$Y(z_1, z_2)$
$ax(n_1, n_2)+by(n_1, n_2)$	$aX(z_1, z_2)+bY(z_1, z_2)$
$x(n_1+n_{10}, n_2+n_{20})$	$z_1^{n_{10}}z_2^{n_{20}}X(z_1, z_2)$
$a^{n_1}b^{n_2}x(n_1, n_2)$	$X(a^{-1}z_1, b^{-1}z_2)$
$n_1n_2x(n_1, n_2)$	$z_1z_2\cdot\dfrac{d^2X(z_1, z_2)}{dz_1\,dz_2}$
$x^*(n_1, n_2)$	$X^*(z_1^*, z_2^*)$
$x(-n_1, -n_2)$	$X(z_1^{-1}, z_2^{-1})$
$\mathrm{Re}\,[x(n_1, n_2)]$	$\frac{1}{2}[X(z_1, z_2)+X^*(z_1^*, z_2^*)]$
$\mathrm{Im}\,[x(n_1, n_2)]$	$\frac{1}{2}[X(z_1, z_2)-X^*(z_1^*, z_2^*)]$
$x(n_1, n_2)*y(n_1, n_2)$	$X(z_1, z_2)Y(z_1, z_2)$
$x(n_1, n_2)y(n_1, n_2)$	$\left(\dfrac{1}{2\pi j}\right)^2\oint_{C_1}\oint_{C_2} X\left(\dfrac{z_1}{v_1}, \dfrac{z_2}{v_2}\right) Y(v_1, v_2)v_1^{-1}v_2^{-1}\,dv_1\,dv_2$

TABLE 1.1. The z-transform of some sequences.

Since linearity is valid (for absolutely convergent systems) for the sum of an infinite number of terms, the linear system can be uniquely determined by its response to unit-sample sequence defined as the sequence

$$\begin{aligned}\delta(n)&=1, \qquad n=0\\ &=0, \qquad n\neq 0\end{aligned} \tag{1.37}$$

Thus we can obtain another input–output relation as follows. By observing that the generic sample $x(n)$ of the input signal can be written in the form

$$x(n)=\sum_{k=-\infty}^{\infty} x(k)\delta(n-k) \tag{1.38}$$

the input–output relation can be written as

$$y(n)=L\left[\left\{\sum_{k=-\infty}^{\infty} x(k)\delta(n-k)\right\}\right] \tag{1.39}$$

and from linearity

$$y(n)=\sum_{k=-\infty}^{\infty} x(k)L[\{\delta(n-k)\}]=\sum_{k=-\infty}^{\infty} x(k)h(k, n) \tag{1.40}$$

where $\{h(k, n)\}$ is the response to the unit sample sequence which in general is a function of k and n.

A more useful relation is obtainable for a subclass of linear systems, the *shift-invariant* linear systems. These systems are characterized by the property that if $\{y(n)\}$ is the response to $\{x(n)\}$, then $\{y(n-k)\}$ is the response to $\{x(n-k)\}$: in other words this implies that the output of the system to an input signal is independent of the instant of its application to the system.

In this case the response of the system to the impulse $\{\delta(n-k)\}$ is $\{h(n-k)\}$ and the input–output relation can be written as

$$y(n)=\sum_{k=-\infty}^{\infty} x(k)h(n-k) \tag{1.41}$$

This relation, which is commonly written also in the symbolic form

$$\{y(n)\}=\{x(n)*h(n)\} \tag{1.42}$$

is called the convolution sum of $\{x(n)\}$ with $\{h(n)\}$. The sequence $\{h(n)\}$ is called the *impulse response* of the system. With a change of variable this equation can be written also in the form

$$y(n)=\sum_{k=-\infty}^{\infty} h(k)x(n-k)=h(n)*x(n) \tag{1.43}$$

that is as a convolution of $\{h(n)\}$ with $\{x(n)\}$. This is equivalent to saying that the result of convolution is independent of the order of summation of the two sequences.

It is important to observe that the cascade of two linear shift invariant systems is again a linear shift invariant system whose impulse response is the convolution of the two impulse responses.

In the z-transform domain, the convolution equation reduces to the form

$$Y(z)=H(z)X(z) \tag{1.44}$$

where

$$H(z)=\sum_{k=-\infty}^{\infty} h(k)z^{-k} \tag{1.45}$$

is the z-transform of the unit-impulse response. $H(z)$ is called the z-transform function of the system and characterizes completely the linear shift invariant system.

In Section IN the z-transform function $H(z)$ of a very important class of digital linear shift invariant filters will be derived and its properties investigated.

IH Fundamental properties of 2-D digital systems

The same definitions as previously can be given in the 2-D case.

A 2-D system can be characterized by an operator transforming an input 2-D sequence $\{x(n_1, n_2)\}$ to an output 2-D sequence $\{y(n_1, n_2)\}$.

The system is linear if and only if the principle of superposition holds, that is if

$$L[\{ax_1(n_1, n_2)+bx_2(n_1, n_2)\}] = aL\{x_1(n_1, n_2)\}+bL\{x_2(n_1, n_2)\} \quad (1.46)$$

If the 2-D linear system is shift invariant, that is to the input $\{x(n_1-k, n_2-i)\}$ there corresponds the output $\{y(n_1-k, n_2-i)\}$, and the output of the system is independent of the position of the input, it is possible to define a 2-D input–output convolution relation as follows.

If $\{h(n_1-k, n_2-i)\}$ is the output of the system to the unit pulse $\{\delta(n_1-k, n_2-i)\}$ defined from the following relations

$$\begin{aligned} \delta(n_1, n_2) &= 1, \qquad n_1, n_2 = 0 \\ &= 0, \qquad n_1, n_2 \neq 0 \end{aligned} \quad (1.47)$$

the convolution sum can be obtained by writing the output signal in the form

$$\begin{aligned} y(n_1, n_2) &= L\left[\sum_{k=-\infty}^{\infty} \sum_{i=-\infty}^{\infty} x(k, i)\delta(n_1-k, n_2-i)\right] \\ &= \sum_{k=-\infty}^{\infty} \sum_{i=-\infty}^{\infty} x(k, i)L[\delta(n_1-k, n_2-i)] \\ &= \sum_{k=-\infty}^{\infty} \sum_{i=-\infty}^{\infty} x(k, i)h(n_1-k, n_2-i) \end{aligned} \quad (1.48)$$

In the 2-D z domain the convolution operator reduces to a multiplication

$$Y(z_1, z_2) = H(z_1, z_2)X(z_1, z_2) \quad (1.49)$$

where the function

$$H(z_1, z_2) = \sum_{k=-\infty}^{\infty} \sum_{i=-\infty}^{\infty} h(k, i)z_1^{-k}z_2^{-i} \quad (1.50)$$

is called the *z-transfer function of the system.* Obviously this function is the 2-D z-transform of the impulse response $h(k, i)$.

II Frequency domain representation

A very important representation of discrete time systems can be obtained in terms of sinusoidal or complex-exponential signals.[13]

This is possible because the linear shift invariant systems have the interesting property that the response to a sinusoidal or complex-exponential input is a sinusoid or complex exponential having the same frequency and an amplitude and a phase which are characteristic of the system under study.

Let us consider a discrete sequence

$$\{x(n)\} = \{e^{jn\omega}\} \tag{1.51}$$

defined for $n = -\infty, \infty$, that is a sampled complex exponential with radian frequency ω. The output of the system having impulse response $\{h(k)\}$ can be written as

$$y(n) = \sum_{k=-\infty}^{\infty} h(k)\, e^{j(n-k)\omega} = e^{jn\omega} \sum_{k=-\infty}^{\infty} h(k)\, e^{-jk\omega} \tag{1.52}$$

and defining

$$H(e^{j\omega}) \triangleq \sum_{k=-\infty}^{\infty} h(k)\, e^{-jk\omega} \tag{1.53}$$

we can write

$$y(n) = e^{jn\omega} H(e^{j\omega}) \tag{1.54}$$

The function $H(e^{j\omega})$, which is called the *frequency response* of the system, describes the change in phase and amplitude of the input exponential, provided that the series for $H(e^{j\omega})$ converges. $H(e^{j\omega})$ is in general a complex number and two representations are possible, that is the representation in terms of its real and imaginary part

$$H(e^{j\omega}) = H_R(e^{j\omega}) + jH_I(e^{j\omega}) \tag{1.55}$$

or in terms of its magnitude and phase

$$H(e^{j\omega}) = |H(e^{j\omega})|\, e^{j \arg H(e^{j\omega})} \tag{1.56}$$

In this representation the phase can be substituted by the *group delay*, where the group delay of course is defined as the negative of the first derivative of the phase with respect to ω.

It can be observed that $H(e^{j\omega})$ is a continuous function of ω and further that it is a periodic function of ω with period 2π. This is a direct consequence of the sampled signal spectrum of eq. (1.12). This is an important point implying that the system behaves in the same way for all frequencies with multiples of 2π added to the basic frequency. Since $H(e^{j\omega})$ is periodic with period 2π it is then possible to interpret the expression (1.53) as the Fourier series of $H(e^{j\omega})$, in which the coefficients are the samples of the impulse response. This interpretation allows the possibility of representing an infinite sequence in terms of an infinite sum of exponentials, in the form of the expression for the coefficients of the Fourier series

$$h(n)=\frac{1}{2\pi}\int_{-\pi}^{+\pi} H(e^{j\omega})\,e^{jn\omega}\,d\omega \tag{1.57}$$

This representation is valid not only for the impulse response of linear systems but also for any sequence for which the series converges.

In this way a frequency representation of the signal can be constructed which is particularly useful for the practical analysis of the behaviour of linear systems to any input. Indeed the behaviour of the system can be described in terms of its frequency response as it will be shown in the following section.

IJ Some properties of the frequency representation

In the preceding section a frequency representation of linear systems and sequences was developed in terms of an infinite sum of complex exponentials.

Since this representation, which is known as the *Fourier transform representation*, is very useful, it is convenient to summarize some of its properties.

It is convenient to consider some definitions, to be used in further development. A sequence $\{x_e(n)\}$ is defined to be a *conjugate symmetric* sequence if the following relation is valid

$$x_e(n)=x_e^*(-n) \tag{1.58}$$

whilst a sequence $\{x_0(n)\}$ is defined to be *conjugate antisymmetric* sequence if

$$x_o(n)=-x_o^*(-n) \tag{1.59}$$

These two types of sequences are important because any sequence can be expressed as a sum of a conjugate symmetric part and a conjugate antisymmetric part, in the form

$$x(n)=x_e(n)+x_o(n) \tag{1.60}$$

where

$$\begin{aligned} x_e(n)&=\tfrac{1}{2}[x(n)+x^*(-n)]\\ x_o(n)&=\tfrac{1}{2}[x(n)-x^*(-n)] \end{aligned} \tag{1.61}$$

as it can be verified substituting (1.61) relations in (1.60).

A real sequence, which is conjugate symmetric, that is $x(n)=x(-n)$, is called *even sequence*, while a real sequence which is conjugate antisymmetric is called *odd sequence.*

The same definitions can be given for the Fourier transforms $X(e^{j\omega})$, which can be divided in a conjugate symmetric part $X_e(e^{j\omega})$ and a conjugate antisymmetric part $X_o(e^{j\omega})$.

In Table 1.2 we give a summary of the properties of the Fourier transform: these properties are readily obtainable. In the same table also are given the symmetry properties for the Fourier transform of a $\{x(n)\}$ real sequence.

Sequence	Fourier Transform
$x(n)$	$X(e^{j\omega})$
$ax(n)+by(n)$	$aX(e^{j\omega})+bY(e^{j\omega})$ (linearity)
$x(n-k)$	$e^{-jk\omega}X(e^{j\omega})$ (shifting theorem)
$x^*(n)$	$X^*(e^{-j\omega})$
$x^*(-n)$	$X^*(e^{j\omega})$
$\mathrm{Re}\,[x(n)]$	$X_e(e^{j\omega})$
$j\,\mathrm{Im}\,[x(n)]$	$X_o(e^{j\omega})$
$x_e(n)$	$\mathrm{Re}\,[X(e^{j\omega})]$
$x_o(n)$	$j\,\mathrm{Im}\,[X(e^{j\omega})]$
for $x(n)$ real	$X(e^{j\omega})=X^*(e^{-j\omega})$
	$\mathrm{Re}\,[X(e^{j\omega})]=\mathrm{Re}\,[X(e^{-j\omega})]$
	$\mathrm{Im}\,[X(e^{j\omega})]=-\mathrm{Im}\,[X(e^{-j\omega})]$
	$\lvert X(e^{j\omega})\rvert=\lvert X(e^{-j\omega})\rvert$
	$\arg\,[X(e^{j\omega})]=-\arg\,[X(e^{-j\omega})]$

TABLE 1.2. Properties of the Fourier transform.

A further important property is the convolution theorem, which can be stated as follows: "If $\{x(n)\}$ and $\{h(n)\}$ have Fourier transforms $X(e^{j\omega})$ and $H(e^{j\omega})$ respectively, the sequence $\{y(n)\}$ defined as

$$y(n)=\sum_{k=-\infty}^{\infty} h(n-k)x(k) \tag{1.62}$$

has as Fourier transform the function $Y(e^{j\omega})$ for which

$$Y(e^{j\omega})=H(e^{j\omega})X(e^{j\omega})\text{"} \tag{1.63}$$

This relation is very important from the point of view of the theory of linear systems: it shows that the behaviour of the system can be described in terms of its frequency response. The output sequence of a linear shift invariant system can be obtained first by computing its Fourier transform as the product of the input sequence Fourier transform by the frequency response of the system, and then by inverting the Fourier transform.

Finally it is useful to point out the relation between the Fourier transform representation and the z-transform.

If we consider the z-transform expression and evaluate it along the unit circle, provided that the unit circle is in the convergence region, we get

$$X(z)|_{z=e^{j\omega}} = \sum_{n=-\infty}^{\infty} x(n)\, e^{-jn\omega} \tag{1.64}$$

which is the same expression as the Fourier transform of the sequence $\{x(n)\}$ defined in the preceding section. The Fourier transform results therefore equal to the z-transform evaluated along the unit circle in the z-plane.

IK 2-D frequency representation

A frequency representation of 2-D linear systems and sequences[13] can be defined by considering the output of 2-D linear shift invariant systems to a complex exponential sequence $\{x(n_1, n_2)\}$ of the form

$$x(n_1, n_2) = e^{jn_1\omega_1}\, e^{jn_2\omega_2} \tag{1.65}$$

The output can be expressed by means of the convolution sum

$$\begin{aligned} y(n_1, n_2) &= \sum_{m_1=-\infty}^{\infty} \sum_{m_2=-\infty}^{\infty} h(m_1, m_2)\, e^{j(n_1-m_1)\omega_1}\, e^{j(n_2-m_2)\omega_2} \\ &= e^{jn_1\omega_1}\, e^{jn_2\omega_2} H(e^{j\omega_1}, e^{j\omega_2}) \end{aligned} \tag{1.66}$$

where

$$H(e^{j\omega_1}, e^{j\omega_2}) = \sum_{m_1=-\infty}^{\infty} \sum_{m_2=-\infty}^{\infty} h(m_1, m_2)\, e^{-jm_1\omega_1}\, e^{-jm_2\omega_2} \tag{1.67}$$

is, by definition, the frequency response of the 2-D system. The function $H(e^{j\omega_1}, e^{j\omega_2})$ is a continuous function of ω_1, ω_2 and it is periodic in both directions ω_1, ω_2 with period 2π, that is

$$H(e^{j\omega_1}, e^{j\omega_2}) = H(e^{j\omega_1+j2k\pi}, e^{j\omega_2+j2i\pi}) \tag{1.68}$$

with $k, i = -\infty, \infty$. This means that $H(e^{j\omega_1}, e^{j\omega_2})$ can be expressed as a

Fourier series. Therefore the inversion relation for $\{h(n_1, n_2)\}$ can be obtained from the definition of Fourier coefficients as

$$h(n_1, n_2) = \frac{1}{4\pi^2} \int_{-\pi}^{+\pi} \int_{-\pi}^{+\pi} H(e^{j\omega_1}, e^{j\omega_2})\, e^{jn_1\omega_1}\, e^{jn_2\omega_2}\, d\omega_1\, d\omega_2 \qquad (1.69)$$

The same relations are valid for any sequence $\{x(n_1, n_2)\}$ which is absolutely summable, so that any sequence of this type has a representation in the frequency domain in the form

$$X(e^{j\omega_1}, e^{j\omega_2}) = \sum_{n_1=-\infty}^{\infty} \sum_{n_2=-\infty}^{\infty} x(n_1, n_2)\, e^{-jn_1\omega_1}\, e^{-jn_2\omega_2} \qquad (1.70)$$

with the inverse relation

$$x(n_1, n_2) = \frac{1}{4\pi^2} \int_{-\pi}^{+\pi} \int_{-\pi}^{+\pi} X(e^{j\omega_1}, e^{j\omega_2})\, e^{jn_1\omega_1}\, e^{jn_2\omega_2}\, d\omega_1\, d\omega_2 \qquad (1.71)$$

The convolution theorem and some symmetry relationships are among the properties of the 2-D Fourier transform of immense importance. The convolution theorem can be expressed in the following form: "If $\{x(n_1, n_2)\}$ and $\{h(n_1, n_2)\}$ have as Fourier transforms $X(e^{j\omega_1}, e^{j\omega_2})$ and $H(e^{j\omega_1}, e^{j\omega_2})$ respectively, the sequence $\{y(n_1, n_2)\}$ defined as

$$y(n_1, n_2) = \sum_{m_1=-\infty}^{\infty} \sum_{m_2=-\infty}^{\infty} h(n_1 - m_1, n_2 - m_2) x(m_1, m_2) \qquad (1.72)$$

has as Fourier transform the function $Y(e^{j\omega_1}, e^{j\omega_2})$ for which the relation

$$Y(e^{j\omega_1}, e^{j\omega_2}) = H(e^{j\omega_1}, e^{j\omega_2}) X(e^{j\omega_1}, e^{j\omega_2}) \qquad (1.73)$$

is valid."

If $h(n_1, n_2)$ is a real sequence, the frequency response satisfies the constraint

$$H(e^{j\omega_1}, e^{j\omega_2}) = H^*(e^{-j\omega_1}, e^{-j\omega_2}) \qquad (1.74)$$

which means that the knowledge of the Fourier transform in the first quadrant implies the knowledge of the values of the Fourier transform in the third quadrant and vice versa. The same is true for the second and fourth quadrant. Therefore the Fourier transform is uniquely determined by its values in a half-plane.

Finally if $\{h(n_1, n_2)\}$ is separable, that is if

$$h(n_1, n_2) = h_1(n_1) h_2(n_2) \qquad (1.75)$$

then $H(e^{j\omega_1}, e^{j\omega_2})$ is separable too and can be expressed in the form

$$H(e^{j\omega_1}, e^{j\omega_2}) = H_1(e^{j\omega_1}) H_2(e^{j\omega_2}) \qquad (1.76)$$

where $H_1(e^{j\omega_1})$, $H_2(e^{j\omega_2})$ are the 1-D Fourier transforms of $\{h_1(n_1)\}$ and $\{h_2(n_2)\}$ respectively.

IL 1-D discrete Fourier transform

In the preceding sections the description of sequences in terms of z-transforms and of Fourier transforms was developed. In particular a relation between these transforms was found, where the Fourier transform is equal to the z-transform, evaluated along the unit circle, that is for $z = e^{j\omega}$.

Let us now consider the representation of a finite sequence of N points in terms of N samples of its Fourier transform.[13,15,16]

If we have a sequence $\{x(n)\}$ with N terms, its z-transform can be written as

$$X(z) = \sum_{n=0}^{N-1} x(n) z^{-n} \tag{1.77}$$

For the values of the z-transform on N equidistant points along the unit circle the following expression can be obtained

$$X(k) = \sum_{n=0}^{N-1} x(n) e^{-j(2\pi kn/N)}, \qquad k = 0, \ldots N-1 \tag{1.78}$$

This expression is known as the *discrete Fourier transform* (DFT) of the sequence $\{x(n)\}$, and can be inverted to yield

$$x(n) = \frac{1}{N} \sum_{k=0}^{N-1} X(k) e^{j(2\pi kn/N)}, \quad n = 0, \ldots N-1 \tag{1.79}$$

This can be verified directly by substituting for $X(k)$ in (1.79) the expression (1.78) and using the orthogonality relations for the functions $e^{j(2\pi kn/N)}$.

The first observation to be made is that the exponential functions are periodic with period N, that is for integer k, n (k, $n = -\infty, +\infty$)

$$e^{j(2\pi kn/N)} = e^{j[2\pi k(n+N)/N]} = e^{j[2\pi n(k+N)/N]} \tag{1.80}$$

This means that the sequences $\{x(n)\}$ and $\{X(k)\}$ as defined by the transformation relations are *periodic sequences.*

We summarize in the following table the main properties of the DFT, without going in detail into the theory since this is not the purpose of this book and such details can be found in several specialized books and publications.

It is necessary, however, to consider in some detail the *discrete convolution theorem* due to its implications in linear filtering applications.

This theorem can be formulated in the following way. If two sequences $\{x_1(n)\}$, $n = 0, \ldots N-1$ and $\{x_2(n)\}$, $n = 0, \ldots N-1$ have DFTs $X_1(k)$,

Finite length sequence (N)	DFT
$x(n)$	$X(k)$
$y(n)$	$Y(k)$
$ax(n)+by(n)$	$aX(k)+bY(k)$ (linearity)
$x(n+m)$	$e^{j(2\pi km/N)}X(k)$
$x^*(n)$	$X^*(-k)$
$x^*(-n)$	$X^*(k)$
$x_e(n)$	$\text{Re}\,[X(k)]$
$x_o(n)$	$j\,\text{Im}\,[X(k)]$
$x(n)$ real	$X(k)=X^*(-k)$
	$\text{Re}\,[X(k)]=\text{Re}\,[X(-k)]$
	$\text{Im}\,[X(k)]=-\text{Im}\,[X(-k)]$
	$\lvert X(k)\rvert=\lvert X(-k)\rvert$
	$\arg\,[X(k)]=-\arg\,[X(-k)]$

TABLE 1.3. Properties of the DFT.

$k=0,\ldots N-1$, $X_2(k)$, $k=0,\ldots N-1$, then for the sequence $\{y(n)\}$ given by

$$y(n)=\frac{1}{N}\sum_{i=0}^{N-1} x_1(i)x_2(n-i)=\frac{1}{N}\sum_{i=0}^{N-1} x_1(n-i)x_2(i) \tag{1.81}$$

we shall have the DFT $\{Y(k)\}$, $k=0,\ldots N-1$, given by

$$Y(k)=X_1(k)X_2(k) \tag{1.82}$$

and for the sequence $\{y_1(n)\}$ defined by

$$y_1(n)=x_1(n)x_2(n)$$

we shall have its DFT $\{Y_1(k)\}$ given by

$$Y_1(k)=\sum_{m=0}^{N-1} X_1(m)X_2(k-m)=\sum_{m=0}^{N-1} X_1(k-m)X_2(m) \tag{1.83}$$

This theorem is very similar to the corresponding theorems for continuous Fourier transforms. However it is important to point out a fundamental difference between them. This difference is caused by the fact that the sequences (1.78) and (1.79) are defined as periodic sequences and consequently the convolutions of (1.81) and (1.83) are also defined as periodic convolutions. This is illustrated in Fig. 1.2.

Therefore it is not possible to compute directly the conventional convolution of two finite sequences by means of the preceding theorem. In fact when one sequence is shifted beyond the last point of the other sequence, it

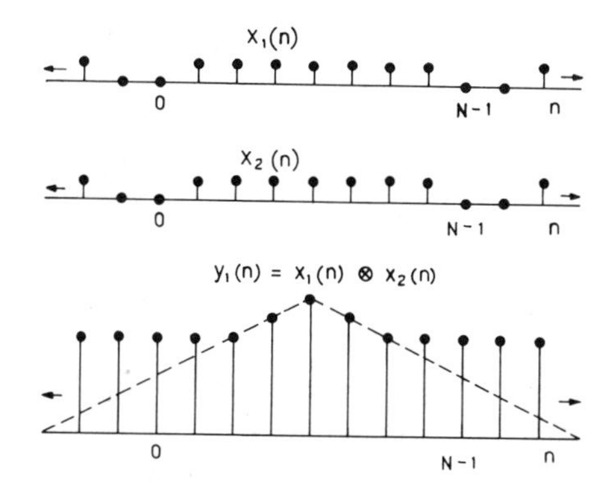

FIG. 1.2. Periodic convolution of two sequences $x_1(n)$ and $x_2(n)$.

will not find zero values as it must, but the periodic repetition of the sequence. Thus to force the discrete convolution to have the same numerical values as the linear convolution of two finite sequences it is necessary to start with two artificial sequences having guard spaces at the end of zero samples to avoid such errors from the periodicity.

If the two sequences to convolve have length N_1 and N_2, respectively, it is obvious that the linear convolution will have a length at most of $N_c = N_1+N_2-1$. Therefore this is the length which has to be used in the computation of DFTs in order to avoid the undesirable results of the circular convolution.

The DFT is one of the most important tools in digital signal processing and in digital filtering, both in the design of digital filters and in the implementation of filtering operations. In fact a very efficient algorithm exists for the evaluation of DFTs, which for example makes the filtering through the convolution theorem far more efficient computationally than direct convolution. This algorithm, which has the name of *Fast Fourier Transform* (FFT), is presented in Appendix 2.

IM 2-D discrete Fourier transform

A 2-D DFT can be defined as the sequence of equidistant values of the z-transform of the finite duration sequence $\{x(n_1, n_2)\}$, $n_1=0, \ldots N_1-1$, $n_2=0, \ldots N_2-1$, along the two circles $z_1=e^{j\omega_1}$, $z_2=e^{j\omega_2}$, that is[9,12,13]

$$X(k_1, k_2)=X(z_1, z_2)|_{z_1=\exp[j(2\pi k_1/N_1)],\ z_2=\exp[j(2\pi k_2/N_2)]}$$

$$= \sum_{n_1=0}^{N_1-1} \sum_{n_2=0}^{N_2-1} x(n_1, n_2) \exp[-j(2\pi k_1 n_1/N_1)] \exp[-j(2\pi k_2 n_2/N_2)]$$

$$k_1=0, \ldots N_1-1,\ k_2=0, \ldots N_2-1 \tag{1.84}$$

The sequence $\{X(k_1, k_2)\}$ represents uniquely the sequence $\{x(n_1, n_2)\}$, which can be recovered from the $\{X(k_1, k_2)\}$ through the relation

$$x(n_1, n_2)=\frac{1}{N_1}\frac{1}{N_2}\sum_{k_1=0}^{N_1-1}\sum_{k_2=0}^{N_2-1} X(k_1, k_2)\exp\left(j\frac{2\pi}{N_1}k_1n_1\right)\exp\left(j\frac{2\pi}{N_2}k_2n_2\right) \tag{1.85}$$

This can be shown in the same way as in the 1-D case.

Most of the properties we considered in the 1-D case are still valid, as, for example, the linearity of the transform and the shifting theorem, which in the 2-D case can be formulated by saying that if $\{X(k_1, k_2)\}$, $k_1 = 0, \ldots N_1-1$, $k_2=0, \ldots N_2-1$ is the DFT of $\{x(n_1, n_2)\}$, $n_1=0, \ldots N_1-1$, $n_2=0, \ldots N_2-1$, then to the sequence $\{x(n_1-m_1, n_2-m_2)\}$ there corresponds the DFT $\{e^{-j(2\pi k_1m_1/N_1)}e^{-j(2\pi k_2m_2/N_2)}X(k_1, k_2)\}$, $k_1=0, \ldots N_1-1$, $k_2=0, \ldots N_2-1$.

Finally the circular convolution theorem is still valid. Thus if $\{X_1(k_1, k_2)\}$ and $\{X_2(k_1, k_2)\}$, $k_1=0, \ldots N_1-1$, $k_2=0, \ldots N_2-1$ are the DFTs of the two sequences $\{x_1(n_1, n_2)\}$, $\{x_2(n_1, n_2)\}$, $n_1=0, \ldots N_1-1$, $n_2= 0, \ldots N_2-1$, then the DFT given below as a product

$$X_3(k_1, k_2)=X_1(k_1, k_2)X_2(k_1, k_2) \tag{1.86}$$

corresponds to the sequence $\{x_3(n_1, n_2)\}$, $n_1=0, \ldots N_1-1$, $n_2= 0, \ldots N_2-1$ given by

$$\begin{aligned} x_3(n_1, n_2) &= \sum_{m_1=0}^{N_1-1}\sum_{m_2=0}^{N_2-1} x_1(m_1, m_2)x_2(n_1-m_1, n_2-m_2) \\ &= \sum_{m_1=0}^{N_1-1}\sum_{m_2=0}^{N_2-1} x_1(n_1-m_1, n_2-m_2)x_2(m_1, m_2) \end{aligned} \tag{1.87}$$

If a new sequence is formed according to eq. (1.88)

$$x_3(n_1, n_2)=x_1(n_1, n_2)x_2(n_1, n_2) \tag{1.88}$$

then it will have a DFT $\{X_3(k_1, k_2)\}$, $k_1=0, \ldots N_1-1$, $k_2=0, \ldots N_2-1$, given by

$$\begin{aligned} X_3(k_1, k_2) &= \sum_{e_1=0}^{N_1-1}\sum_{e_2=0}^{N_2-1} X_1(e_1, e_2)X_2(k_1-e_1, k_2-e_2) \\ &= \sum_{e_1=0}^{N_1-1}\sum_{e_2=0}^{N_2-1} X_1(k_1-e_1, k_2-e_2)X_2(e_1, e_2) \end{aligned} \tag{1.89}$$

The relations (1.87) and (1.89) of course represent circular convolutions. A linear convolution can be obtained by transforming a lengthened sequence with dimensions greater than N_1+N_2-1, N_1+N_2-1 as discussed in the 1-D case.

The most important symmetry properties of 2-D DFTs are summarized in Table 1.4.

Sequence	DFT
$x(n_1, n_2)$	$X(k_1, k_2)$
$x^*(n_1, n_2)$	$X^*(-k_1, -k_2)$
$x_e(n_1, n_2)$	$\mathrm{Re}\,[X(k_1, k_2)]$
$x_o(n_1, n_2)$	$j\,\mathrm{Im}\,[X(k_1, k_2)]$
$\mathrm{Re}\,[x(n_1, n_2)]$	$X_e(k_1, k_2)$
$\mathrm{Im}\,[x(n_1, n_2)]$	$X_o(k_1, k_2)$
$x(n_1, n_2)$ real	$\mathrm{Re}\,[X(k, l)] = \mathrm{Re}\,[X(-k, -l)]$
	$\mathrm{Im}\,[X(k, l)] = -\mathrm{Im}\,[X(k, l)]$
	$\lvert X(k, l)\rvert = \lvert X(-k, -l)\rvert$

TABLE 1.4. Properties of the 2-D DFT.

Of particular interest are the above symmetry properties of the DFT, where it is possible to show, as in the continuous Fourier transform case, that the real sequence $\{x(n_1, n_2)\}$ is specified by the values of DFT in an half plane.

IN 1-D linear systems described by difference equations

A very important class of linear shift invariant systems is the one described by the following equation

$$\sum_{k=0}^{N-1} a(k)x(n-k) = \sum_{k=0}^{M-1} b(k)y(n-k) \tag{1.90}$$

where $x(n)$ are the samples of the input sequence, $y(n)$ are the samples of the output sequence and $a(k)$ and $b(k)$ are the coefficients which define the system or indeed filter.[13,15,16]

In general, of course, a digital filter is not uniquely specified by the difference equation (1.90). That is, to any solution of (1.90) we can add a component which satisfies the homogeneous difference equation (that is, the difference equation with the left hand side equal to zero) so that the overall sum will still satisfy eq. (1.90). It is possible to prove that the general solution of the homogeneous equation is of the type ka^n, where k is an arbitrary constant (i.e. a^n is an eigenfunction of eq. (1.90)). Thus if (1.90) is satisfied by $y_1(n)$ for $x(n) = x_1(n)$, then any solution of the type $y(n) = y_1(n) + ka^n$ satisfies the same equation.

Therefore, as with differential equations in the continuous time case, it is necessary to specify the initial conditions of the system. These initial conditions must be such that the system is *linear* and *recursive*.

For the *recursivity* of the system, it is necessary that any output sample is *computable* from the knowledge of previously computed samples or from the initial conditions. If, for example, the system is causal, we have to specify initial conditions so that if $x(n)=0$ for $n<n_0$, then $y(n)=0$ for $n<n_0$. In this case the linear difference equation provides an explicit relationship between the input and the output, in the form

$$y(n)=\sum_{k=0}^{N-1}\frac{a(k)}{b(0)}x(n-k)-\sum_{k=1}^{M-1}\frac{b(k)}{b(0)}y(n-k) \tag{1.91}$$

with initial conditions $y(-1), y(-2), \ldots, y(-M+1)=0$.

It is very useful to sketch graphically the operation described by the relation (1.91). In Fig. 1.3 the input and output sequences are shown, together with the operation performed, for the particular case $N=M=3$.

It is clear by the above considerations that under the initial conditions $y(-1), \ldots y(-M+1)=0$, the equation (1.91) is recursive. Indeed any output sample $y(n)$ can be obtained from the previously computed samples $y(i)$, $i<n$, and from the initial conditions if $n<(M-1)$.

In the z-domain the system (1.90) can be represented by its transfer function $H(z)$, which in this case has a very simple form.

If we take the z-transform of the two sides of eq. (1.90) we obtain

$$\sum_{n=-\infty}^{\infty}\left[\sum_{k=0}^{N-1}a(k)x(n-k)\right]z^{-n}=\sum_{n=-\infty}^{\infty}\left[\sum_{k=0}^{M-1}b(k)y(n-k)\right]z^{-n}$$

$$\sum_{k=0}^{N-1}a(k)\sum_{n=-\infty}^{\infty}x(n-k)z^{-n}=\sum_{k=0}^{M-1}b(k)\sum_{n=-\infty}^{\infty}y(n-k)z^{-n}$$

$$X(z)\sum_{k=0}^{N-1}a(k)z^{-k}=Y(z)\sum_{k=0}^{M-1}b(k)z^{-k}$$

from which it is possible to obtain an input–output relation in the z-domain in the form

$$Y(z)=H(z)X(z) \tag{1.92}$$

where

$$H(z)=\frac{\sum_{k=0}^{N-1}a(k)z^{-k}}{\sum_{k=0}^{M-1}b(k)z^{-k}} \tag{1.93}$$

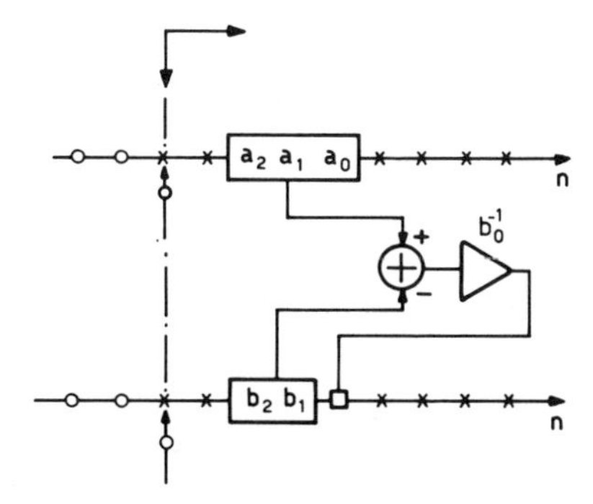

FIG. 1.3. Example of recursive operation (1.91), for the particular case $N=M=3$.

Therefore, in the case of linear shift invariant systems described by a difference equation of the type (1.90), it is very easy to find the transfer function in the z-domain. This is a rational function, the coefficients of which are those of the difference equation.

Two cases can now be considered, leading to two fundamental classes of filters. If the coefficients satisfy the following conditions

$$\begin{aligned} &b(0)=1 \\ &b(k)=0, \qquad k \neq 0 \end{aligned} \tag{1.94}$$

the difference equation (1.90) reduces to

$$y(n)=\sum_{k=0}^{N-1} a(k)x(n-k) \tag{1.95}$$

and hence the transfer function reduces to

$$H(z)=\sum_{k=0}^{N-1} a(k)z^{-k} \tag{1.96}$$

In this case the output samples of the filter depend only on the input samples without any feedback of the past output samples on the current output sample.

These filters have obviously finite responses of length N to the unit sample sequence $\delta(n)$. For this reason, they are known as *finite impulse response filters* (FIR) as opposed to *infinite impulse response filters* (IIR), which are described by (1.90) when at least one (in addition to $b(0)$) of $b(k)$ coefficients with $k \neq 0$ is different from zero.

Transfer functions of the type (1.93) and (1.96) are uniquely determined within a constant factor by the roots of denominator and numerator polynomials, together with the causality condition in the IIR case. The roots of numerator polynomials are generally called zeros of the filter, while the roots of denominator polynomials are known as poles of the filter. Thus FIR filters have only finite zeros, while IIR filters have both zeros and poles.

As far as the distribution of zeros and poles is concerned on the z-plane, we consider the numerator and denominator polynomials as having real coefficients and hence their roots are either simple and real or they occur in complex conjugate pairs. This leads to a typical pattern of zeros and poles as the one shown in Fig. 1.4.

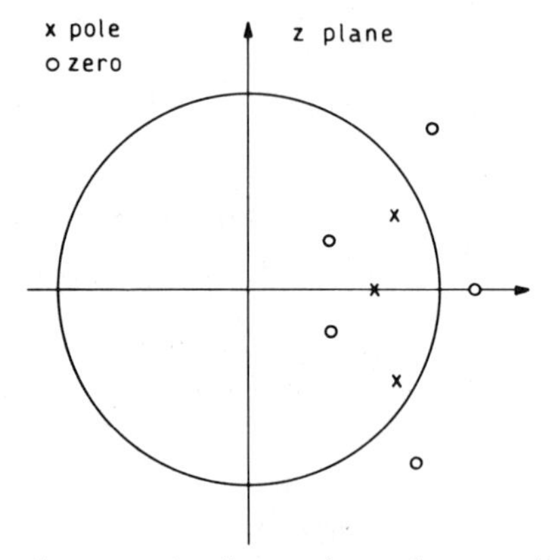

FIG. 1.4. Typical pattern of zeros and poles on the z-plane considering numerator and denominator polynomials with real coefficients.

It is important to observe that the form of eq. (1.91) is not the only recursive form of 1-D difference equations of eq. (1.90). Another form corresponds to a noncausal impulse response, that is to a system for which $h(n)$ is zero for $n>0$ and whose output is zero for $n>n_0$, if $x(n)$ is zero for $n>n_0$. The corresponding difference equation can be obtained by pulling out in eq. (1.91) the term $y(n-N+1)$ instead of the term $y(n)$, obtaining in the case $N=M$

$$b(N-1)y(n-N+1)=\sum_{k=0}^{N-1} a(k)x(n-k)-\sum_{k=0}^{N-2} b(k)y(n-k) \tag{1.97}$$

With the change of variable $m=n-N+1$, (1.97) becomes

$$b(N-1)y(m)=\sum_{k=0}^{N-1} a(k)x(m+N-1-k)-\sum_{k=0}^{N-2} b(k)y(m+N-1-k) \tag{1.98}$$

and with the coefficients rearranged according to

$$\begin{aligned} \tilde{a}(k)&=a(N-1-k), \qquad k=0,\ldots N-1 \\ \tilde{b}(k)&=b(N-1-k), \qquad k=1,\ldots N-1 \end{aligned} \tag{1.99}$$

we obtain the noncausal recursive equation for (1.91) as follows

$$b(N-1)y(m)=\sum_{k=0}^{N-1}\tilde{a}(k)x(m+k)-\sum_{k=1}^{N-1}\tilde{b}(k)y(m+k) \qquad (1.100)$$

with initial conditions $y(1), \ldots, y(N-1)=0$

10 2-D linear systems described by linear difference equations

The 2-D difference equation which defines a *linear shift invariant filter* has the form

$$\sum_{n_1=0}^{N_1-1}\sum_{n_2=0}^{N_2-1} a(n_1, n_2)x(m_1-n_1, m_2-n_2)$$

$$=\sum_{n_1=0}^{M_1-1}\sum_{n_2=0}^{M_2-1} b(n_1, n_2)y(m_1-n_1, m_2-n_2) \qquad (1.101)$$

where $\{x(n_1, n_2)\}$ is the input matrix, $\{y(n_1, n_2)\}$ is the output matrix, and $\{a(n_1, n_2)\}$, $\{b(n_1, n_2)\}$ are the coefficient matrices which define the transfer function of the filter.

In the (z_1, z_2) plane the filter (1.101) is described by the transfer function

$$H(z_1, z_2)=\frac{\sum_{n_1=0}^{N_1-1}\sum_{n_2=0}^{N_2-1} a(n_1, n_2)z_1^{-n_1}z_2^{-n_2}}{\sum_{n_1=0}^{M_1-1}\sum_{n_2=0}^{M_2-1} b(n_1, n_2)z_1^{-n_1}z_2^{-n_2}} \qquad (1.102)$$

which has the form of a ratio of two bivariate polynomials. If $b(0, 0)=1$ and $b(n_1, n_2)=0$ for n_1 and/or $n_2 \neq 0$, the (1.102) reduces to

$$H(z_1, z_2)=\sum_{n_1=0}^{N_1-1}\sum_{n_2=0}^{N_2-1} a(n_1, n_2)z_1^{-n_1}z_2^{-n_2} \qquad (1.103)$$

with a corresponding difference equation in the form

$$y(m_1, m_2)=\sum_{n_1=0}^{N_1-1}\sum_{n_2=0}^{N_2-1} a(n_1, n_2)x(m_1-n_1, m_2-n_2) \qquad (1.104)$$

As in the 1-D case, equations (1.103) and (1.104) define FIR systems, while equations (1.101) and (1.102) represent IIR systems.

In the 2-D case it is necessary to discuss with particular care the choice of suitable initial conditions and recursivity problems, which are of greater importance due to the increase of dimensionality.[5,12] In the discussion of Section IN the key point was the specification of zones along the axis where the impulse response was zero. In this way it was possible to find zones in which the output was definitely equal to zero. This in turn provided a set of initial conditions for two different formulations of recursive equations corresponding to the causal and the noncausal case.

In the 2-D case the problem is more involved because the possible number of different recursive relations with zero initial conditions is increased. We shall consider here the case of *quadrant recursive filters* and *half plane recursive filters*.

The first class consists of the filters whose impulse response is defined on a quadrant of the (n_1, n_2) plane. Obviously four different quadrant filters can be defined. If we consider the condition $y(n_1, n_2)=0$ for $n_1<0$ or $n_2<0$, if $x(n_1, n_2)=0$ for $n_1<0$ or $n_2<0$, it is possible to write an input–output recursive equation of the type for $b(0, 0)=1$

$$y(m_1, m_2)=\sum_{n_1=0}^{N_1-1}\sum_{n_2=0}^{N_2-1} a(n_1, n_2)x(m_1-n_1, m_2-n_2) - \sum_{\substack{n_1=0 \\ n_1+n_2\neq 0}}^{M_1-1}\sum_{n_2=0}^{M_2-1} b(n_1, n_2)y(m_1-n_1, m_2-n_2) \quad (1.105)$$

which corresponds to the 2-D quadrant causal system. Graphically the relation (1.105) can be represented as in Fig. 1.5 for $N_1=N_2=M_1=M_2=3$.

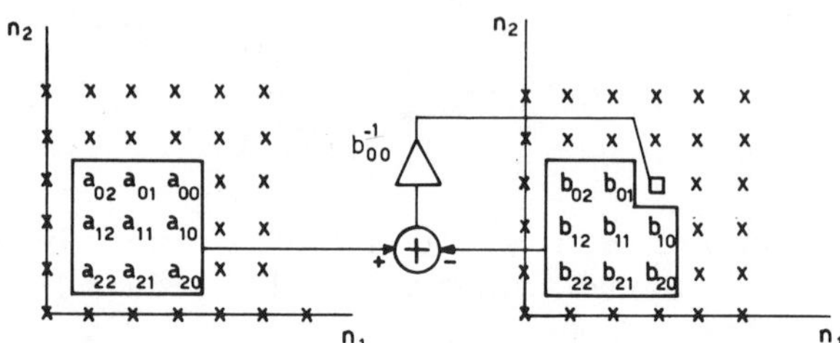

FIG. 1.5. Example of 2-D recursive operation (1.105) for $N_1=N_2=M_1=M_2=3$.

The system described by the (1.105) relation is recursive if its initial conditions are defined as in Fig. 1.6. With these initial conditions every output point can be obtained by means of (1.105) from the previously computed samples and from the initial conditions. The recursion can be performed both along the columns and along the rows of the arrays.

The matrix $\{b(n_1, n_2)\}$ which defines the recursive part of the filter is different from zero only in the region $n_1\geq 0$ and $n_2\geq 0$. This circumstance can be indicated using the symbol

$$\{{}^1b(n_1, n_2)\}, \qquad n_1\geq 0 \qquad n_2\geq 0 \qquad (1.106)$$

$\{{}^1b(n_1, n_2)\}$ is called a *first quadrant sequence*, while the corresponding filter is called a *first quadrant filter*. Filters with impulse responses defined on the

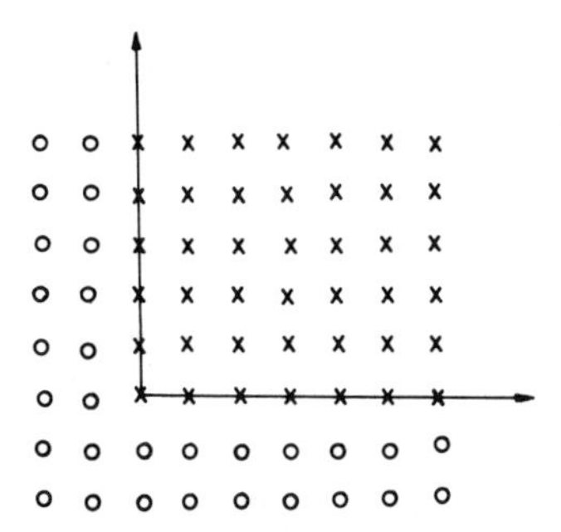

FIG. 1.6. Initial conditions for a recursive system described by eq. (1.105).

other quadrants as shown in Fig. 1.7 can also be defined. Their coefficient matrices are then *second, third* and *fourth quadrant functions* and can be defined as follows

$$\begin{aligned}
&\{{}^2b(n_1, n_2)\} && \begin{matrix} n_1 \geq 0 \\ n_2 \leq 0 \end{matrix} \\
&\{{}^3b(n_1, n_2)\} && \begin{matrix} n_1 \leq 0 \\ n_2 \leq 0 \end{matrix} \\
&\{{}^4b(n_1, n_2)\} && \begin{matrix} n_1 \leq 0 \\ n_2 \geq 0 \end{matrix}
\end{aligned} \tag{1.107}$$

These filters are called *noncausal filters.* The corresponding recursive relations can be easily obtained from (1.101), by pulling out of the expression the suitable output sample.

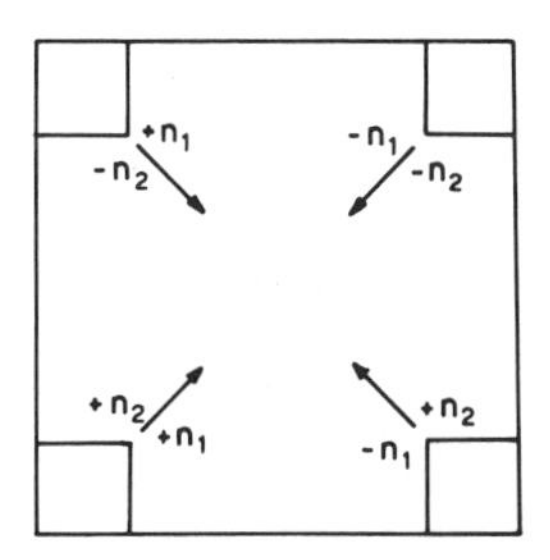

FIG. 1.7. Filters with impulse responses defined on the four quadrants.

In the previous discussion we showed that in the case of quadrant filters it is possible to define a difference equation and a set of initial conditions to obtain a linear shift invariant filter described by a recursive algorithm. However this quadrant form is not the only one in which a recursive realization can be found. This can be seen by considering the following form of difference equation

$$\sum_{n_1=0}^{N_1-1}\sum_{n_2=0}^{N_2-1} a(n_1, n_2)x(m_1-n_1, m_2-n_2)$$
$$= \sum_{n_1=1}^{M_1-1}\sum_{n_2=-(M_2-1)}^{M_2-1} b(n_1, n_2)y(m_1-n_1, m_2-n_2)$$
$$+ \sum_{n_2=0}^{M_2-1} b(0, n_2)y(m_1, m_2-n_2) \tag{1.108}$$

in which the recursive coefficients $b(n_1, n_2)$ are non zero only in the region

$$\begin{aligned} &1\leq n_1\leq M_1-1 \qquad -(M_2-1)\leq n_2\leq (M_2-1) \\ &n_1=0 \qquad\qquad\qquad 0\leq n_2\leq M_2-1 \end{aligned} \tag{1.109}$$

as shown in Fig. 1.8.

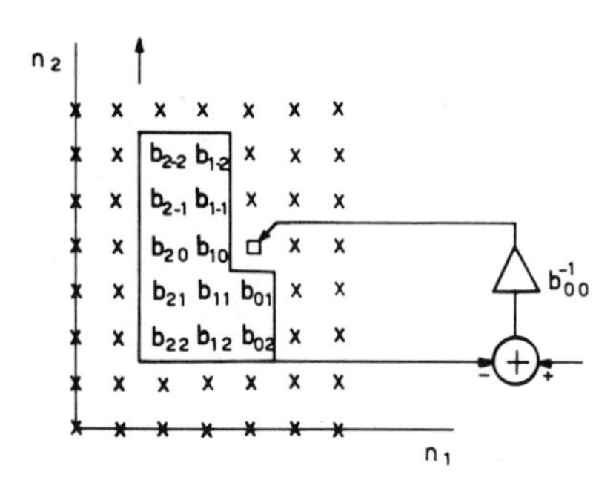

FIG. 1.8. Example of recursive system with coefficients $b(n_1, n_2)$ non-zero only in the region specified by the conditions (1.109).

By considering in eq. (1.108) the point $y(m_1, m_2)$ an input–output relation can be obtained in the form $(b(0, 0)=1)$

$$y(m_1, m_2)= \sum_{n_1=0}^{N_1-1}\sum_{n_2=0}^{N_2-1} a(n_1, n_2)x(m_1-n_1, m_2-n_2)$$
$$- \sum_{n_1=1}^{M_1-1}\sum_{n_2=-(M_2-1)}^{M_2-1} b(n_1, n_2)y(m_1-n_1, m_2-n_2)$$
$$- \sum_{n_2=1}^{M_2-1} b(0, n_2)y(m_1, m_2-n_2) \tag{1.110}$$

It is obvious, by inspection, that this difference equation is recursive if we move in the upward direction along the columns starting from the left lowest point of the input matrix. As far as the initial conditions are concerned, the situation is somewhat involved.

It is, however, possible to see that in general it is necessary to define a zero initial condition zone of the type shown in Fig. 1.9 for $M_1 = M_2 = 3$.

The filter described by the difference equation (1.108) is known as an *unsymmetrical half-plane filter.* Eight different equations of this type can be written corresponding to filters which differ in the half-plane on which the coefficient matrix is defined and in the direction of recursion.

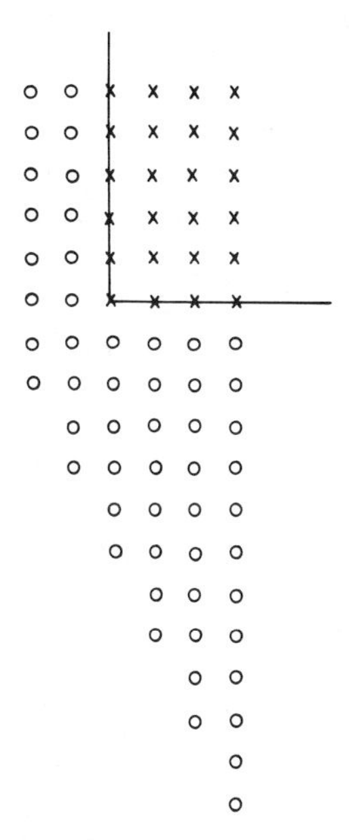

FIG. 1.9. Initial conditions for a recursive system described by eq. (1.110) with $M_1 = M_2 = 3$.

IP 1-D digital filter stability

To be practically usable, a digital filter must be *stable* or in any case we must know the *stability performance* of the filter.[21]

Several definitions can be used for defining the stability. The most useful one is the BIBO (bounded input, bounded output) *stability criterion.* This can be expressed by saying that a digital filter is stable if its response to a limited input is also limited. It is possible to show that linear shift-invariant filters are BIBO-stable, if and only if

$$\sum_{n=-\infty}^{\infty} |h(n)| < \infty \tag{1.111}$$

where $\{h(n)\}$ is the impulse sequence of the filter.

It is useful to observe that obviously this stability criterion is always satisfied if $\{h(n)\}$ has a finite number of terms.

The above stability criterion can be related to the analytic regions of the z-transform function of the system.

In Section IN we considered the form of the z-transform transfer function and we found that this has the form of a rational function

$$H(z)=\frac{A(z)}{B(z)} \tag{1.112}$$

where $A(z)$ and $B(z)$ are polynomials in z.

The filter $H(z)$ can be considered as the cascade of two filters $H(z)=H_1(z)H_2(z)$, where $H_1(z)=A(z)$ and $H_2(z)=1/B(z)$, which correspond to the cascade of a FIR filter $H_1(z)$ with a IIR filter $H_2(z)=1/B(z)$.

According to our previous considerations, the stability of the filter $H(z)$ is only related to the stability of $H_2(z)$, since $H_1(z)$ is always stable. If we indicate with $\{h_2(n)\}$ the impulse response of the $H_2(z)$ filter, for the definition of impulse response and z-transform function, the following relation can be written

$$\sum_{n=0}^{\infty} h_2(n)z^{-n}=\frac{1}{B(z)} \tag{1.113}$$

and $h_2(n)$ can be obtained inverting $B(z)=\sum_{n=0}^{N-1} b(n)z^{-n}$.

By assuming $b(0)=1$, let us consider first the case $B(z)=1+kz^{-1}$; we can write

$$\frac{1}{B(z)}=\frac{1}{1+kz^{-1}}=1-kz^{-1}+k^2z^{-2}-k^3z^{-3}+\cdots$$

$$=\sum_{n=0}^{\infty}(-1)^n(kz^{-1})^n \tag{1.114}$$

and by equating the coefficients of equal powers in the (1.114) $\{h_2(n)\}$ can be obtained in the form

$$h_2(n)=(-k)^{+n} \tag{1.115}$$

The series (1.111) converges if $|k|<1$, but it does not converge for $|k|>1$.

If we define the case $|k|<1$ as the *minimum phase* case and the case $|k|>1$ as the *maximum phase* case, the first order filter $1/(1+kz^{-1})$ is stable if it is minimum phase, while it is unstable if it is maximum phase. It is useful to note that if the first order filter is maximum phase, it is possible to write

$$\frac{1}{1+kz^{-1}}=\frac{k^{-1}z}{1+k^{-1}z} \tag{1.116}$$

and

$$\frac{1}{B(z)}=\sum_{n=0}^{\infty}(-1)^n(k^{-1}z)^n=\sum_{n=0}^{\infty}h_2(-n)z^{+n} \qquad (1.117)$$

where the series $\sum_{n=0}^{\infty}|h_2(-n)|$ is convergent. Thus a maximum phase filter is unstable if used in a filtering operation toward increasing n, while it is stable if it is used in the decreasing n direction. This is the type of filter which in Section IN was called *noncausal filter.* This result can be formulated by saying that the causal first order cell is stable if $|k|<1$, while the noncausal first order cell is stable if $|k|>1$.

In the general case, $B(z)$ is an N order polynomial, but it can be decomposed in first order terms of the form $1/(1+k_iz^{-1})$, where k_i are in general complex. However if $B(z)$ is a real coefficient polynomial, the complex k_i can appear only in a complex conjugate pair and the two factors will have the same phase, that is to say they are both minimum phase or both maximum phase.

In the general case three different possibilities exist:

(1) All factors are of minimum phase. $B(z)$ is consequently of minimum phase and the filter is stable, being the cascade of stable filters; $h_2(n)$ is a causal sequence.

(2) All factors are of maximum phase and hence $B(z)$ is maximum phase so that $1/B(z)$ is an unstable filter. However the filter having as impulse response the time-reversed version $h_2(-n)$ of this unstable filter, that is the noncausal filter, is itself stable.

(3) Some factors are of minimum phase and some are of maximum phase. In this case it is possible to decompose the filter in two filters, one of which is stable going in the positive n direction, while the other is stable going in the negative n direction.

It is very simple to relate the previous considerations to the position of poles of $H(z)$, that is of the zeros of $B(z)$, in the z-plane. If we consider a first order factor of the form $1/(1+kz^{-1})$, its denominator has a zero for $z=-k$. Now if $|k|<1$, then it is stable, while if $|k|>1$, it is unstable. This can be expressed saying that if the pole of this factor is inside the unit circle in the z-plane, then it is stable, while if the pole is outside the unit circle, it is unstable.

Since any $H(z)$ can be expressed in a cascade of first order factors, this means that a causal filter is stable if and only if all the poles are inside the unit circle in the z-plane, while it is unstable if its poles are outside the unit circle. In this case the noncausal filter is stable.

Due to this last result the stability test is very simple and efficient. It is in fact a direct evaluation of pole positions in the z-plane and very efficient computational methods exist for finding the roots of polynomials.

The discussion on causal and noncausal systems and their stability allows us to discuss the possibility of having IIR filtering operations without phase distortion.

A causal IIR filter cannot have a linear phase. In fact a linear phase filter must have a real frequency response with a linear phase factor. This implies that impulse response must be symmetrical with respect to a general point n_0.

This condition obviously cannot be satisfied by a causal filter, whose impulse response is of infinite extent in one direction only.

However when the required filtering does not have to be done as a *real-time operation*, but an operation on stored data of finite extent is performed, a zero phase filter can be implemented as the cascade connection of the causal version of the stable filter $H(z)$ and the noncausal filter $H(z^{-1})$. According to the preceding discussion in fact if $H(z)$ has all the poles inside the unit circle and is therefore stable in the causal form, $H(z^{-1})$ has all the poles outside the unit circle and is therefore stable only if implemented in its noncausal form. Now on the unit circle $H(z^{-1})|_{z=e^{j\omega}} = H^*(z)|_{z=e^{j\omega}}$. Thus the cascade of the two filters has a frequency response

$$H(e^{j\omega})H^*(e^{j\omega}) = |H(e^{j\omega})|^2 \tag{1.118}$$

that is a zero phase frequency response equal to the squared amplitude response of the filter $H(z)$.

The same result can be obtained using only the filter $H(z)$ and inverting the sequence to be filtered, according to the scheme of Fig. 1.10. In this case the following relations can be written

$$\begin{aligned} Y_1(z) &= X(z^{-1}) \\ Y_2(z) &= X(z^{-1})H(z) \\ Y_3(z) &= X(z)H(z^{-1}) \\ Y_4(z) &= X(z)H(z)H(z^{-1}) \end{aligned} \tag{1.119}$$

from which it is seen that the overall frequency response of the system is indeed $|H(e^{j\omega})|^2$.

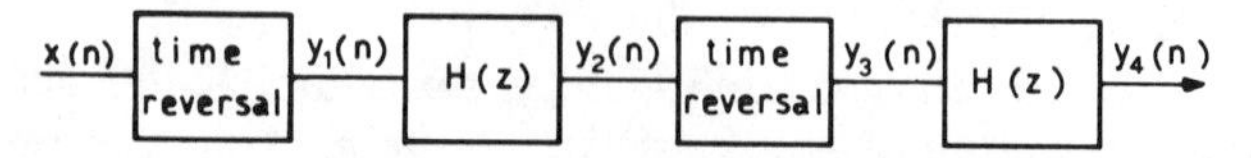

FIG. 1.10. Implementation of a zero phase filter by inverting the sequence of data to be filtered.

IQ 1-D stabilization methods

In the preceding section the general problem of stability was considered and a stability test was obtained based on the pole position evaluation.

Unfortunately it is not generally easy to control the stability of filters during the design procedure. Moreover it is often convenient, when phase is not important, to design filters defined through their squared magnitude transfer function $|H(e^{j\omega})|^2$, which has an expression very simple and suitable for several approximation procedures. In this last case the relation (1.119) shows that this function will contain all the poles of the causal filter and all the poles of the noncausal filter.

Thus the problem of factorization of a squared magnitude function in the stable causal and noncausal component has to be considered, together with the problem of obtaining from an unstable filter a stable causal filter having the same magnitude response.

Several methods of stabilization and factorization are in principle possible. However, due to the very efficient methods for finding the roots of polynomials the most convenient approach is the one based on the pole position evaluation.

At this point the factorization problem can be easily solved. The function $H(z)$ can be obtained by choosing the roots inside the unit circle and the $H(z^{-1})$ using the poles outside the unit circle. Further the stabilization problem can be solved by substituting the poles in the instability region with poles in conjugate symmetric positions with respect to the unit circle (zeros of $H(z)$ can be chosen arbitrarily or on minimum phase basis).

This last stabilization procedure can be proved by starting from the knowledge of *all-pass systems*, that is systems whose frequency response is equal to 1 for all ω. These systems can be constructed by means of a cascade of factors of form

$$\frac{z^{-1}-a^*}{1-az^{-1}} \tag{1.120}$$

having poles and zeros in conjugate reciprocal locations. The all-pass characteristics of such a factor can be easily verified by evaluating its modulus along the unit circle. Let us now consider a non-minimum phase system $H(z)$ with, for example, two poles outside the unit circle in the positions z_0 and z_0^* with $|z_0|>1$.

$H(z)$ can be written as

$$H(z)=H_m(z)\frac{1}{1-z_0z^{-1}}\frac{1}{1-z_0^*z^{-1}} \tag{1.121}$$

where $H_m(z)$ is the minimum phase part. But $H(z)$ can be expressed in the form

$$H(z)=H_m(z)\frac{1}{(1-z_0z^{-1})}\frac{1}{(1-z_0^*z^{-1})}\frac{(z^{-1}-z_0)}{(z^{-1}-z_0)}\frac{(z^{-1}-z_0^*)}{(z^{-1}-z_0^*)}$$

$$=H_m(z)\frac{1}{(z^{-1}-z_0)}\frac{1}{(z^{-1}-z_0^*)}\frac{(z^{-1}-z_0)}{(1-z_0^*z^{-1})}\frac{(z^{-1}-z_0^*)}{(1-z_0z^{-1})}$$

$$=H_{\min}(z)\frac{(z^{-1}-z_0)(z^{-1}-z_0^*)}{(1-z_0^*z^{-1})(1-z_0z^{-1})} \quad (1.122)$$

where $H(z)$ has been expressed as a cascade of a minimum phase function $H_{\min}(z)$ with two all-pass sections. By examining $H_{\min}(z)$ it is seen that this is a minimum phase function having the same amplitude response as $H(z)$ along the unit circle, and it has been obtained by replacing the poles outside the unit circle with poles in conjugate symmetric positions.

IR 2-D digital filter stability

In 2-D the BIBO (bounded input, bounded output) stability criterion for quadrant causal filters can be expressed in the form

$$\sum_{n_1=0}^{\infty}\sum_{n_2=0}^{\infty}|h(n_1, n_2)|<\infty \quad (1.123)$$

where $\{h(n_1, n_2)\}$ is the impulse response of the filter. If the filter is a FIR, the condition (1.123) is always satisfied. Thus FIR filters do not present stability problems. For IIR filters, in 1-D the BIBO stability condition can be related to the position of the poles of the z-transfer function, that is to the region of analyticity for the function $H(z)$. A similar theorem relating the stability of the filter to the position of the zeros of the denominator bivariate polynomial can also be formulated in the 2-D case. This theorem, first presented by Shanks,[19] can be expressed in the following form.

If $B(z_1, z_2)$ is a polynomial in z_1 and z_2, the expansion of $1/B(z_1, z_2)$ in positive powers of z_1 and z_2 converges absolutely if and only if $B(z_1, z_2)$ is not zero for $|z_1|$ and $|z_2|$ simultaneously less or equal to 1.†

This theorem has the same general form as the 1-D theorem, relating the stability of the filter to the position of the zeros of the denominator of the filter. However the stability test is more involved, because in the 2-D case

† It must be noted that the *Shanks theorem*, which we present in its original form, assumes an expansion in positive powers of z_1 and z_2, which is in contrast with the definition of the z-transform we assumed. Obviously with the definition in terms of negative powers, the denominator has to be not zero for $|z_1|\geq 1$ and $|z_2|\geq 1$.

it is not possible in general to factorize the polynomial $B(z_1, z_2)$, due to the lack of an appropriate factorization theorem of algebra. The Shanks stability theorem is very important from a theoretical point of view, but it is not very useful for establishing a directly usable stability test. In fact the theorem requires the mapping of the unit disk $D_1 \equiv \{z_1 : |z_1| \leq 1\}$ in the z_2 plane according to the relation $z_2 = f(z_1)$, obtained from the $B(z_1, z_2) = 0$ equation. If the locus of mapped points obtained from the mapping has no intersection with the disk $D_2 \equiv \{z_2 : |z_2| \leq 1\}$, then the equation $B(z_1, z_2)$ has no solution for $|z_1| \leq 1$ and $|z_2| \leq 1$ and the filter is stable. Unfortunately this technique involves, in principle, the solution of the equation $B(z_1, z_2) = 0$, for an infinite number of points $|z_1| \leq 1$.

Some modified forms of the *Shanks theorem* have been presented as for example the *Huang theorem* a statement of which is as follows.[10]

A causal recursive filter having a z-transform

$$H(z_1, z_2) = \frac{A(z_1, z_2)}{B(z_1, z_2)} \tag{1.124}$$

is stable if and only if

(a) the mapping of $\delta_1 \equiv \{z_1 : |z_1| = 1\}$ in the z_2 plane according to the relation $B(z_1, z_2) = 0$ does not intersect the circle $D_2 \equiv \{z_2 : |z_2| \leq 1\}$.

(b) no point of $D_1 \equiv \{z_1 : |z_1| \leq 1\}$ is mapped to the $z_2 = 0$ point by the $B(z_1, z_2) = 0$ relation.

A test therefore based on the above will involve the mapping of only the circumference δ_1 and the solution of the equation $B(z_1, 0)$, to test the absence of solutions for $|z_1| \leq 1$. However it needs in principle an infinite number of operations. But by means of the *bilinear transformation*

$$p_1 = \frac{1 - z_1}{1 + z_1}, \qquad p_2 = \frac{1 - z_2}{1 + z_2} \tag{1.125}$$

the preceding theorem can be related to a lemma by Ansell, obtaining conditions which can be tested in a finite number of operations. The bilinear transformation maps the $z_2 = 0$ point in the $p_2 = 1$ point and the unit circle $|z_1| \leq 1$ in the $\mathrm{Re}p_1 \geq 0$ region, whilst the function $1/B(z_1, z_2)$ is transformed in a function $E(p_1, p_2)/F(p_1, p_2)$. So the condition (b) $B(z_1, 0) \neq 0$ for $|z_1| \leq 1$ corresponds to the condition $F(p_1, 1) \neq 0$ for $\mathrm{Re}p_1 \geq 0$ and the condition (a) $B(z_1, z_2) \neq 0$ for $|z_1| = 1$ and $|z_2| \leq 1$ corresponds to $F(p_1, p_2) \neq 0$ for $\mathrm{Re}p_1 = 0$ and $\mathrm{Re}p_2 \geq 0$.

By using the bilinear transformation, the Huang theorem can be re-expressed as follows:

"The causal recursive filter $H(z_1, z_2)$ is stable if and only if

(a) for every ω real and finite, the complex polynomial in p_2 $F(j\omega, p_2)$ has no zeros in $\mathrm{Re}p_2 \geq 0$;

(b) the polynomial $F(p_1, 1)$ has no zeros in $\text{Re}p_1 \geq 0$."

The second condition is straightforward in the test. The main problem is now the test of the first condition. This can be done in a finite number of steps with a method proposed by Ansell[2] for a similar test for continuous systems. In the following we describe the procedure to perform the test without the proof, as this can be found in the original Ansell work.

First we express $F(j\omega, j\Omega)$ in the form

$$F(j\omega, j\Omega) = b_0(\omega)\Omega^n + b_1(\omega)\Omega^{n-1} + \ldots + b_n(\omega)$$
$$+ j[a_0(\omega)\Omega^n + a_1(\omega)\Omega^{n-1} + a_n(\omega)] \tag{1.126}$$

where the $a_i(\omega)$ and $b_i(\omega)$ are real polynomials and $a_0(\omega)$ and $b_0(\omega)$ are not identically zero. We define

$$e_{rs}(\omega) = a_r(\omega)b_s(\omega) - a_s(\omega)b_r(\omega) \qquad 0 \leq r, s \leq n \tag{1.127}$$

and we call $D(\omega)$ the symmetric polynomial matrix whose b_{ij} element is obtained by summing all the $e_{rs}(\omega)$ for which $s + r = i + j - 1$ and $s - r > |i - j|$.

It is possible to prove that if all the successive principal minors of $D(\omega)$ are positive for any real ω, the condition (a) of Ansell lemma is satisfied. To test if any minor of $D(\omega)$ is positive in a finite number of steps, the Sturm method can be used. This method can be stated in the following way: "Given a polynomial $f_0(\omega)$ and its derivative $f_1(\omega)$, if we construct the polynomial sequence $f_2(\omega), f_3(\omega) \ldots$, where the polynomial $f_{k+1}(\omega)$ is defined as the remainder of the quotient $f_{k-1}(\omega)/f_k(\omega)$, with the sign inverted, then the number of zeros of $f_0(\omega)$ in the interval (a, b) is given by $n_a - n_b$, where n_a and n_b are the sign changes of the polynomial sequence in the a and b points, respectively. A change in the point a occurs when in a the sign of the k polynomial is different from the sign of the $k+1$ polynomial."

It is obvious that this method can be used to test if any minor $D(\omega)$ is positive definite on the real axis. In fact it is sufficient to compute the value of the minor in any point of the real axis, for example $\omega = 0$ and to construct the sequence of Sturm polynomials. If at the chosen point the minor is positive and the number of variations from $-\infty$ is equal to the number of variations to $+\infty$, then the minor has no zero in the real axis and therefore its sign is >0 everywhere on the real axis. Starting from Huang theorem several methods have been proposed for the stability test, which differ in the algorithms used to test the conditions of the theorem.[1,11]

As an example we present here a very efficient method proposed by Anderson and Jury, which avoids the necessity of the bilinear transformation and is based on the properties of the *Shur–Cohn matrices*.

Given a polynomial of degree n, $\sum_{i=0}^{n} a_i z^{-i}$ $(a_n \neq 0)$ a Hermitian matrix $n \times n$ can be constructed, the Shur–Cohn matrix, with the elements e_{ij} given by

$$e_{ij} = \sum_{k=1}^{i} (a_{n-i+k} a_{n-j+k}^{*} - a_{i-k} a_{j-k}^{*}) \qquad i \leq j \tag{1.128}$$

This matrix has the property that the number of zeros of the polynomial with modulus less than 1 and whose inverse is not a zero, is equal to the number of positive eigenvalues of the matrix. The number of zeros having modulus greater than 1 and whose inverse is not a zero, is equal to the number of negative eigenvalues. The number of zeros having modulus equal to 1 or whose inverse is a zero is equal to the number of zero eigenvalues.

This property of the Shur–Cohn matrices can be used to test the stability conditions of the Huang theorem, since all the eigenvalues have to be negative if all the poles are outside the region $|z_1| \leq 1, |z_2| \leq 1$. However, from the Jacobi theorem we have that the number of negative eigenvalues is equal to the number of sign changes in the sequence $1, D_1, D_2 \ldots D_n$, where all the D_i are the determinants of the principal minors of increasing order. In our case the first determinant has to be negative and the signs of subsequent determinants have to alternate at every step. Thus the odd order determinants must be negative and the even order determinants must be positive.

This test is very easy for the second Huang condition because the matrix will be real, since the coefficient of $B(z_1, 0)$ are real.

The problem is more difficult for the first condition however. In fact $B(z_1, z_2)$ in this case is a polynomial in z_2, the coefficients of which are polynomials in z_1. A Shur–Cohn matrix can then be constructed, whose elements are polynomials in z_1 and z_1^{*} with real coefficients. The condition on the sign of the minors must be valid in this case for all the values of z_1 with $|z_1| = 1$. It can be noted that, since the matrix is Hermitian, its principal minors have the property that the coefficients of the same powers of z_1 and z_1^{*} are equal. Moreover, since the polynomials are evaluated on the unit circle we shall have z_1^{*} equal to z_1^{-1}. Thus the determinants of the principal minors are autoinverse or mirror image polynomials (MIP) and the test consists in verifying a MIP polynomial $f(z_1) = \sum_{k=0}^{N} c_k (z_1^{k} + z_1^{-k})$ that is positive definite along the unit circle.

The problem can be solved by the substitution $z_1 = \mathrm{e}^{j\theta}$, from which we obtain

$$f(\mathrm{e}^{j\theta}) = \sum_{k=0}^{N} 2c_k \cos k\theta, \qquad \theta \in [-\pi, \pi] \tag{1.129}$$

and with $\cos\theta = x$

$$\sum_{k=0}^{N} 2c_k T_k(x) = \sum_{k=0}^{N} d_k x^k \qquad x \in [-1, +1] \tag{1.130}$$

being $T_k(x)$ a Chebyshev polynomial of order k.

The positiveness of (1.130) on $[-1, 1]$ can be verified with the Sturm method considered earlier.

The Shanks theorem is valid only for causal quadrant filters with a finite $\{^1b(n_1, n_2)\}$ coefficient matrix. However it is possible to remove these limitations.

Let us now consider a very important case from a theoretical point of view, of causal filters with the denominator of the transfer function represented by an infinite series. It has been shown by Bednar and Farmer[3] that a causal recursive filter of the form

$$\frac{1}{B(z_1, z_2)} = \frac{1}{\sum_{n_1=0}^{\infty} \sum_{n_2=0}^{\infty} {}^1b(n_1, n_2) z_1^{n_1} z_2^{n_2}} \tag{1.131}$$

with the following condition

$$\sum_{n_1=0}^{\infty} \sum_{n_2=0}^{\infty} |{}^1b(n_1, n_2)| < \infty \tag{1.132}$$

is stable if and only if $B(z_1, z_2) \neq 0$ for both $|z_1|$ and $|z_2|$ less than or equal to 1.

Causal sequences with this analytic region are normally called *minimum phase sequences.*

As far as the causality condition is concerned, we observed in Section IO that four different recursive implementations of the same linear difference equation can be defined, leading to filters defined on the four quadrants. However any noncausal algorithm can be transformed to a causal algorithm with a rotation performed on the input, output and coefficient matrices. Let us consider for example the case of a filtering operation which starts from the right upper corner of the input matrix. The coefficient matrix will be a third quadrant matrix $\{^3b(n_1, n_2)\}$, while the input matrix will be of the form $\{^3x(n_1, n_2)\}$. If we indicate with ${}^3B(z_1, z_2)$ and ${}^3X(z_1, z_2)$ the z-transform of $\{^3b(n_1, n_2)\}$ and $\{^3x(n_1, n_2)\}$ we can write

$${}^3Y(z_1, z_2) = \frac{{}^3X(z_1, z_2)}{{}^3B(z_1, z_2)} \tag{1.133}$$

If we now rotate $\{^3b(n_1, n_2)\}$ and $\{^3x(n_1, n_2)\}$ to get an algorithm recursive in the first quadrant, this algorithm will produce an output

$$^1Y(z_1, z_2) = \frac{^1X(z_1, z_2)}{^1B(z_1, z_2)} = \frac{^3X(z_1^{-1}, z_2^{-1})}{^3B(z_1^{-1}, z_2^{-1})} = {^3Y}(z_1^{-1}, z_2^{-1}) \quad (1.134)$$

and it is evident that the inverse rotation of $^1Y(z_1, z_2)$ will produce the same output as the recursive filter which is defined on the third quadrant.

This means that a noncausal system will be stable if the causal system obtained by rotation is stable. This allows us to use the same stability test for the noncausal systems too.

Finally it can be observed that only one of the four filters obtained starting from the same difference equation can be stable.

Let us now consider the problem of performing IIR 2-D filtering operations with linear phase. This is highly desirable when visual images are to be processed.[9] This is so since the shape of an object in a given image is directly connected to the phase information. Thus a phase error can modify significantly the shape of the objects to be processed in the image. Then in general in 2-D, the signals to be processed are of finite extent and completely stored. This means that the causality conditions are not so important as in most of the 1-D applications, which in general are real time applications.

It is possible to obtain a zero phase filter using a cascade of quadrant IIR filters.

For example if we filter an input matrix with a first quadrant filter $H(z_1, z_2)$ and then in cascade with a third quadrant filter with the same coefficient matrix and consequently with a transfer function $H(z_1^{-1}, z_2^{-1})$ the overall filter has a transfer function

$$H_c(z_1, z_2) = H(z_1, z_2)H(z_1^{-1}, z_2^{-1}) \quad (1.135)$$

In the frequency domain eq. (1.135) can be written

$$H_c(e^{j\omega_1}, e^{j\omega_2}) = H(e^{j\omega_1}, e^{j\omega_2})H(e^{-j\omega_1}, e^{-j\omega_2})$$

and since $H(e^{-j\omega_1}, e^{-j\omega_2}) = H^*(e^{j\omega_1}, e^{j\omega_2})$, the frequency transfer function becomes

$$H_c(e^{j\omega_1}, e^{j\omega_2}) = |H(e^{j\omega_1}, e^{j\omega_2})|^2 \quad (1.136)$$

which is, of course, zero phase.

A zero phase filter can be obtained also by filtering the input matrix with the two filters considered before and by adding the results. In this case the frequency response is given by

$$H_c(e^{j\omega_1}, e^{j\omega_2}) = H(e^{j\omega_1}, e^{j\omega_2}) + H^*(e^{j\omega_1}, e^{j\omega_2}) = 2\,\mathrm{Re}\,[H(e^{j\omega_1}, e^{j\omega_2})] \quad (1.137)$$

These two filtering operations are zero-phase but not symmetric with respect to both axis. If we need a filtering operation which is symmetric with respect to both axis (for example circularly symmetric filters can be obtained in this way), two alternative forms can be used, either

$$H_c(z_1, z_2) = H(z_1, z_2)H(z_1, z_2^{-1})H(z_1^{-1}, z_2)H(z_1^{-1}, z_2^{-1}) \quad (1.138)$$

or

$$H_c(z_1, z_2) = H(z_1, z_2) + H(z_1, z_2^{-1}) + H(z_1^{-1}, z_2) + H(z_1^{-1}, z_2^{-1}) \quad (1.139)$$

The same result can be obtained with suitable half-plane filters as defined in Section IO. In this case however the stability test theory has not yet been developed and only one test has been proposed. This will be considered in the following section for the study of the factorization methods.

IS 2-D stabilization procedures

The same considerations made in the 1-D case about the stabilization or factorization methods can be applied to the 2-D case also. However it is necessary to observe that in the 2-D case the stability tests are much more involved than in the 1-D case. Thus it is more difficult to control the stability of filters in the design procedures and it can be more useful than in the 1-D case not to consider the stability problem in the design stage but to try to stabilize the eventually unstable filters.

Unfortunately also the stabilization procedures are more complicated in the 2-D case. We observed in the study of stability tests that due to the lack of an appropriate factorization theorem of algebra (Section IR), it is impossible to test the stability in 2-D by finding the roots of the denominator polynomials and so determining the analytic regions of the z-transfer function. This fact has a fundamental consequence in the stabilization problem also. In fact it is not possible to stabilize filters simply by substituting the poles in the instability region with suitable poles in the stability region. But it will be necessary to use different techniques based either on the properties of the complex *cepstrum* of minimum phase sequences or on the Hilbert transform relation between magnitude and phase of the Fourier transform of this type of sequences.

In the 1-D case it was shown that an unstable filter can be decomposed in two filters, one inherently stable and the other considered stable when realized in the noncausal form. Moreover when the transfer function was defined as a squared magnitude function, the two filters obtained from it had transfer functions $H(z)$ and $H(z^{-1})$, having the same magnitude transfer function. The problem now is to find if it is possible to obtain the same result in 2-D, that is to divide an unstable filter into several (usually four) filters each stable if realized in a suitable causal or noncausal algorithm.

This is equivalent to a deconvolution problem. In general we want to divide a filter $H(z_1, z_2)$ in a product of filters of the type ${}^1H_1(z_1, z_2)$, ${}^2H_2(z_1, z_2^{-1})$, ${}^3H_3(z_1^{-1}, z_2^{-1})$, ${}^4H_4(z_1^{-1}, z_2)$, each of which is stable if realized through a recursion in the suitable quadrant. This relation in the z domain corresponds in the space domain to the convolution of the four sequences that are the inverse z-transform of the above four transfer functions. Thus the problem corresponds to the problem of finding the four sequences starting from their convolution, with the additional constraint that these sequences are defined on different portions of the (n_1, n_2) plane, which have an intersection only at the boundaries, i.e. on the axis of the (n_1, n_2) plane and that their z-transforms have different analytic regions.

Two approaches of the above are considered, one of which is based on the properties of the *complex cepstrum* of causal minimum phase sequences[7,14] and noncausal sequences, whilst the other is based on the relation between the logarithm of the magnitude and the phase of the causal minimum phase sequence through a suitable two-dimensional *Hilbert transform.*[18]

These problems are very involved and proofs are concerned with properties necessary for the derivation of the procedures for factorization. The complete mathematical theory is however outside the scope of this book. Therefore an outline for a practical use of these will be presented, since the rather involved details of the derivation can be found in specialized literature.[7,14,18]

The complex cepstrum is defined as the inverse z-transform of the complex logarithm of the z-transform of a sequence. That is if $x(n_1, n_2)$ is a 2-D sequence, its *cepstrum* is defined as[6]

$$\hat{x}(n_1, n_2) = Z^{-1}[\ln [Z[x(n_1, n_2)]]] \tag{1.140}$$

Before we proceed to the explanation of how the complex cepstrum can be used in factorization problems, it is necessary to state the conditions for the existence of cepstrum and also its main properties.

If we consider the subclass of sequences whose z-transform is a ratio of polynomials, as the ones in which we are interested, it is possible to show[6] that they have a well defined complex cepstrum provided that (i) the Fourier transform is not equal to zero or infinity at any frequency and (ii) any linear phase component has been eliminated through an appropriate shift of the original sequence. These conditions are necessary to ensure the analyticity of the logarithm along the unit circles.

Obviously if the coefficient matrix on which the above operation is performed is the coefficient matrix of a squared magnitude transfer function the condition (ii) is automatically satisfied, and in most of the design procedures it is possible to introduce the constraint of positiveness in the squared magnitude transfer function.

The main property of the cepstrum of minimum phase quadrant signals is that such sequences have a cepstrum which is quadrant causal, that is according to our definition of causality a cepstrum which is different from zero in the region $n_1 \geq 0$, $n_2 \geq 0$. The same result can be obtained for the noncausal sequences too, provided that they have z-transforms with suitable regions of analyticity. Thus the four quadrant sequences

$$\{{}^1b(n_1, n_2)\} \text{ with } {}^1B(z_1, z_2) \text{ analytic for } |z_1| \leq 1, |z_2| \leq 1$$

$$\{{}^2b(n_1, n_2)\} \text{ with } {}^2B(z_1, z_2) \text{ analytic for } |z_1| \leq 1, |z_2| \geq 1$$

$$\{{}^3b(n_1, n_2)\} \text{ with } {}^3B(z_1, z_2) \text{ analytic for } |z_1| \geq 1, |z_2| \geq 1$$

$$\{{}^4b(n_1, n_2)\} \text{ with } {}^4B(z_1, z_2) \text{ analytic for } |z_1| \geq 1, |z_2| \leq 1$$

have cepstra which are one quadrant sequences.

Let us now consider the case of a squared magnitude transfer function and let us assume that the function is everywhere different from zero. In this case the matrix $\{b(n_1, n_2)\}$ which defines the transfer function can be expressed in the form

$$\{b(n_1, n_2)\} = \{{}^1b(n_1, n_2)\} * \{{}^2b(n_1, n_2)\} * \{{}^3b(n_1, n_2)\} * \{{}^4b(n_1, n_2)\} \quad (1.141)$$

By using the fundamental property of the cepstrum which states that if two sequences convolve, their cepstra, if they exist, sum, it is possible to write for the cepstra the relation

$$\{\hat{b}(n_1, n_2)\} = \{{}^1\hat{b}(n_1, n_2)\} + \{{}^2\hat{b}(n_1, n_2)\} + \{{}^3\hat{b}(n_1, n_2)\} + \{{}^4\hat{b}(n_1, n_2)\} \quad (1.142)$$

Now due to the previous property of cepstra, $\{{}^1\hat{b}(n_1, n_2)\}$, $\{{}^2\hat{b}(n_1, n_2)\}$, $\{{}^3\hat{b}(n_1, n_2)\}$, and $\{{}^4\hat{b}(n_1, n_2)\}$ are one-quadrant and hence it is possible to separate the cepstra corresponding to the different factors and so to reconstruct the original sequences by means of the inverse transformation (from cepstra to sequences). Some indeterminacy in the decomposition can arise at the two axes where two cepstra add and at the origin where four cepstra add. A simple solution to this problem is the following: we let ${}^1\hat{b}(0, 0) = {}^2\hat{b}(0, 0) = {}^3\hat{b}(0, 0) = {}^4\hat{b}(0, 0) = \hat{b}(0, 0)/4$, $\{{}^1b(n_1, 0)\} = \{{}^2b(n_1, 0)\} = \{b(n_1, 0)/2\}$, $\{{}^1b(0, n_2)\} = \{{}^4b(0, n_2)\} = \{b(0, n_2)/2\}$, $\{{}^4b(-n_1, 0)\} = \{{}^3b(-n_1, 0)\} = \{b(-n_1, 0)/2\}$, $\{{}^3b(0, -n_2)\} = \{{}^2b(0, -n_2)\} = \{b(0, -n_2)/2\}$.

The problem now is to compute the complex cepstrum of a sequence $\{b(n_1, n_2)\}$ with $n_1, n_2 = 0, \ldots N-1$. An approximation can be constructed through the 2-D DFT as follows. First the DFT of $\{b(n_1, n_2)\}$ is constructed, then the complex logarithm is computed by taking particular care in the numerical computation to obtain a continuous phase (without jumps at multiples of 2π). At this point the cepstrum can be obtained by means of the inverse DFT to produce the sequence $\{\hat{b}_a(n_1, n_2)\}$. The main problem in this procedure is that the cepstrum is not constrained to be zero outside the region $n_1, n_2 = 0, \ldots N-1$. Thus in reality $\{\hat{b}_a(n_1, n_2)\}$ is equal to the sum of shifted version of $\{\hat{b}(n_1, n_2)\}$, that is we can write

$$\{\hat{b}_a(n_1, n_2)\} = \sum_{i,j}\sum \hat{b}(n_1 + iN, n_2 + jN) \tag{1.143}$$

and the approximation $\{\hat{b}_a(n_1, n_2)\} \simeq \{\hat{b}(n_1, n_2)\}$ is good only if $\{\hat{b}(n_1, n_2)\}$ goes to zero fast enough when n_1 and n_2 approach to N.

The possible non-finiteness of the cepstra and therefore of the sequences which are involved in the convolution is not only a computational problem as seen before, but it is also a more fundamental drawback. In fact to obtain usable recursive filters, some truncation will be necessary. This truncation must be done with particular care: firstly because this truncation modifies the form of the magnitude of the transfer function; secondly because in general it is not possible to guarantee that truncating a minimum phase infinite sequence, the resulting sequence is still minimum phase. Therefore a stability test will be in general necessary after the truncation of the matrix.

The same considerations can be applied also to the decomposition in two unsymmetrical half-plane filters, that is the given coefficient matrix can be considered as the convolution of two half-plane sequences with transfer functions having the analytic properties to be used for a stable recursive operation. In particular, in this case the property of the conservation of the region of definition is valid in going from the sequence to its cepstrum.

We observe that this last property can also be used for the stability test.[7] In fact, for example, if the computed cepstrum of one-quadrant causal sequence is quadrant causal, then the sequence is minimum phase and the corresponding filter is stable. This type of stability test is the only one which has been proposed for half-plane filters.

A second method of factorization is based on the properties of the Fourier transform of causal minimum phase sequences, for which a relation between the logarithm of the amplitude and the phase can be found, as a Hilbert transform relation. A relation of this type can be found also for finite extent sequences, if we define as causal a finite sequence $\{x(n_1, n_2)\}$, $n_1 = 0, \ldots N_1 - 1$, $n_2 = 0, \ldots N_2 - 1$, when

$$x(n_1, n_2) = 0 \qquad \text{for } n_1 \geq N_1/2, \qquad n_2 \geq N_2/2 \tag{1.144}$$

To derive this relation, we start from the definition of the odd and the even parts of the sequence (1.144)

$$x_e(n_1, n_2) = \tfrac{1}{2}[x(n_1, n_2) + x(N_1 - n_1, N_2 - n_2)] \tag{1.145}$$

$$x_o(n_1, n_2)=\tfrac{1}{2}[x(n_1, n_2)-x(N_1-n_1, N_2-n_2)] \tag{1.146}$$

The odd and even parts of the causal sequence are related by

$$x_o(n_1, n_2)=[\text{sgn }(n_1+n_2)+\delta(n_1, n_2)]x_e(n_1, n_2) \tag{1.147}$$

where the 2-D sgn function is defined as

$$\text{sgn }(n_1, n_2)=\begin{cases} 1 & 0<n_1<N_1/2, \quad 0<n_2<N_2/2 \\ -1 & N_1/2<n_1<N_1, \quad N_2/2<n_1<N_2 \\ 0 & \text{elsewhere} \end{cases} \tag{1.148}$$

and the $\delta(n_1, n_2)$ is defined as

$$\delta(n_1, n_2)=\begin{cases} 1 & n_2=0 \quad \text{and} \quad 0<n_1<N_1/2 \\ -1 & n_2=0 \quad \text{and} \quad N_1/2<n_1<N_1 \\ 1 & n_1=0 \quad \text{and} \quad 0<n_2<N_2/2 \\ -1 & n_1=0 \quad \text{and} \quad N_2/2<n_2<N_2 \\ 0 & \text{elsewhere} \end{cases} \tag{1.149}$$

Now using the properties of the 2-D DFT it is possible to write

$$X(k_1, k_2)=\text{DFT }[x(n_1, n_2)]=\text{DFT }[x_e(n_1, n_2)]+\text{DFT }[x_o(n_1, n_2)] \tag{1.150}$$

and in terms of the real and imaginary parts of the DFT,

$$\text{Re }[X(k_1, k_2)]=\text{DFT }[x_e(n_1, n_2)] \tag{1.151}$$

$$\text{Im }[X(k_1, k_2)]=-j \text{ DFT }[x_o(n_1, n_2)] \tag{1.152}$$

Taking the inverse transform of the (1.151) and substituting in (1.147) we get

$$x_o(n_1, n_2)=[\text{sgn }(n_1, n_2)+\delta(n_1, n_2)] \text{ IDFT }[\text{Re } X(k_1, k_2)] \tag{1.153}$$

and transforming the two sides we obtain

$$\text{Im }[X(k_1, k_2)]=-j \text{ DFT }[[\text{sgn }(n_1, n_2)+\delta(n_1, n_2)] \text{ IDFT }[\text{Re }[X(k_1, k_2)]]] \tag{1.154}$$

which is a relation between the real and imaginary part of the DFT of a causal sequence.

If the sequence is not only causal but also minimum phase, the same relation holds for magnitude and phase of the Fourier transform. So that the phase $\{\phi(k_1, k_2)\}$ of the transform of a minimum phase causal sequence can be obtained from the amplitude $\{A(k_1, k_2)\}$ in the form

$$\phi(k_1, k_2)=-j \text{ DFT }[[\text{sgn }(n_1, n_2)+\delta(n_1, n_2)] \text{ IDFT }[\log [A(k_1, k_2)]]] \tag{1.155}$$

Thus given a non-minimum phase finite sequence, it can be augmented with zeros to satisfy the causality conditions and through the DFT the logarithm of the amplitude spectrum can be obtained. Then applying the (1.155), the phase of the spectrum of the minimum phase sequence can be computed by means of an IDFT taken on the obtained spectrum.

The described procedure is very simple and fast. Unfortunately the same problems are present as in the method based on the cepstrum decomposition. Mainly particular care must be taken because of the aliasing which results from the use of DFTs of continuous Fourier transforms and because of the truncation which is necessary to obtain a finite sequence.

IT 1-D recursive filter structures

As shown in Section IN, infinite response linear shift invariant systems can be described by finite difference equations of the type

$$y_n = \sum_{k=0}^{N-1} a_k x(n-k) - \sum_{k=1}^{M-1} b_k y(n-k) \tag{1.156}$$

where $a_k, b_k = a(k), b(k)$ of (1.90) and they have rational z-transforms

$$H(z) = \frac{\sum_{k=0}^{N-1} a_k z^{-k}}{\sum_{k=0}^{M-1} b_k z^{-k}} = \frac{A(z)}{B(z)} \tag{1.157}$$

Let us now consider briefly the problem of synthesis of the filter, that is of the algorithms required to perform the filtering operation.[16]

It is evident that the problem is simpler than in the analogue case, where the knowledge of the system differential equation or its pole and zero pattern is only the first of many steps in the design.

In the digital case the problem is in principle solved from the knowledge of the coefficients a_k and b_k because eq. (1.156) can be directly used to synthesize the system using a digital computer or a special purpose hardware system.

Drawing a block diagram of the difference equation we get Fig. 1.11.† It is evident from the figure that to synthesize the filter we need, in the case $M = N$, $2N$ memory cells and $2N$ products and sums. We have mentioned before that it is possible in principle to synthesize the filter in this way. However, normally other forms are used which differ from this direct method in that a reduction in memory can be achieved or some form of decomposition of $H(z)$ in terms of lower order may be used.

† In Fig. 1.11 and following figs $N = n$ and $M = m$.

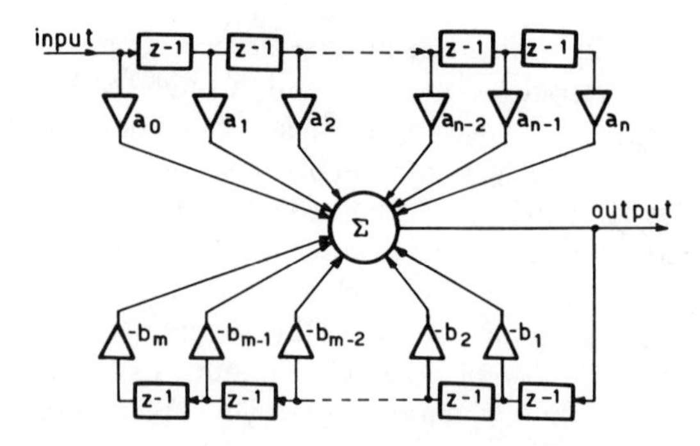

FIG. 1.11. Block diagram of the direct synthesis structure.

Referring to Chapter 4 for a discussion of errors in digital filter synthesis, we remark that these forms may be at times more convenient from the point of view of error behaviour.

Before discussing the decompositions of $H(z)$, we wish to describe a second form of direct synthesis, which is known as the *canonical direct form.* This structure is *canonic* in the sense that it requires a minimum number of delays.

To get this canonic form, we observe that $H(z)$ can be obtained through the cascade of two filters

$$H(z) = H_1(z)H_2(z) = \left(\frac{1}{B(z)}\right) A(z) \tag{1.158}$$

Denoting by $w(n)$ the output of the first filter the two difference equations corresponding to the (1.158) for $M = N$ are the following

$$w(n) = x(n) - \sum_{k=0}^{N-1} b_k w(n-k) \tag{1.159}$$

$$y(n) = \sum_{k=0}^{N-1} a_k w(n-k) \tag{1.160}$$

and it is evident that in this case only N delays are necessary for the quantities $w(n)$, which are normally called the *states* of the system.

The block diagram of the canonic direct synthesis method is shown in Fig. 1.12.

It can be observed now that the $H(z)$ function can be written, if p_i and z_i are its poles and zeros, in the form

$$H(z) = \frac{(z - z_1)(z - z_1^*)\ldots}{(z - p_1)(z - p_1^*)\ldots} \tag{1.161}$$

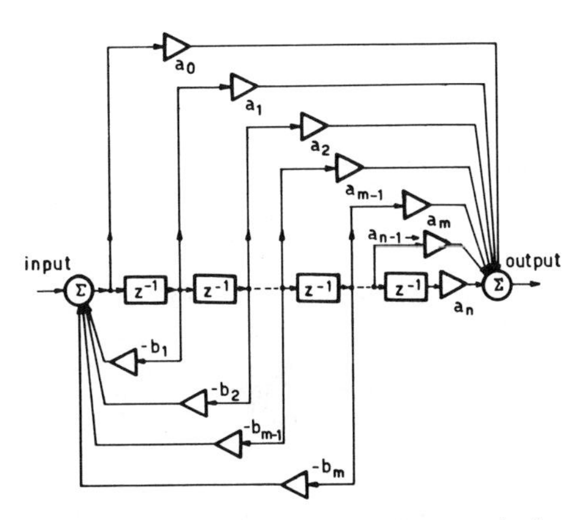

FIG. 1.12. Block diagram of the canonical direct synthesis structure.

and by considering pairs of complex conjugate zeros and poles it is possible to write the $H(z)$ as the cascade of first and second order sections

$$H(z)=A\frac{\prod_i(\alpha_{0i}+\alpha_{1i}z^{-1}+\alpha_{2i}z^{-2})\prod_i(\gamma_{0i}+\gamma_{1i}z^{-1})}{\prod_i(\beta_{0i}+\beta_{1i}z^{-1}+\beta_{2i}z^{-2})\prod_i(\delta_{0i}+\delta_{1i}z^{-1})} \tag{1.162}$$

Each section of the decomposition can be synthesized both in the direct and in the canonic form, obtaining a block diagram as shown in Fig. 1.13.

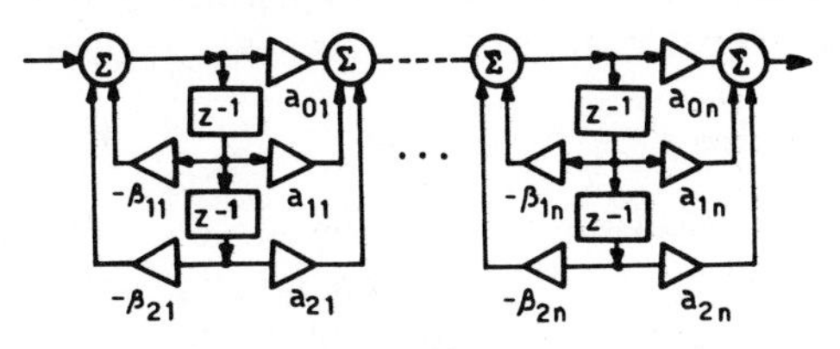

FIG. 1.13. Block diagram of the cascade synthesis structure.

This structure may be convenient from the point of view of error performance even though particular care must be taken in pairing of zeros and poles (see Chapter 4).

A second decomposition is the partial fraction expansion in the form

$$H(z)=c+\sum_{i=1}^{N}H_i(z) \tag{1.163}$$

where $H_i(z)$ are the second order sections of the form

$$H_i(z)=\frac{\gamma_{0i}+\gamma_{1i}z^{-1}}{\beta_{0i}+\beta_{1i}z^{-1}+\beta_{2i}z^{-2}} \tag{1.164}$$

or first order sections of the form

$$H_i(z)=\frac{\gamma_{0i}}{\delta_{0i}+\delta_{1i}z^{-1}} \tag{1.165}$$

from which a synthesis structure of the form shown in Fig. 1.14 is possible.

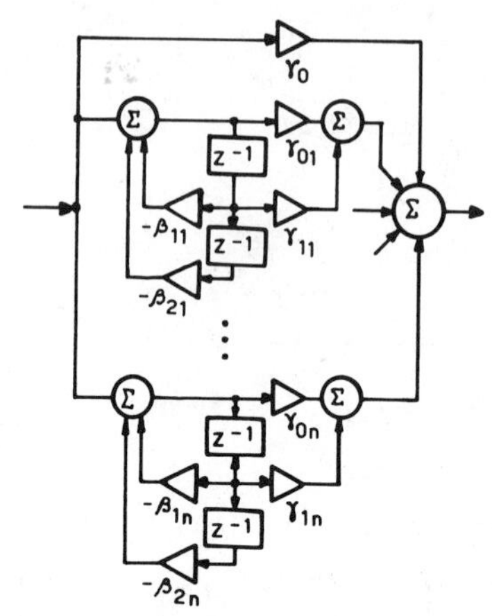

FIG. 1.14. Block diagram of the parallel synthesis structure.

The direct or canonic direct form are not the only structures possible for the synthesis of second order sections. Many other forms have been found but the preceding ones are more commonly used. A very important class of realization structures that possess very low sensitivity is that of *wave digital filters* and associated realizations. These are examined in Chapter 3.

IU 1-D nonrecursive filter structures

We have seen in the previous section how structures for recursive filters may be obtained by expressing a given function in different algebraic forms. The same is true for nonrecursive filters.

The most widely known and used structure corresponds to a *tapped delay line* form as shown in Fig. 1.15. The transfer function is, of course, given by (see eq. (1.96))

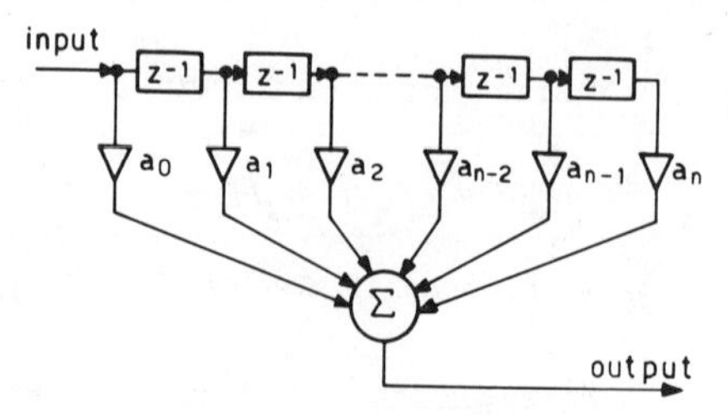

FIG. 1.15. Block diagram of 1-D nonrecursive filter structure, tapped delay-line form.

$$H(z) = a_0 + a_1 z^{-1} + a_2 z^{-2} + \ldots + a_{N-1} z^{-(N-1)} \tag{1.166}$$

where $a_i = a(i)$ of eq. (1.96).

Equation (1.166) can be re-expressed in many forms each of which is likely to produce different structures for $H(z)$.

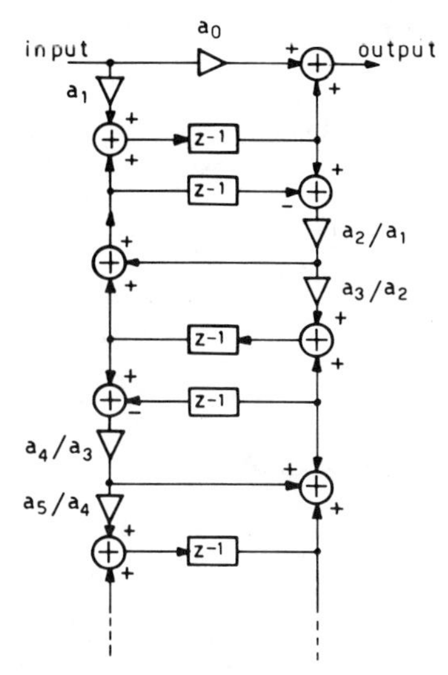

FIG. 1.16. Block diagram of nonrecursive filter structure, defined by using continued fraction expansion (eq. (1.167)).

Interesting structures have also been presented by using interpolation formulas[20] and continued fractions.[4] An example is shown in Fig. 1.16 corresponding to an expression of $H(z)$ given by[4]

$$H(z) = a_0 + \cfrac{a_1 z^{-1}}{1 - \cfrac{z^{-1}}{z^{-1} + \cfrac{\frac{a_1}{a_2}}{1 + \cfrac{\frac{a_3 z^{-1}}{a_2}}{1 - \cfrac{z^{-1}}{z^{-1} + \cfrac{a_3}{a_4} \; \cdots \qquad 1 - \cfrac{z^{-1}}{z^{-1} + \cfrac{a_{N-2}}{a_{N-1}}}}}}}} \tag{1.167}$$

References

1. Anderson, B. D. O. and Jury, E. I. (1973). Stability test of two-dimensional recursive filters. *I.E.E.E. Trans. Audio Electroacoustics* **AU-21**, **4**, 366–72.
2. Ansell, H. G. (1964). On certain two-variable generalizations of circuit theory, with applications to networks of transmission lines and lumped reactances. *I.E.E.E. Trans. Circuit Theory* **CT-11**, **2**, 214–23.
3. Bednar, J. B. and Farmer, C. (1972). Stability of spatial digital filters. *Math. Biosc.* **14**, 113–19.
4. Constantinides, A. G. and Dasgupta, S. (1974). Realization structures for non-recursive digital filters. Paper presented at "Digitale Systeme zur Signalverarbeitung", Erlangen, Germany.
5. Dudgeon, D. E. (1974). Two-dimensional recursive filtering. Sc.D. Dissertation, M.I.T.
6. Dudgeon, D. E. (1975). The existence of cepstra for two-dimensional rational polynomials. *I.E.E.E. Trans. Acoustics, Speech and Signal Processing* **ASSP-23**, **2**, 242–3.
7. Ekstrom, M. P. and Woods, J. W. (1976). Two-dimensional spectral factorization with applications in recursive digital filtering. *I.E.E.E. Trans. Acoustics, Speech and Signal Processing* **ASSP-24**, **2**, 115–28.
8. Freeman, H. (1965). "Discrete Time Systems". Wiley, New York.
9. Huang, T. S., Schreiber, W. F. and Tretiak, O. (1971). Image processing. *I.E.E.E. Proceedings* **59**, **11**, 1586–1609.
10. Huang, T. S. (1972). Stability of two-dimensional recursive filters. *I.E.E.E. Trans. Audio Electroacoustics* **AU-20**, **2**, 158–63.
11. Maria, G. A. and Fahmi, M. M. (1973). On the stability of two dimensional digital filters. *I.E.E.E. Trans. Audio Electroacoustics* **AU-21**, **5**, 470–72.
12. Mersereau, R. M. and Dudgeon, D. E. (1975). Two-dimensional digital filtering. *I.E.E.E. Proceedings* **63**, **4**, 610–23.
13. Oppenheim A. V. and Shafer, R. W. (1975). "Digital Signal Processing". Prentice Hall, Englewood Cliffs, New Jersey.
14. Pistor, P. (1974). Stability criterion for recursive filters. *IBM J. Res. Div.* **18**, **1**, 59–71.
15. Rabiner, L. R. and Gold, B. (1975). "Theory and Application of Digital Signal Processing". Prentice Hall, Englewood Cliffs, New Jersey.
16. Rader, C. M. and Gold, B. (1969). "Digital Processing of Signals". McGraw Hill, New York.
17. Ragazzini, J. R. and Franklin, G. F. (1958). "Sampled Data Control Systems". McGraw-Hill, New York.
18. Read, R. R. and Treitel, S. (1973). The stabilization of two-dimensional recursive filters via the discrete Hilbert transform. *I.E.E.E. Trans. Geoscience Elec.* **GE-11**, **3**, 153–60.
19. Shanks, J. L., Treitel, S. and Justice, J. H. (1972). Stability and synthesis of two-dimensional recursive filters. *I.E.E.E. Trans. Audio Electroacoustics* **AU-20**, **2**, 115–28.
20. Schüssler, W. (1972). On structures for non-recursive digital filters. *AEÜ*, **26**,, 255–8.
21. Treitel, S. and Anderson, E. A. (1964). The stability of digital filters. *I.E.E.E. Trans. Geoscience Elec.* **GE-12**, 6–18.

Chapter 2

Design Methods of Finite Impulse Response Digital Filters

IIA Introduction

In this chapter the main design methods of finite impulse response (FIR) digital filters are presented. Firstly the general transfer functions of 1-D and 2-D linear phase FIR filters are described. Then different design methods of 1-D and 2-D FIR digital filters are presented in detail: filters using windows, filters with frequency sampling, optimum filters, equiripple filters. Finally the methods using transformations from 1-D filters to 2-D filters are considered.

Some typical design examples of the above FIR digital filters are also presented as a guideline for their practical use.

IIB Transfer functions of 1-D linear phase FIR filters

One of the main characteristics of FIR digital filters (viz. eqs (1.95), (1.96)) is that they can be designed to have linear phase. This means that their frequency response can be considered as the product of a pure delay term and a term which is either real or completely imaginary. Let us now consider the conditions for the real impulse response $\{h(n)\}$ to produce linear phase filters.

In Chapter 1 it was seen that: (1) a delay in the impulse response corresponds to a linear phase in the frequency response; (2) a noncausal sequence symmetrical with respect to the origin has a real transfer function whilst a noncausal sequence antisymmetrical with respect to the origin has a completely imaginary transfer function. That is to say that the first case corresponds to a noncausal zero-phase filter, whereas the second case corresponds to a noncausal filter with a constant phase shift of $\pi/2$ radiants.

Consequently the conditions on the impulse response to produce a linear phase causal filter can be studied firstly by considering the conditions necessary to produce a noncausal filter with real or imaginary frequency response and secondly by evaluating the amount of delay that has to be introduced in the noncausal filter to obtain a causal filter. This delay is obviously giving the linear phase of the causal filter.

Four different cases have to be studied corresponding to the number of points N of the impulse response and the type of symmetry (even or odd). These cases are the following:

(1) N odd, even symmetry;
(2) N odd, odd symmetry;
(3) N even, even symmetry;
(4) N even, odd symmetry.

In the following the symmetry conditions on the impulse response of the noncausal and causal filters will be obtained and the form of the frequency response for the causal representation of the four filters will be derived.

IIB1 *N* odd, even symmetry

In this case the zero-phase symmetry for the noncausal filter can be written as

$$h_{nc}(-n) = h_{nc}(n) \qquad n = 0, 1, \ldots, \left(\frac{N-1}{2}\right) \tag{2.1}$$

as shown in Fig. 2.1(a).

The frequency response can be written as

$$H(e^{j\omega}) = h_{nc}(0) + \sum_{n=1}^{(N-1)/2} [h_{nc}(n)\,e^{-j\omega n} + h_{nc}(-n)\,e^{j\omega n}] \tag{2.2}$$

and using the (2.1) we obtain

$$\begin{aligned} H(e^{j\omega}) &= h_{nc}(0) + \sum_{n=1}^{(N-1)/2} h_{nc}(n)(e^{-j\omega n} + e^{j\omega n}) \\ &= h_{nc}(0) + \sum_{n=1}^{(N-1)/2} 2h_{nc}(n) \cos n\omega \\ &= \sum_{n=0}^{(N-1)/2} a(n) \cos n\omega \end{aligned} \tag{2.3}$$

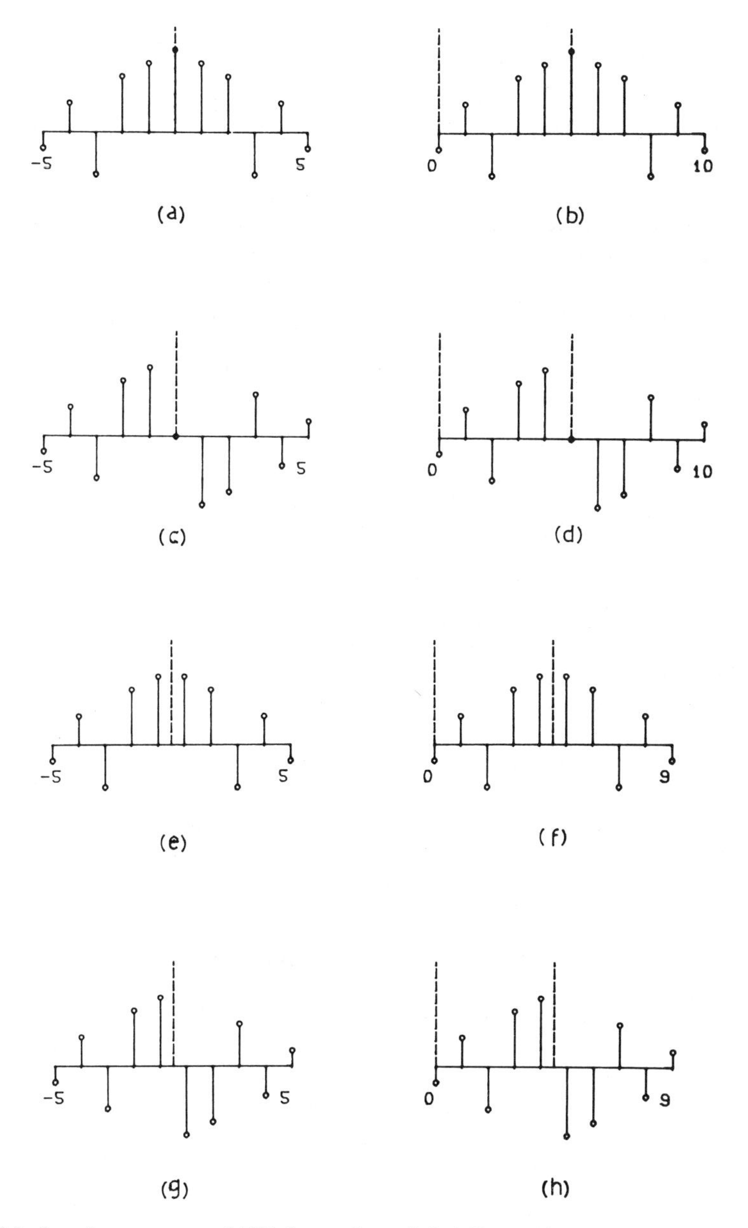

FIG. 2.1. Impulse responses of FIR linear phase digital filters: (a) noncausal filter, N odd, even symmetry; (b) causal filter, N odd, even symmetry; (c) noncausal filter, N odd, odd symmetry; (d) causal filter, N odd, odd symmetry; (e) noncausal filter, N even, even symmetry; (f) causal filter, N even, even symmetry; (g) noncausal filter, N even, odd symmetry; (h) causal filter, N even, odd symmetry.

where

$$a(0) = h_{nc}(0)$$

$$a(n) = 2h_{nc}(n), \qquad n = 1, \ldots, \left(\frac{N-1}{2}\right) \tag{2.4}$$

To obtain the causal impulse response, it is necessary to delay the noncausal impulse response for an interval which corresponds to $[(N-1)/2]$ samples, as can be obtained by observing the Fig. 2.1(a). This delay corresponds to a phase shift of $e^{-j\omega[(N-1)/2]}$ and hence the transfer function of the causal filter can be written as

$$H(e^{j\omega}) = \exp\left[-j\omega\left(\frac{N-1}{2}\right)\right] \sum_{n=0}^{(N-1)/2} a(n)\cos n\omega \tag{2.5}$$

while the symmetry conditions for the impulse as results from Fig. 2.1(b) are

$$h(n) = h(N-1-n) \qquad n = 0, \ldots, (N-1)/2 \tag{2.6}$$

IIB2 *N* odd, odd symmetry

In this case linear phase filters with an imaginary frequency response are obtained. This is possible if the noncausal filter has an impulse response with the symmetry

$$h_{nc}(-n) = -h_{nc}(n) \qquad n = 0, \ldots, (N-1)/2 \tag{2.7}$$

which also implies that

$$h_{nc}(0) = 0 \tag{2.8}$$

as shown in Fig. 2.1(c).

The frequency response using the (2.7) and (2.8) can be written as

$$H(e^{j\omega}) = \sum_{n=1}^{(N-1)/2} [h_{nc}(n)\, e^{-j\omega n} + h_{nc}(-n)\, e^{+j\omega n}]$$

$$= -j \sum_{n=1}^{(N-1)/2} b(n) \sin n\omega \tag{2.9}$$

where

$$b(n) = 2h_{nc}(n) \qquad n = 1, \ldots, (N-1)/2 \tag{2.10}$$

Obviously filters of this type, having a frequency response which is completely imaginary for all frequencies, can be used to design differentiators and Hilbert transformers. However it is important to point out that the frequency response is zero in the points $\omega = 0$ and $\omega = \pi$ independently of the coefficients. Hence filters of this type may not be useful for full-band differentiators or Hilbert transformers.

By introducing the same amount of delay in the impulse response as in the preceding case, the causal filter frequency response can be written in the form

$$H(e^{j\omega}) = -\exp\left[-j\left(\frac{N-1}{2}\right)\omega\right] j \sum_{n=1}^{(N-1)/2} b(n) \sin n\omega \tag{2.11}$$

whilst the causal impulse response will have the symmetry conditions

$$h(n) = -h(N-1-n), \qquad n = 0, 1, \ldots, (N-1)/2 \tag{2.12}$$

with

$$h\left(\frac{N-1}{2}\right) = 0 \tag{2.13}$$

as shown in Fig. 2.1(d).

IIB3 *N* even, even symmetry

It is evident that in this case it is not possible to satisfy the zero or $\pi/2$ phase condition since we need to have a sample of the noncausal impulse response at $n = 0$. It is necessary however to have the origin located between the $N/2$ and $(N/2+1)$ samples as shown in Fig. 2.1(e). In such a case the impulse response with even symmetry will be

$$h_{nc}(-n) = h_{nc}(n) \qquad n = 1, \ldots, N/2 \tag{2.14}$$

and the frequency response has the form

$$H(e^{j\omega}) = \sum_{n=1}^{N/2} [h_{nc}(n)\, e^{-j(n-\frac{1}{2})\omega} + h_{nc}(-n)\, e^{j(n-\frac{1}{2})\omega}] \tag{2.15}$$

and using the (2.14)

$$H(e^{j\omega}) = \sum_{n=1}^{N/2} c(n) \cos[(n-\tfrac{1}{2})\omega] \tag{2.16}$$

where

$$c(n) = 2h_{nc}(n) \qquad n = 1, \ldots, N/2 \tag{2.17}$$

To obtain a causal filter, it is necessary to delay the impulse response of a period $N/2-1/2=(N-1)/2$, which corresponds to a phase factor of $e^{-j[(N-1)/2]\omega}$. Therefore the causal filter frequency response is given by

$$H(e^{j\omega}) = \exp\left[-j\left(\frac{N-1}{2}\right)\omega\right] \sum_{n=1}^{N/2} c(n)\cos\left[(n-\tfrac{1}{2})\omega\right] \tag{2.18}$$

with a causal impulse response having the symmetry

$$h(n)=h(N-1-n) \qquad n=1,\ldots,N/2 \tag{2.19}$$

as shown in Fig. 2.1(f).

In this case the shift $(N-1)/2$, with N even, corresponds to a delay of an integer number $N/2$ of sample intervals and in addition of an interval equal to one half of the distance between two successive samples. This is equivalent to saying that the output signal is not sampled at the same points as the input signal, but it is sampled midway between the samples of the input signal.

IIB4 *N* even, odd symmetry

In the case of an odd symmetry as in Fig. 2.1(g)

$$h_{nc}(-n)=-h_{nc}(n) \qquad n=1,\ldots,N/2 \tag{2.20}$$

the frequency response of the noncausal filter is given by

$$H(e^{j\omega}) = \sum_{n=1}^{N/2} \left[h_{nc}(n)\, e^{-j(n-\frac{1}{2})\omega} + h_{nc}(-n)\, e^{j(n-\frac{1}{2})\omega}\right]$$

$$= -j \sum_{n=1}^{N/2} d(n)\sin\left[(n-\tfrac{1}{2})\omega\right] \tag{2.21}$$

where

$$d(n)=2h_{nc}(n) \qquad n=1,\ldots,N/2 \tag{2.22}$$

By using the same delay as in the preceding case, the causal filter frequency response is obtained in the form

$$H(e^{j\omega}) = -j\exp\left[-j\left(\frac{N-1}{2}\right)\omega\right] \sum_{n=1}^{N/2} d(n)\sin\left[(n-\tfrac{1}{2})\omega\right] \tag{2.23}$$

with an impulse response having the symmetry

$$h(n)=-h(N-1-n) \qquad n=1,\ldots,N/2 \tag{2.24}$$

as shown in Fig. 2.1(h).

The results obtained are summarized in Table 2.1 for the causal case.

Symmetry in $h(n)$	Frequency response
N odd, even symmetry $h(n)=h(N-1-n)$ $n-0,\ldots,\frac{N-1}{2}$	$H(e^{j\omega})=\exp\left[-j\left(\frac{N-1}{2}\right)\omega\right]\sum_{n=0}^{(N-1)/2} a(n)\cos n\omega$ $a(0)=h_{nc}(0)$ $a(n)=2h_{nc}(n) \quad n=1,\ldots\frac{N-1}{2}$
N odd, odd symmetry $h(n)=-h(N-1-n)$ $n=1,\ldots,\frac{N-1}{2}$ $h(N-1)/2=0$	$H(e^{j\omega})=-j\exp\left[-j\left(\frac{N-1}{2}\right)\omega\right]\sum_{n=1}^{(N-1)/2} b(n)\sin n\omega$ $b(n)=2h_{nc}(n) \quad n=1,\ldots\frac{N-1}{2}$
N even, even symmetry $h(n)=h(N-1-n)$ $n=1,\ldots,\frac{N}{2}$	$H(e^{j\omega})=\exp\left[-j\left(\frac{N-1}{2}\right)\omega\right]\sum_{n=1}^{N/2} c(n)\cos[(n-\frac{1}{2})\omega]$ $c(n)=2h_{nc}(n), n=1,\ldots,\frac{N}{2}$
N even, odd symmetry $h(n)=-h(N-1-n)$ $n=1,\ldots,\frac{N}{2}$	$H(e^{j\omega})=-j\exp\left[-j\left(\frac{N-1}{2}\right)\omega\right]\sum_{n=1}^{N/2} d(n)\sin[(n-\frac{1}{2})\omega]$ $d(n)=2h_{nc}(n) \quad n=1,\ldots,\frac{N}{2}$

TABLE 2.1. Frequency responses for different symmetry impulse responses.

IIC Transfer functions of 2-D linear phase FIR filters

As in the 1-D case so in the 2-D case, before embarking on the details of the design methods of FIR digital filters, it is convenient to study in general the form of the frequency response and the symmetry conditions on the impulse response, to produce filters with completely real or completely imaginary frequency responses modified by a linear phase factor. In a manner similar to the 1-D case, to simplify the derivation of the form of the frequency response, we consider a noncausal filter, observing that the causal filter can be obtained from this through a translation that introduces only linear phase term.

IIC1 N_1 and N_2 odd

In this case the frequency response of the noncausal FIR filter with N_1 and N_2 odd can be written in the form

$$H(e^{j\omega_1}, e^{j\omega_2}) = \sum_{n_1=-(N_1-1)/2}^{(N_1-1)/2} \sum_{n_2=-(N_2-1)/2}^{(N_2-1)/2} h_{nc}(n_1, n_2)\, e^{-jn_1\omega_1} e^{-jn_2\omega_2} \quad (2.25)$$

The condition on the coefficients $\{h(n_1, n_2)\}$ to yield a noncausal filter with a real transfer function, derivable from the properties of the 2-D Fourier transform, is the following

$$h_{nc}(-n_1, -n_2) = h_{nc}(n_1, n_2) \tag{2.26}$$

This condition corresponds to a symmetry with respect to the origin of the first and the third quadrant and/or the second and fourth quadrant. On considering the case of an impulse response defined in the first and third quadrant the real transfer function can be written in the following form

$$\begin{aligned} H(e^{j\omega_1}, e^{j\omega_2}) = h_{nc}(0,0) &+ \sum_{\substack{n_1=-(N_1-1)/2 \\ n_1 \neq 0}}^{(N_1-1)/2} h_{nc}(n_1, 0)\, e^{-jn_1\omega_1} \\ &+ \sum_{\substack{n_2=-(N_2-1)/2 \\ n_2 \neq 0}}^{(N_2-1)/2} h_{nc}(0, n_2)\, e^{-jn_2\omega_2} \\ &+ \sum_{n_1=-(N_1-1)/2}^{(N_1-1)/2} \sum_{\substack{n_2=-(N_2-1)/2 \\ n_1+n_2 \neq 0 \\ n_1 \neq 0,\, n_2 \neq 0}}^{(N_2-1)/2} h_{nc}(n_1, n_2)\, e^{-jn_1\omega_1}\, e^{-jn_2\omega_2} \end{aligned} \tag{2.27}$$

and, on using the symmetry conditions (2.26), we have

$$H(e^{j\omega_1}, e^{j\omega_2}) = \sum_{n_1=0}^{(N_1-1)/2} \sum_{n_2=0}^{(N_2-1)/2} \tilde{a}(n_1, n_2) \cos(n_1\omega_1 + n_2\omega_2) \tag{2.28}$$

where

$$\begin{aligned} \tilde{a}(0,0) &= h_{nc}(0,0) \\ \tilde{a}(n_1, n_2) &= 2h_{nc}(n_1, n_2) \end{aligned} \tag{2.29}$$

$$n_1 = 0, \ldots, (N_1 - 1)/2; \; n_2 = 0, \ldots, (N_2 - 1)/2$$

$$n_1 + n_2 \neq 0$$

The function $\cos(n_1\omega_1 + n_2\omega_2)$ becomes $\cos(n_1\omega_1 - n_2\omega_2)$ if the impulse response is defined on the second and fourth quadrants.

It can be observed at this point that the frequency response in this case is real and not symmetrical with respect to the ω_1, ω_2 axes, but only with respect to the origin of the axes, i.e.

$$H(-\omega_1, -\omega_2) = H(\omega_1, \omega_2) \tag{2.30}$$

If a symmetry is necessary with respect to the axes too (as, for example, in the case of circularly symmetric filters) another symmetry condition must be present in the impulse response, namely

$$h_{nc}(n_1, n_2) = h_{nc}(n_1, -n_2) = h_{nc}(-n_1, -n_2) = h_{nc}(-n_1, n_2) \quad (2.31)$$

In this case the transfer function of eq. (2.25) can be written in the form

$$H(e^{j\omega_1}, e^{j\omega_2}) = \sum_{n_1=0}^{(N_1-1)/2} \sum_{n_2=0}^{(N_2-1)/2} a(n_1, n_2) \cos n_1\omega_1 \cos n_2\omega_2 \quad (2.32)$$

where

$$\begin{aligned} a(0, 0) &= h_{nc}(0, 0) \\ a(n_1, 0) &= 2h_{nc}(n_1, 0) \quad n_1 = 1, \ldots, (N_1-1)/2 \\ a(0, n_2) &= 2h_{nc}(0, n_2) \quad n_2 = 1, \ldots, (N_2-1)/2 \\ a(n_1, n_2) &= 4h_{nc}(n_1, n_2) \quad n_1 = 1, \ldots, (N_1-1)/2;\ n_2 = 1, \ldots, (N_2-1)/2 \end{aligned} \quad (2.33)$$

Our assumption concerning the symmetry with respect to the axes is evident from the obtained function. In fact a change in sign of either ω_1 or ω_2 does not change the value of the frequency response due to the symmetry of the cosine function.

To obtain a causal impulse response, the noncausal $h_{nc}(n_1, n_2)$ has to be shifted by $(N_1-1)/2$ intervals in the positive n_1 direction and by $(N_2-1)/2$ points in the positive n_2 direction. This corresponds to a linear phase term

$$\exp\left[-j\left(\frac{N_1-1}{2}\right)\omega_1\right] \exp\left[-j\left(\frac{N_2-1}{2}\right)\omega_2\right] \quad (2.34)$$

and the expressions of eqs (2.28) and (2.32) have to be multiplied by this quantity to produce the frequency response of the causal filter.

Finally the symmetry relations for the impulse response of the causal filter can be written in the form

$$h(n_1, n_2) = h(N_1-1-n_1, N_2-1-n_2) \quad (2.35)$$
$$n_1 = 0, \ldots, (N_1-1)/2; \quad n_2 = 0, \ldots, (N_2-1)/2$$

in the two quadrant case, whilst in the four quadrant case they can be written as

$$\begin{aligned} h(n_1, n_2) &= h(n_1, N_2-1-n_2) = h(N_1-1-n_1, N_2-1-n_2) \\ &= h(N_1-1-n_1, n_2) \end{aligned} \quad (2.36)$$
$$n_1 = 0, \ldots, (N_1-1)/2; \quad n_2 = 0, \ldots, (N_2-1)/2$$

To obtain a completely imaginary frequency response, the symmetry conditions for the noncausal impulse response are of the form

$$h_{nc}(-n_1, -n_2) = -h_{nc}(n_1, n_2) \tag{2.37}$$

which of course imply the condition $h_{nc}(0, 0) = 0$.

The impulse response in the two quadrant case can be written in the form

$$H(e^{j\omega_1}, e^{j\omega_2}) = -j \sum_{n_1=0}^{(N_1-1)/2} \sum_{\substack{n_2=0 \\ n_1+n_2\neq 0}}^{(N_2-1)/2} \tilde{b}(n_1, n_2) \sin(n_1\omega_1 + n_2\omega_2) \tag{2.38}$$

with

$$\tilde{b}(n_1, n_2) = 2h_{nc}(n_1, n_2) \qquad n_1 = 0, \ldots, (N_1-1)/2;\ n_2 = 0, \ldots, (N_2-1)/2 \quad n_1 + n_2 \neq 0 \tag{2.39}$$

If we impose, however, the four quadrant symmetry conditions

$$h_{nc}(n_1, n_2) = -h_{nc}(n_1, -n_2) = -h_{nc}(-n_1, -n_2) = h_{nc}(-n_1, +n_2) \tag{2.40}$$

the following additional constraints are obtained

$$\begin{aligned} &h_{nc}(0, 0) = 0 \\ &h_{nc}(n_1, 0) = 0 \qquad n_1 = -(N_1-1)/2, \ldots, (N_1-1)/2 \end{aligned} \tag{2.41}$$

from which we obtain the frequency response

$$H(e^{j\omega_1}, e^{j\omega_2}) = -j \sum_{n_1=0}^{(N_1-1)/2} \sum_{n_2=1}^{(N_2-1)/2} b(n_1, n_2) \cos n_1\omega_1 \sin n_2\omega_2 \tag{2.42}$$

where

$$\begin{aligned} &b(0, n_2) = 2h_{nc}(0, n_2) \qquad n_2 = 1, \ldots, (N_2-1)/2 \\ &b(n_1, n_2) = 4h_{nc}(n_1, n_2) \quad n_1 = 1, \ldots, (N_1-1)/2;\ n_2 = 1, \ldots, (N_2-1)/2 \end{aligned} \tag{2.43}$$

The function of eq. (2.42) is zero on the ω_1 axis and it changes sign on either side of the ω_2 axis.

As in the previous case, the causal filter frequency response can be obtained by multiplying the expressions (2.38) and (2.42) by the phase term of eq. (2.35).

The causal impulse response has the symmetry in the two quadrant case

$$h(n_1, n_2) = -h(N_1 - 1 - n_1, N_2 - 1 - n_2) \tag{2.44}$$

$$n_1 = 0, \ldots, (N_1-1)/2;\ n_2 = 0, \ldots, (N_2-1)/2$$

with $h[(N_1-1)/2, (N_2-1)/2]=0$, whereas for the four quadrant case we have

$$\begin{aligned} h(n_1, n_2) &= -h(n_1, N_2-1-n_2) = -h(N_1-1-n_1, N_2-1-n_2) \\ &= h(N_1-1-n_1, n_2) \qquad (2.45) \\ n_1 &= 0, \dots, (N_1-1)/2; \qquad n_2 = 0, \dots, (N_2-1)/2-1 \end{aligned}$$

with $h[n_1, (N_2-1)/2]=0$, $n_1=0, \dots, (N_1-1)$.

Obviously this is only one of the four different forms in which the four quadrant symmetry conditions producing a completely imaginary frequency response can be written. For instance, using the symmetry conditions

$$\begin{aligned} h(n_1, n_2) &= h(n_1, N_2-1-n_2) = -h(N_1-1-n_1, N_2-1-n_2) \\ &= -h(N_1-1-n_1, n_2) \qquad (2.46) \\ n_1 &= 0, \dots, (N_1-1)/2-1; \qquad n_2 = 0, \dots, (N_2-1)/2 \end{aligned}$$

with $h[(N_1-1)/2, n_2]=0$, $n_2=0, \dots, N_2-1$, a completely imaginary transfer function which is zero along the ω_2 axis can be obtained.

IIC2 N_1 and N_2 even

Let us now discuss the symmetry conditions necessary to produce linear phase filters for the case when N_1 and N_2 are even. As in the 1-D case, particular care must be taken to define the symmetry conditions. Since N_1 and N_2 are even, it is not possible to satisfy the symmetry conditions of the form considered in the previous section with samples on the n_1 or n_2 grid. The samples of the impulse in fact have to be chosen at the grid points $n_1-\frac{1}{2}$ and $n_2-\frac{1}{2}$.

The derivation of the form of the frequency response with different forms of symmetry conditions is the same as in the case when N_1 and N_2 are odd. We summarize the main results therefore leading directly the symmetry conditions and the consequent transfer functions for the causal filters.

For the two quadrants real linear phase filter the symmetry conditions are the following

$$\begin{aligned} h(n_1, n_2) &= h(N_1-1-n_1, N_2-1-n_2) \qquad (2.47) \\ n_1 &= 0, \dots, (N_1/2-1); \qquad n_2 = 0, \dots, (N_2/2-1) \end{aligned}$$

and the corresponding frequency response is

$$H(e^{j\omega_1}, e^{j\omega_2}) = \exp\left[-j\left(\frac{N_1-1}{2}\right)\omega_1\right] \exp\left[-j\left(\frac{N_2-1}{2}\right)\omega_2\right]$$

$$\times \sum_{n_1=1}^{N_1/2} \sum_{n_2=1}^{N_2/2} \tilde{c}(n_1, n_2) \cos\left[(n_1 - \tfrac{1}{2})\omega_1 + (n_2 - \tfrac{1}{2})\omega_2\right] \qquad (2.48)$$

with

$$\tilde{c}(n_1, n_2) = 2h(N_1/2 - n_1, N_2/2 - n_2) \qquad n_1 = 1, \ldots, N_1/2;\ n_2 = 1, \ldots, N_2/2 \qquad (2.49)$$

Equation (2.48) is of course not symmetrical respect to the axes. However, if this constraint is necessary, the symmetry conditions are

$$h(n_1, n_2) = h(n_1, N_2 - 1 - n_2) = h(N_1 - 1 - n_1, N_2 - 1 - n_2)$$

$$= h(N_1 - 1 - n_1, n_2) \qquad (2.50)$$

$$n_1 = 0, \ldots, N_1/2 - 1; \quad n_2 = 0, \ldots, N_2/2 - 1$$

and can be used to produce a filter whose frequency transfer function is given by

$$H(e^{j\omega_1}, e^{j\omega_2}) = \exp\left[-j\left(\frac{N_1-1}{2}\right)\omega_1\right] \exp\left[-j\left(\frac{N_2-1}{2}\right)\omega_2\right]$$

$$\sum_{n_1=1}^{N_1/2} \sum_{n_2=1}^{N_2/2} c(n_1, n_2) \cos\left[(n_1 - \tfrac{1}{2})\omega_1\right] \cos\left[(n_2 - \tfrac{1}{2})\omega_2\right] \qquad (2.51)$$

where

$$c(n_1, n_2) = 4h(N_1/2 - n_1, N_2/2 - n_2) \qquad (2.52)$$

$$n_1 = 1, \ldots, N_1/2; \quad n_2 = 1, \ldots, N_2/2$$

If a linear phase completely imaginary transfer function is required, the following symmetry conditions for the impulse response can be used in the two quadrant case

$$h(n_1, n_2) = -h(N_1 - 1 - n_1, N_2 - 1 - n_2) \qquad (2.53)$$

$$n_1 = 0, \ldots, N_1/2 - 1; \quad n_2 = 0, \ldots, N_2/2 - 1$$

and the corresponding frequency response becomes

$$H(e^{j\omega_1}, e^{j\omega_2}) = -j \exp\left[-j\left(\frac{N_1-1}{2}\right)\omega_1\right] \exp\left[-j\left(\frac{N_2-1}{2}\right)\omega_2\right]$$

$$\times \sum_{n_1=1}^{N_1/2} \sum_{n_2=1}^{N_2/2} \tilde{d}(n_1, n_2) \sin\left[(n_1 - \tfrac{1}{2})\omega_1 + (n_2 - \tfrac{1}{2})\omega_2\right] \qquad (2.54)$$

where

$$\tilde{d}(n_1, n_2) = 2h(N_1/2 - n_1, N_2/2 - n_2) \qquad n_1 = 1, \ldots, N_1/2;\ n_2 = 1, \ldots, N_2/2 \tag{2.55}$$

Finally for the four quadrant case a linear phase completely imaginary frequency response can be obtained with a causal impulse response satisfying the conditions

$$\begin{aligned} h(n_1, n_2) &= -h(n_1, N_2 - 1 - n_2) = -h(N_1 - 1 - n_1, N_2 - 1 - n_2) \\ &= h(N_1 - 1 - n_1, n_2) \\ n_1 &= 0, \ldots, N_1/2 - 1; \qquad n_2 = 0, \ldots, N_2/2 - 1 \end{aligned} \tag{2.56}$$

The frequency response will then be

$$H(e^{j\omega_1}, e^{j\omega_2}) = -j \exp\left[-j\left(\frac{N_1 - 1}{2}\right)\omega_1\right] \exp\left[-j\left(\frac{N_2 - 1}{2}\right)\omega_2\right] \times \sum_{n_1=1}^{N_1/2} \sum_{n_2=1}^{N_2/2} d(n_1, n_2) \cos\left[(n_1 - \tfrac{1}{2})\omega_1\right] \sin\left[(n_2 - \tfrac{1}{2})\omega_2\right] \tag{2.57}$$

where

$$d(n_1, n_2) = 4h(N_1/2 - n_1, N_2/2 - n_2) \qquad n_1 = 1, \ldots, N_1/2; \quad n_2 = 1, \ldots, N_2/2 \tag{2.58}$$

We can conclude by observing that the causal filter, having N_1 and N_2 even, has an impulse response delayed by a non integer number of samples with respect to the noncausal response. This means that as in the corresponding 1-D case so in the 2-D case the output for this class of filters is not sampled at the same points as the input signal.

IID 1-D digital filter design: the window method

As a first design method of 1-D FIR digital filters (viz. eq. (1.95)) we describe the Fourier series technique. In this method given a frequency function $H(e^{j\omega})$ periodic with period 2π, a FIR digital filter with a transfer function corresponding to $H(e^{j\omega})$ can be obtained simply computing the coefficients of the Fourier series of $H(e^{j\omega})$ and using these coefficients as the impulse response of the filter.[25]

The main problem is that in general the Fourier series so obtained has an infinite number of coefficients, resulting in a filter which of course cannot be implemented. It is necessary therefore in this case to truncate in some way the Fourier series so obtained. This truncation unfortunately leads to the Gibbs phenomenon which causes the appearance of an overshoot in the approximated frequency response near a discontinuity.

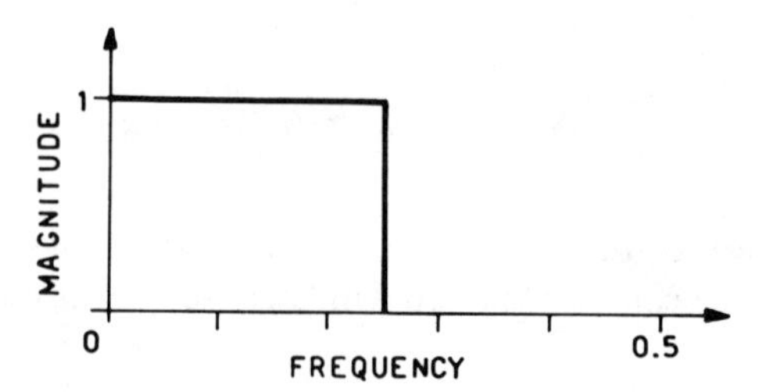

FIG. 2.2. Frequency response of the ideal low-pass filter.

To illustrate this point let us consider the design of a low-pass filter, whose ideal frequency response is shown in Fig. 2.2. The magnitude of this filter is defined equal to unity in the interval $0 \div \omega_c$ and equal to zero in the interval $\omega_c \div \pi$. The impulse response $\{h(n)\}$ can be obtained by computing the coefficients of the Fourier series for $H(e^{j\omega})$ which are given by

$$h(n) = \frac{1}{2\pi} \int_{-\omega_c}^{\omega_c} e^{j\omega n} \, d\omega = \frac{\sin(n\omega_c)}{\pi n} \tag{2.59}$$

This response has the form shown in Fig. 2.3 for $N = 101$.

If now we consider the frequency transfer function of the filter resulting from a truncated Fourier series whose coefficients are as in Fig. 2.3, that is

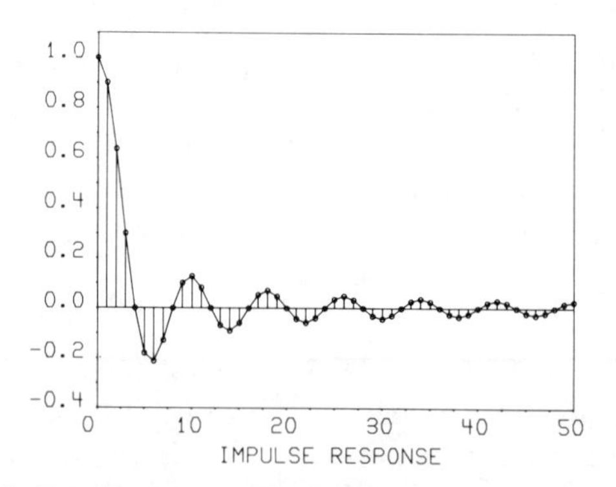

FIG. 2.3. Impulse response of the ideal low-pass filter ($N = 101$).

expression (2.59), we obtain the result of Fig. 2.4. This figure shows the amount of the error due to the Gibbs phenomenon and the fact that a finite transition band substitutes for the discontinuity in the ideal frequency response.

The error introduced by the truncation of the Fourier series, which in the case of an ideal low-pass or band-pass filter is about 9% of the value of the discontinuity, is independent from the length of the impulse response.

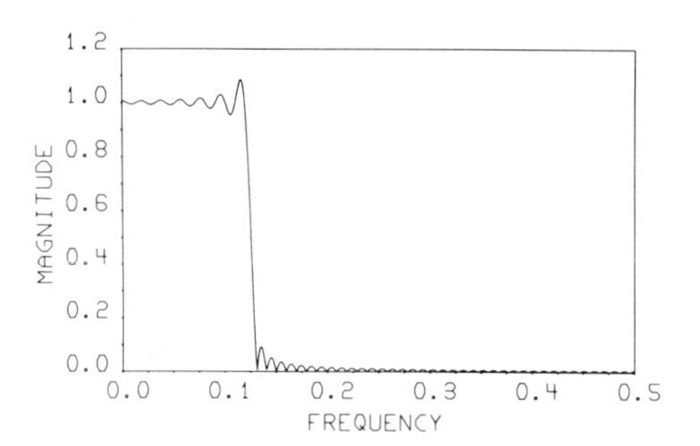

FIG. 2.4. Frequency response of a rectangular-window low-pass filter ($N = 101$).

When the impulse response is lengthened, the maximum error is the same but the oscillations are confined to a smaller frequency range, as can be seen in Fig. 2.5, where the transfer function of the low-pass filter with $N = 201$ is shown.

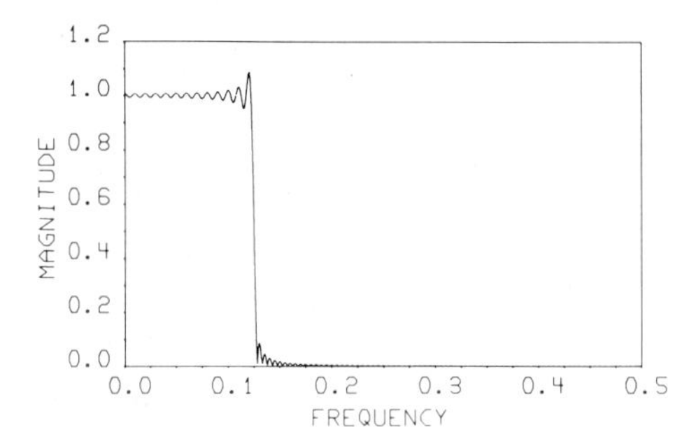

FIG. 2.5. Frequency response of a rectangular-window low-pass filter ($N = 201$).

This effect can be easily appreciated through the convolution theorem. Indeed the truncation of the Fourier series corresponds to a multiplication of the Fourier series with a window defined as†

$$w_N(n) = \begin{cases} 1 & -\left(\dfrac{N-1}{2}\right) \leq n \leq \left(\dfrac{N-1}{2}\right) \\ 0 & \text{otherwise} \end{cases} \tag{2.60}$$

which has a frequency response of the form

$$W_N(\omega) = \frac{\sin(N\omega/2)}{\sin(\omega)} \tag{2.61}$$

shown in Fig. 2.6 for $N = 41$. This function has a main lobe with a width of

† We consider the noncausal form of the filter with N odd.

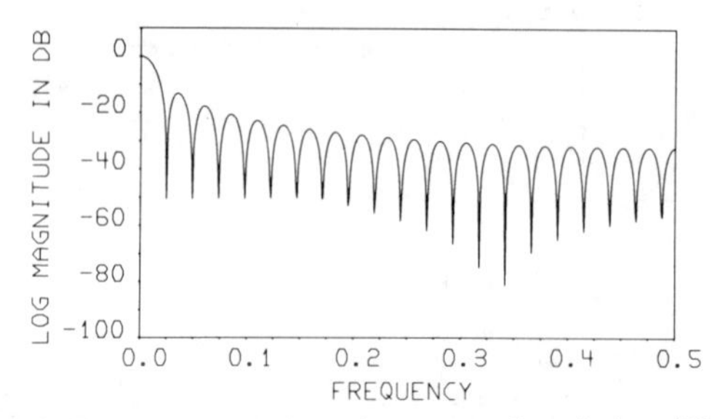

FIG. 2.6. Frequency response of a rectangular window ($N=41$).

$(4\pi)/N$ and a number of sidelobes. When N increases the distance between the zeros decreases but the area under each lobe remains constant.

The transfer function of the truncated filter can be obtained by means of the convolution of the transfer function of the infinite length filter (ideal low-pass filter) and the frequency response of the window. When the point is far from the discontinuity the contribution of the two parts of the frequency window in the convolution integral is roughly equal, giving a very low error. Near the discontinuity two effects are produced by the convolution:

(i) an error in the frequency response due to the unequal contribution of the two parts of the frequency window.

(ii) a smoothing of the discontinuity in a finite transition band.

Obviously the width of the transition band is a function of the width of the frequency window, whilst the ripple is a function of the magnitude of the sidelobes. Recalling the preceding considerations concerning the form of the frequency window corresponding to the rectangular window, it is evident that the error in the frequency response thus obtained is independent from the number N, since it depends on the areas of the sidelobes.

From the above qualitative discussion it is evident that the optimum window should have:[25,37]

(i) a minimum width for the main lobe;

(ii) a minimum area under the sidelobes.

Unfortunately these requirements are not compatible and a trade-off is necessary between the two requirements.

IIE Description of some windows

Several windows $w(n)$ have been proposed and used for the design of FIR digital filters.†

Some of these windows have been first proposed in the spectrum analysis of short periodograms, as the ones listed below.[1,25]

† The coefficients $a(k)$ in (1.95) (1.96) will now become $h(n)w(n)$.

(a) Hanning window

$$w(n)=\frac{1}{2}\left[1+\cos\left(\frac{2\pi n}{N-1}\right)\right] \qquad n=-\frac{N-1}{2},\ldots,\frac{N-1}{2} \tag{2.62}$$

(b) Hamming window

$$w(n)=0.54+0.46\cos\left(\frac{2\pi n}{N-1}\right) \qquad n=-\frac{N-1}{2},\ldots,\frac{N-1}{2} \tag{2.63}$$

(c) Blackman window

$$w(n)=0.42+0.5\cos\left(\frac{2\pi n}{N-1}\right)+0.08\cos\left(\frac{4\pi n}{N-1}\right) \tag{2.64}$$

$$n=-\frac{N-1}{2},\ldots,\frac{N-1}{2}$$

In general these windows, as the above listed ones, are very simple from the point of view of computation, but are not very efficient for the FIR filter design.

Other windows can be defined from the classical works on the Fourier series summation, where some methods have been proposed to speed up the convergence of the Fourier series. It can be shown that some of these correspond to a multiplication of the original coefficients by suitable weights, that is to say they are equivalent to using a window.

The Fejer method for the sum of the Fourier series leads to a window (triangular or Fejer window) of the form[25]

$$w(n)=\left(1-\left|\frac{2n}{N-1}\right|\right) \qquad n=-\frac{N-1}{2},\ldots,\frac{N-1}{2} \tag{2.65}$$

while the Lanczos method of summation leads to a window of the form[7]†

$$w_L(n)=\left(\frac{\sin\dfrac{2n\pi}{N-1}}{\dfrac{2n\pi}{N-1}}\right)^L \tag{2.66}$$

where L is an integer ($L>0$).

This last Lanczos window was extended by Cappellini, by using at the place of L any positive number m.[8] The resulting window $w_m(n)$ is very interesting for its simplicity and for the fact that the parameter m can be easily used to control the trade-off between the width of the main lobe of the transfer function and the area under the sidelobes. Increasing m decreases the amplitude of the sidelobes and the width of the main lobe

† The same limits for n as in (2.65) are valid for this and the following windows.

increases. Hence for any value of N different transition bands and sidelobes amplitudes can be obtained.

Other windows have been defined according to some optimality criterion. Examples of this kind include Dolph–Chebyshev window, Kaiser window, Weber type (Cappellini) window, Papoulis window.

Dolph–Chebyshev window, as defined by Helms,[15] has a frequency expression given by

$$W_D(\omega) = \frac{\cos\,[(N-1)\cos^{-1}(\alpha\cos(\omega T/2))]}{\cosh\,[(N-1)\cosh^{-1}(\alpha)]} \tag{2.67}$$

where α is a positive number and T is the time sampling period.

Kaiser window and Weber type (Cappellini) windows are derived by minimizing uncertainty relations in suitable forms.

As it is known, the classical Heisemberg uncertainty relation is defined as[11]

$$\Delta t \Delta \omega \geq \tfrac{1}{2} \tag{2.68}$$

where

$$\Delta t^2 = \frac{\int_{-\infty}^{\infty} (t-t_0)^2 |x(t)|^2\, \mathrm{d}t}{\int_{-\infty}^{\infty} |x(t)|^2\, \mathrm{d}t} \tag{2.69}$$

$$\Delta \omega^2 = \frac{\int_{-\infty}^{\infty} (\omega-\omega_0)^2 |X(\omega)|^2\, \mathrm{d}\omega}{\int_{-\infty}^{\infty} |X(\omega)|^2\, \mathrm{d}\omega} \tag{2.70}$$

where t_0 and ω_0 are the first order moments of $|x(t)|^2$ and $|X(\omega)|^2$. The equality in the relation (2.68) is true if $x(t)$ (and hence $X(\omega)$) is a Gaussian function.

Landau, Slepian and Pollack have defined the measure of the dispersion of $x(t)$ and $X(\omega)$ in finite time and frequency domains of more practical interest as[32,38]

$$\alpha^2 = \frac{\int_{-\tau}^{\tau} |x(t)|^2\, \mathrm{d}t}{\int_{-\infty}^{\infty} |x(t)|^2\, \mathrm{d}t} \tag{2.71}$$

$$\beta^2 = \frac{\int_{-\Omega}^{\Omega} |X(\omega)|^2\, \mathrm{d}\omega}{\int_{-\infty}^{\infty} |X(\omega)|^2\, \mathrm{d}\omega} \tag{2.72}$$

where τ and Ω are constant (assigned time and frequency extension parameters). Landau, Slepian and Pollack have found the function $x_0(t)$ which gives the maximum of $\alpha^2\beta^2$ and in particular the maximum of β: prolate spheroidal functions were obtained.[32,38]

Kaiser has defined a function, which closely approximates the spheroidal functions, expressed as sampled form by[25]

$$w_K(n) = \frac{I_0\{\omega_a[(N-1)/2]T\sqrt{1-[2n/(N-1)]^2}\}}{I_0(\omega_a[(N-1)/2]T)} \tag{2.73}$$

where I_0 is the modified Bessel function of the first kind and zero order and ω_a is a positive number.

Further by considering real even functions $x(t)$ for which the positive frequency behaviour is sufficient to be analysed,[26] Hilberg and Rothe have found the function minimizing the modified uncertainty relation (by using only positive frequencies in (2.70)): the Fourier transform of this function results to be $W(\omega) = W(-\omega) = \Phi(-\omega)$, being $\Phi(\omega)$, defined for $\omega > 0$, the solution of the second order differential equation (of Weber type)[20]

$$\Phi''(\omega) - \lambda\omega(\omega-2)\Phi(\omega) = 0 \tag{2.74}$$

with $\omega > 0$, $\Phi'(0) = 0$, $\Phi(\infty) = 0$ and being λ the least positive eigenvalue of the equation. The solution of the preceding equation results[20]

$$\Phi(\omega) = aD_\nu(r) + bD_\nu(-r) \tag{2.75}$$

where $r = (\omega-1)\sqrt{2\sqrt{\lambda}}$, $D_\nu(r)$ is the Weber function of the argument r, with $\nu = (1/2)(\sqrt{\lambda}-1)$ and a, b (determined from the condition $\Phi'(0) = 0$ and from the normalization $\Phi(0) = 1$) are given by functions of $D'_\nu(\sqrt{2\sqrt{\lambda}})$ and $\Gamma(\nu)$ (being Γ the Eulero function). The expression of the function $w(t)$ is somewhat complicated: it was evaluated in some points as given by Hilberg and Rothe.[20] Cappellini found a very close approximation to this function as a third-order polynomial[2,3,4]

$$w_C(t) = at^3 + bt^2 + ct + d \tag{2.76}$$

where a, b, c and d are given by Table 2.2 (related to a time interval of 1.5). This window having so simple definition has a very good efficiency and results to be a flexible window by considering $w_C(t_1)$ with $t_1 = at$, a being a positive time expansion or concentration factor.[2]

$0 \le t < 0.75$	$0.75 \le t \le 1.5$
$a = 0.828217$	$a = 0.065062$
$b = -1.637363$	$b = 0.372793$
$c = 0.041186$	$c = -1.701521$
$d = 0.99938$	$d = 1.496611$

TABLE 2.2. Weber type (Cappellini) window.

Papoulis window defined as[30]

$$w_P(n)=\frac{1}{\pi}\left|\sin\left(\frac{2\pi n}{N-1}\right)\right|+\left(2\frac{|n|}{N-1}\right)\cos\left(\frac{2\pi n}{N-1}\right) \tag{2.77}$$

is finally an example of optimum window minimizing the bias integral for spectral estimates.[29] Other windows of this type have been defined.[5]

Figure 2.7 shows a comparison of time behaviour of three windows (limited to negative values): w_1, Weber type (Cappellini) window (Table 2.2); w_2, Kaiser window ($\omega_a(N-1)T=12$); w_3, Lanczos window multiplied by an exponential factor.[8]

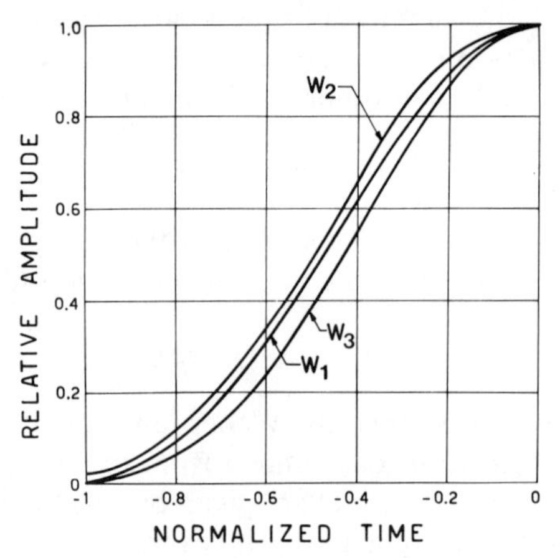

FIG. 2.7. Comparison of time behaviour of three windows (limited to negative values): w_1, Weber type (Cappellini) window (Table 2.2); w_2, Kaiser window ($\omega_a(N-1)T=12$); w_3, Lanczos window multiplied by an exponential factor.

Figure 2.8 shows some examples of frequency response of FIR digital filters using window design technique: (a) with Lanczos window, $L=2$, $N=65$; (b) with Weber type (Cappellini) window (Table 2.2), $N=65$.

IIF 2-D FIR digital filter design: the window method

The window method can also be directly applied to the design of 2-D FIR filters.

If the design of a 2-D filter with a frequency response $H(e^{j\omega_1}, e^{j\omega_2})$ periodic in ω_1 and ω_2 is required, in principle this filter can be synthesized by using as impulse response the sequence $\{h(n_1, n_2)\}$ obtained by computing the coefficients of the 2-D Fourier series of the periodic frequency response.

However to have a practically useful filter it is necessary to truncate the impulse response to obtain a finite matrix, of finite dimensions, for example

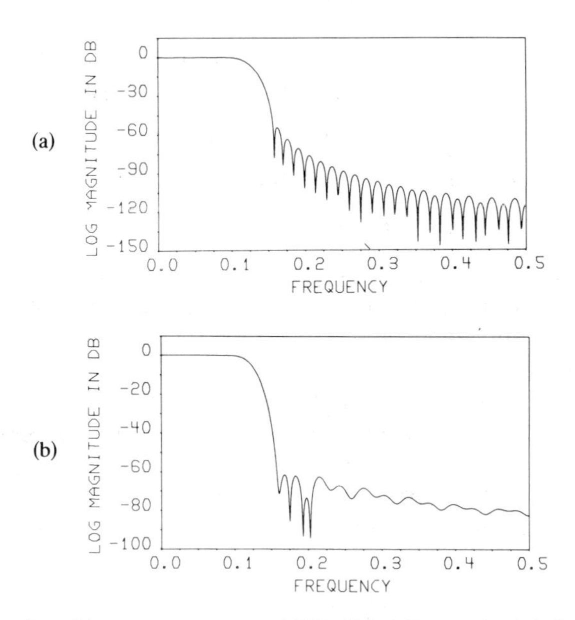

FIG. 2.8. Examples of frequency response of FIR digital filters using window design technique: (a) with Lanczos window, $L=2$, $N=65$; (b) with Weber type (Cappellini) window (Table 2.2.), $N=65$.

$N_1\,N_2$. This can be done by multiplying $h(n_1, n_2)$ by the samples $w(n_1, n_2)$ of a function defined equal to zero outside the region of interest. In this case the impulse response for the noncausal filter with N_1, N_2 odd becomes

$$h_w(n_1, n_2) = h(n_1, n_2)w(n_1, n_2) \tag{2.78}$$

$$n_1 = -\left(\frac{N_1-1}{2}\right), \ldots, \left(\frac{N_1-1}{2}\right); \quad n_2 = -\left(\frac{N_2-1}{2}\right), \ldots, \left(\frac{N_2-1}{2}\right)$$

The frequency response is equal to the convolution of $H(e^{j\omega_1}, e^{j\omega_2})$ and $W(e^{j\omega_1}, e^{j\omega_2})$, where $W(e^{j\omega_1}, e^{j\omega_2})$ is the Fourier transform of $w(n_1, n_2)$.

If we use as window the sequence defined as

$$w(n_1, n_2) = \begin{cases} 1 & n_1 = -\left(\frac{N_1-1}{2}\right), \ldots, \left(\frac{N_1-1}{2}\right); \quad n_2 = -\left(\frac{N_2-1}{2}\right), \ldots, \left(\frac{N_2-1}{2}\right) \\ 0 & \text{otherwise} \end{cases} \tag{2.79}$$

some oscillations will occur in the frequency transfer function of the filter obtained in this manner. It is necessary therefore to find suitable windows as weights to the truncated impulse response to reduce these oscillations.

The choice of the window is straightforward if the filter to design is separable, that is if the frequency response can be written in the form

$$H(e^{j\omega_1}, e^{j\omega_2}) = H_1(e^{j\omega_1})H_2(e^{j\omega_2}) \tag{2.80}$$

and correspondingly in the space domain

$$h(n_1, n_2) = h_1(n_1)h_2(n_2) \tag{2.81}$$

In this case, in fact, the 2-D filter is really the cascade of two 1-D filters, one operating on the rows, and the other on the columns of a given input array. Consequently the same considerations as in 1-D filter design are directly applicable.

A second and very important class of 2-D filters, where the results of 1-D design can be easily generalized, is that of circularly symmetric filters.

In general if $w(t)$ is a good window in the 1-D case and in addition if $W(\omega)$ is narrow with respect to the bands of filters to be designed, then the 2-D circularly symmetric window obtained from a rotation of $w(t)$ in the form[24]

$$w(x, y) = w(\sqrt{x^2 + y^2}) \tag{2.82}$$

is also a good 2-D window.

As an example consider the design of 2-D circularly symmetric low-pass filters having frequency response in the form[6]

$$H(e^{j\omega_1}, e^{j\omega_2}) = \begin{cases} 1 & \sqrt{\omega_1^2 + \omega_2^2} \leq \omega_c \\ 0 & \sqrt{\omega_1^2 + \omega_2^2} > \omega_c \end{cases} \tag{2.83}$$

The coefficients of the Fourier series of this function have the form

$$h(n_1, n_2) = \alpha \frac{J_1(2\pi\alpha\sqrt{n_1^2 + n_2^2})}{\sqrt{n_1^2 + n_2^2}} \tag{2.84}$$

where α is the normalized frequency f_c/f_s and J_1 is the first order Bessel function.

In Fig. 2.9 the frequency response is shown obtained without windowing for $N_1 = N_2 = 16$ and $\alpha = 1/4$. Figure 2.10 shows the result obtained using the Lanczos-extension (Cappellini) window with $m = 1.6$.[6]

IIG 1-D frequency sampling filters

In the previous section we considered the problem of approximation of a continuous periodic frequency response by means of a finite number of coefficients. An approximation through the Fourier series was obtained and some methods for truncating in a useful way the Fourier series were studied.

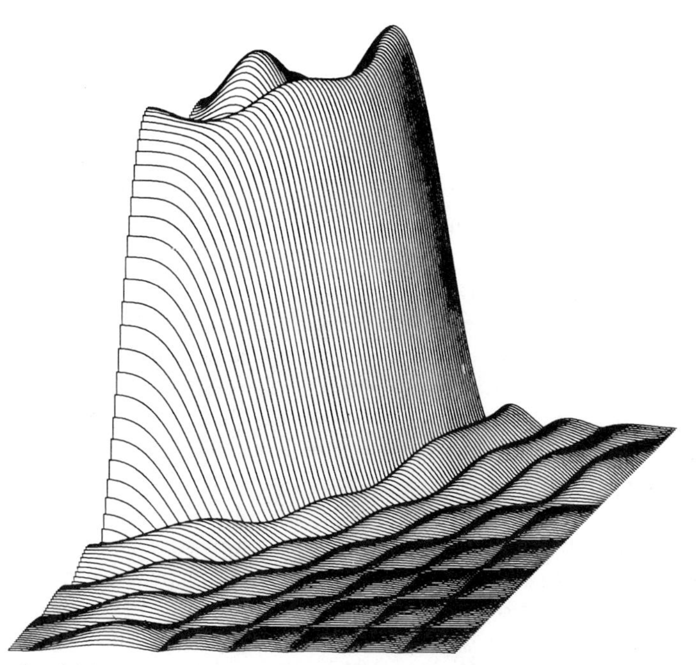

FIG. 2.9. Spatial frequency response (one quadrant) of a circular window 2-D low-pass digital filter ($N_1 = N_2 = 16$).

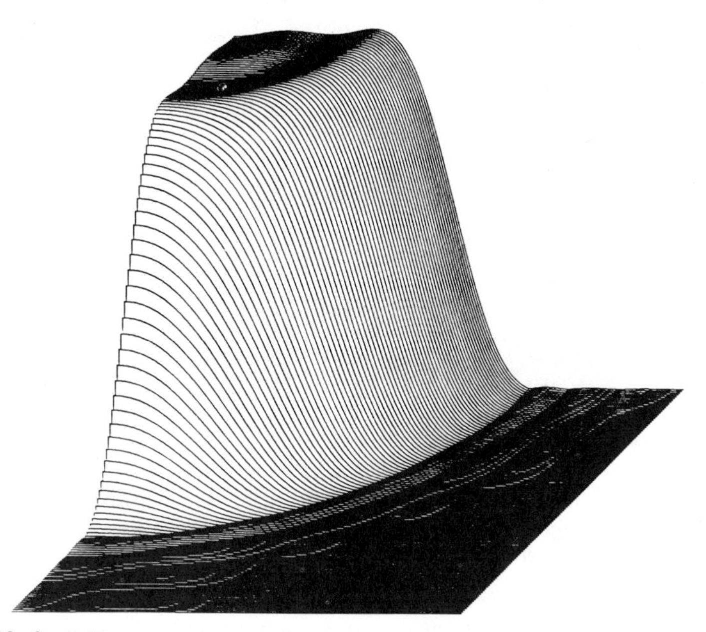

FIG. 2.10. Spatial frequency response (one quadrant) of a circular 2-D low-pass digital filter using Lanczos-extension (Cappellini) window ($N_1 = N_2 = 16$), $m = 1.6$.

With reference to the DFT relations, however, it is evident that these give a direct method for obtaining the impulse response of a filter whose transfer function is identical to N imposed values at N frequency points. In fact if we chose N values of the frequency response $\{H(k)\}$, the finite impulse response of this filter is obtained from the IDFT relation

$$h(n)=\frac{1}{N}\sum_{k=0}^{N-1} H(k)\exp\left(j\frac{2\pi kn}{N}\right) \tag{2.85}$$

This filter naturally has a frequency response which assumes the values $H(k)$ at N equidistant values on the frequency axis and this is in accordance to the invertibility of the discrete Fourier transform (DFT).

The above procedure is known as the *frequency sampling method* and is in principle very simple, reducing to the following basic steps:

(1) choose N equispaced values of the frequency response;

(2) evaluate the IDFT of this frequency samples.

Unfortunately this direct procedure is not of practical interest because it is impossible to predict the frequency response between the chosen samples and moreover the behaviour within these intervals is in most applications not very satisfactory. This effect is produced by the aliasing in the time domain and it can be easily visualized for a low-pass filter design. In Fig. 2.11 N samples of the transfer function are shown and Fig. 2.12 shows the interpolated frequency response.

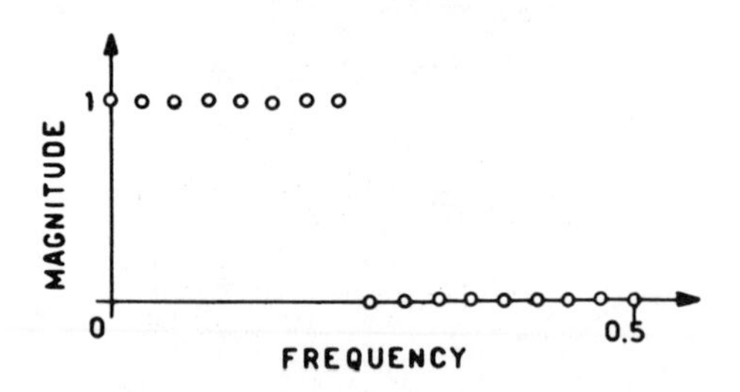

FIG. 2.11. Specification of an ideal low-pass frequency-sampling filter ($N=32$).

Before describing the methods which have been proposed to reduce these ripples it is convenient to write an expression for the interpolated frequency response and to state the conditions which must be satisfied in order to have linear phase filters with a real impulse response.

As mentioned earlier the starting point for the design procedure corresponds to the choice of N frequency samples $\{H(k)\}$, $k=0,\ldots,N-1$, and consequently the N impulse response samples $\{h(n)\}$. The z-transform of the impulse response sequence $\{h(n)\}$, $n=0,\ldots,N-1$, is given by

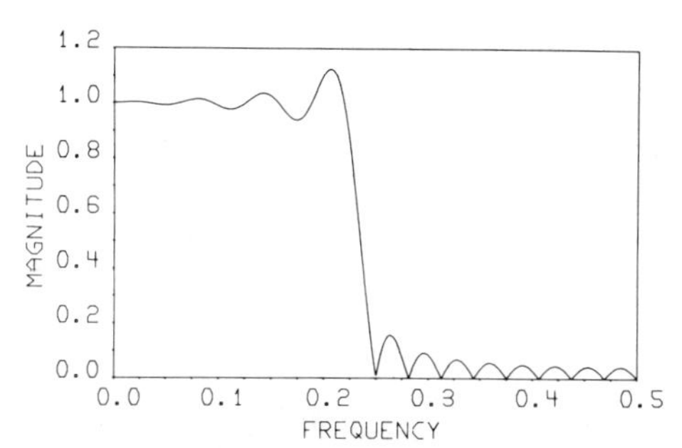

FIG. 2.12. Interpolated frequency response corresponding to the frequency samples of Fig. 2.11.

substituting for $h(n)$ as a function of the $H(k)$ (i.e. the IDFT relationship) to produce

$$H(z)=\sum_{n=0}^{N-1} h(n)z^{-n}=\sum_{n=0}^{N-1}\left[\frac{1}{N}\sum_{k=0}^{N-1} H(k)\exp\left(j\frac{2\pi}{N}kn\right)\right]z^{-n}$$

$$=\sum_{k=0}^{N-1}\frac{H(k)}{N}\sum_{n=0}^{N-1}\left[\exp\left(j\frac{2\pi}{N}k\right)z^{-1}\right]^n=\sum_{k=0}^{N-1}\frac{H(k)}{N}\frac{1-\exp(+j2\pi k)\,z^{-N}}{1-\exp\left(+j\frac{2\pi}{N}k\right)z^{-1}}$$

$$=\frac{(1-z^{-N})}{N}\sum_{k=0}^{N-1}\frac{H(k)}{1-\exp\left(+j\frac{2\pi}{N}k\right)z^{-1}} \tag{2.86}$$

and the frequency response can be obtained by substituting z^{-1} by $e^{-j\omega}$ to yield

$$H(e^{j\omega})=\frac{(1-e^{-jN\omega})}{N}\sum_{k=0}^{N-1}\frac{H(k)}{\left[1-\exp(-j\omega)\exp\left(+j\frac{2\pi}{N}k\right)\right]}$$

$$=\frac{\exp\left(-j\frac{N\omega}{2}\right)\left[\exp\left(j\frac{N\omega}{2}\right)-\exp\left(-j\frac{N\omega}{2}\right)\right]}{N}$$

$$\times\sum_{k=0}^{N-1}\frac{H(k)}{\exp\left(-j\frac{\omega}{2}\right)\exp\left(j\frac{\pi}{N}k\right)\left[\exp\left(j\frac{\omega}{2}\right)\exp\left(-j\frac{\pi}{N}k\right)-\exp\left(-j\frac{\omega}{2}\right)\exp\left(j\frac{\pi}{N}k\right)\right]}$$

$$=\frac{\exp\left(-j\omega\frac{N-1}{2}\right)}{N}\sum_{k=0}^{N-1}\frac{H(k)\exp\left(-j\frac{\pi}{N}k\right)\sin[(N\omega)/2]}{\sin[\omega/2-(\pi k/N)]} \tag{2.87}$$

Equation (2.86) which expresses the z-transform of the filter as a function of the frequency samples is very interesting indeed. It suggests the possibility of a recursive synthesis of a FIR filter.[13,34,35] In fact expression (2.86) corresponds to a cascade of two filters. The first is a nonrecursive one (*comb filter*) with a transfer function

$$H(z)=\frac{1}{N}(1-z^{-N}) \tag{2.88}$$

having N zeros on the unit circle at the positions $z_k=e^{j(2\pi/N)k}$, $k=0,\ldots,N-1$. Its difference equation is given by

$$y(n)=\frac{1}{N}x(n)-\frac{1}{N}x(n-N) \tag{2.89}$$

The second filter in cascade with the above is the parallel combination of N IIR recursive first order sections of the form

$$H(z)=\frac{H(k)}{1-\exp\left(j\frac{2\pi k}{N}\right)z^{-1}} \tag{2.90}$$

These sections have a pole in the point $e^{j(2\pi/N)k}$. If we consider the corresponding section with the pole $z=e^{-j(2\pi/N)k}$, a second order section with real coefficients can be obtained, described by the difference equation

$$y(n)=2A(k)\left\{\cos\theta(k)x(n)-\cos\left[\theta(k)-\frac{2\pi}{N}k\right]x(n-1)\right\}$$
$$+2\cos\left(\frac{2\pi}{N}k\right)y(n-1)-y(n-2) \tag{2.91}$$

where $H(k)=A(k)\,e^{j\theta(k)}$ and $A(k)$, $\theta(k)$ have the symmetry conditions necessary to have a real impulse response. It is very simple to appreciate how this cascade has a finite impulse response. The impulse response of the comb filter is equal to 1 for $n=0$ and equal to -1 for $n=N$ and zero elsewhere. On the other hand the impulse response of the second order section is equal to the periodic repetition of N samples of the cosine function at the points $i\alpha_k$, where $i=0,\ldots,N-1$ and $\alpha_k=(2\pi/N)k$. The second order section is excited at the instant $n=0$ and it begins to oscillate according to the cosinusoidal impulse response. But at the instant N it is excited exactly in the opposite sense. This new excitation then theoretically cancels out the previous oscillations of the second order section. Hence the impulse response of the cascade of the comb filter and any second order section in the structure is finite and it is equal to N samples of a cosine waveform.

Let us now consider the constraints on the frequency samples to produce a real impulse response and a linear phase response. Two different cases for the choice of the frequency samples can be considered. The first choice is equidistant frequency samples at the points

$$f_k = \frac{k}{N} \qquad k = 0, \ldots, N-1 \tag{2.92}$$

where N can be odd or even. This means that a sample will exist at zero frequency. In the odd case the position of the frequency samples is as shown in Fig. 2.13C whereas for N even the position is as in Fig. 2.13A.

In general to have linear phase filters the frequency samples must be symmetrical in magnitude and have a linear antisymmetrical phase within the interval $(-\pi, \pi)$ according to the general properties of the Fourier transform. Due to the nature of the IDFT however it is more convenient to expresss the symmetries in the interval 0–2π rather than in the interval $-\pi$, $+\pi$.

If the frequency samples are written in the form $H(k) = |H(k)|\, e^{j\varphi(k)}$, the symmetries for N odd can be written in the form

$$|H(k)| = |H(N-k)|, \qquad k = 0, \ldots, (N-1)/2$$

$$\varphi(k) = \begin{cases} -\dfrac{2\pi}{N} k \left(\dfrac{N-1}{2}\right) & k = 0, 1, \ldots, \dfrac{N-1}{2} \\[2ex] \dfrac{2\pi}{N}(N-k)\left(\dfrac{N-1}{2}\right) & k = \dfrac{N+1}{2}, \ldots, N-1 \end{cases} \tag{2.93}$$

Introducing these relations in eq. (2.87) we obtain (omitting the linear phase term)

$$\begin{aligned} H(e^{j\omega}) = {} & \frac{|H(0)|}{N} \frac{\sin \dfrac{N\omega}{2}}{\sin \dfrac{\omega}{2}} \\ & + \frac{1}{N} \sum_{k=1}^{(N-1)/2} \frac{|H(k)| \exp\left[-j\dfrac{2\pi}{N} k \left(\dfrac{N-1}{2}\right)\right] \exp\left(-j\dfrac{\pi}{N}k\right) \sin\left(\dfrac{N\omega}{2}\right)}{\sin\left(\dfrac{\omega}{2} - \dfrac{\pi k}{N}\right)} \\ & + \frac{1}{N} \sum_{K=(N+1)/2}^{N-1} \frac{|H(k)| \exp\left[j\dfrac{2\pi}{N}(N-k)\dfrac{N-1}{2}\right] \exp\left(-j\dfrac{\pi}{N}k\right) \sin \dfrac{N\omega}{2}}{\sin\left(\dfrac{\omega}{2} - \dfrac{\pi k}{N}\right)} \end{aligned} \tag{2.94}$$

and, with the substitution $m = N - k$ we have

$$H(e^{j\omega}) = \frac{|H(0)|}{N} \frac{\sin \frac{N\omega}{2}}{\sin \frac{\omega}{2}}$$

$$+ \sum_{k=1}^{(N-1)/2} \frac{|H(k)|}{N} \frac{\exp\left[-j\frac{2\pi}{N}k\left(\frac{N-1}{2}\right)\right] \exp\left(-j\frac{\pi k}{N}\right) \sin \frac{N\omega}{2}}{\sin\left(\frac{\omega}{2} - \frac{\pi k}{N}\right)}$$

$$+ \sum_{m=1}^{(N-1)/2} \frac{|H(N-m)|}{N} \frac{\exp\left[j\frac{2\pi}{N}m\left(\frac{N-1}{2}\right)\right] \exp\left[-j\frac{\pi(N-m)}{N}\right] \sin \frac{N\omega}{2}}{\sin\left(\frac{\omega}{2} - \frac{\pi(N-m)}{N}\right)}$$

$$= \frac{|H(0)|}{N} \frac{\sin \frac{N\omega}{2}}{\sin \frac{\omega}{2}} + \sum_{k=1}^{(N-1)/2} \frac{|H(k)|}{N}$$

$$\times \sin \frac{N\omega}{2} \left[\frac{(-1)^k}{\sin\left(\frac{\omega}{2} - \frac{\pi k}{N}\right)} + \frac{(-1)^k}{\sin\left(\frac{\omega}{2} + \frac{\pi k}{N}\right)} \right] \tag{2.95}$$

Finally, using the trigonometric relations

$$\sin(\alpha + k\pi) = \sin(\alpha - k\pi) = (-1)^k \sin \alpha \tag{2.96}$$

we can obtain

$$H(e^{j\omega}) = \frac{|H(0)|}{N} \frac{\sin \frac{N\omega}{2}}{\sin \frac{\omega}{2}} + \sum_{k=1}^{(N-1)/2} \frac{|H(k)|}{N} \left\{ \frac{\sin\left[N\left(\frac{\omega}{2} - \frac{k\pi}{N}\right)\right]}{\sin\left(\frac{\omega}{2} - \frac{k\pi}{N}\right)} + \frac{\sin\left[N\left(\frac{\omega}{2} + \frac{k\pi}{N}\right)\right]}{\sin\left(\frac{\omega}{2} + \frac{k\pi}{N}\right)} \right\} \tag{2.97}$$

which is the convenient form of the frequency response to be used in the design techniques.

If N is even however the symmetry conditions can be written in the form

$$|H(k)| = |H(N-k)| \qquad k = 0, \ldots, \frac{N}{2} - 1$$

$$\left|H\left(\frac{N}{2}\right)\right| = 0$$

$$\varphi(k) = \begin{cases} -\dfrac{2\pi}{N} k\left(\dfrac{N-1}{2}\right) & k = 0, \ldots, \dfrac{N}{2} - 1 \\ \dfrac{2\pi}{N}(N-k)\left(\dfrac{N-1}{2}\right) & k = \dfrac{N}{2} + 1, \ldots, N-1 \\ 0 & k = \dfrac{N}{2} \end{cases} \tag{2.98}$$

The condition for the $k = N/2$ sample is related to the fact that a linear phase filter with N even must have $H(e^{j\omega}) = 0$ for $\omega = \pi$ as in eq. (2.16).

The corresponding transfer function can be easily written in the form

$$H(e^{j\omega}) = \frac{|H(0)|}{N} \frac{\sin \dfrac{\omega N}{2}}{\sin \dfrac{\omega}{2}} + \sum_{k=1}^{(N/2)-1} \frac{|H(k)|}{N}$$

$$\times \left\{ \frac{\sin\left[N\left(\dfrac{\omega}{2} - \dfrac{\pi k}{N}\right)\right]}{\sin\left(\dfrac{\omega}{2} - \dfrac{\pi k}{N}\right)} + \frac{\sin\left[N\left(\dfrac{\omega}{2} + \dfrac{\pi k}{N}\right)\right]}{\sin\left(\dfrac{\omega}{2} + \dfrac{\pi k}{N}\right)} \right\} \tag{2.99}$$

The choice of the frequency samples at the positions of Fig. 2.13A and Fig. 2.13C (which corresponds to the usual positions of the values of the DFT) is not the only one possible. In fact the samples can be chosen also at the positions

$$f_k = \left(k + \frac{1}{2}\right)\frac{1}{N} \tag{2.100}$$

giving configurations for N odd and N even as shown in Fig. 2.13D and Fig. 2.13B, respectively.

A relation between these frequency samples and the impulse response of the filter can be obtained starting from the definition of the z-transform.

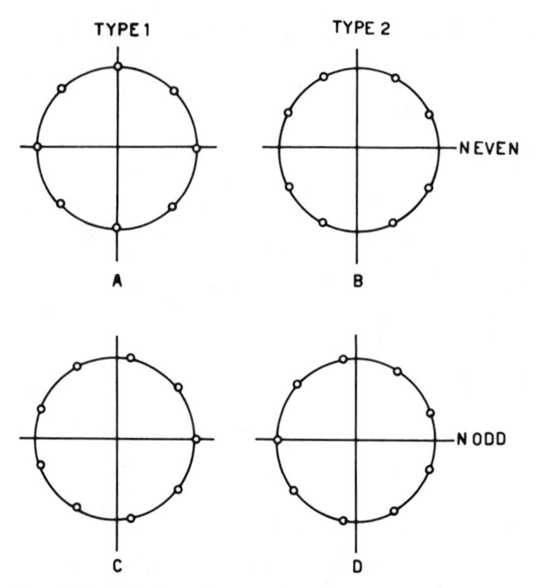

FIG. 2.13. The four different frequency sampling grids.

The frequency samples are the values of $H(z)$ on the unit circle at the points given by eq. (2.100). Hence the expression for $H(k)$ can be written in the form

$$
\begin{aligned}
H(k) &= \sum_{n=0}^{N-1} h(n) \exp\left[-j\frac{2\pi}{N} n\left(k+\frac{1}{2}\right)\right] \\
&= \sum_{n=0}^{N-1} h(n) \exp\left(-j\frac{\pi n}{N}\right) \exp\left(-j\frac{2\pi}{N} nk\right) = \sum_{n=0}^{N-1} g(n) \exp\left(-j\frac{2\pi}{N} nk\right)
\end{aligned}
\tag{2.101}
$$

that is equivalent to saying that the frequency samples can be obtained by means of a DFT taken on the sequence $\{g(n)\}$ defined as

$$g(n) = h(n) \exp\left(-j\frac{\pi n}{N}\right) \qquad n = 0, \ldots, N-1 \tag{2.102}$$

Obviously $g(n)$ is the inverse Fourier transform of the frequency samples $\{H(k)\}$ and therefore we have

$$g(n) = \frac{1}{N} \sum_{k=0}^{N-1} H(k) \exp\left(j\frac{2\pi}{N} nk\right) \qquad n = 0, \ldots, N-1 \tag{2.103}$$

whilst the impulse response $\{h(n)\}$ can be obtained from the relation

$$h(n) = g(n)\exp\left(j\frac{\pi n}{N}\right), \; n=0,\ldots,N-1 \tag{2.104}$$

The z-transform of the filter as a function of the frequency samples, and correspondingly the recursive filter structure, can be obtained by means of eq. (2.104) and eq. (2.103) yielding

$$\begin{aligned}
H(z) &= \sum_{n=0}^{N-1} h(n)z^{-n} = \sum_{n=0}^{N-1} g(n)\exp\left(j\frac{\pi n}{N}\right)z^{-n} \\
&= \sum_{n=0}^{N-1}\left[\frac{1}{N}\sum_{k=0}^{N-1} H(k)\exp\left(j\frac{2\pi}{N}nk\right)\right]\exp\left(j\frac{\pi n}{N}\right)z^{-n} \\
&= \sum_{k=0}^{N-1}\frac{H(k)}{N}\sum_{n=0}^{N-1}\left[\exp\left(j\frac{2\pi}{N}\left(k+\frac{1}{2}\right)\right)z^{-1}\right]^n \\
&= \frac{1-z^{-N}}{N}\sum_{k=0}^{N-1}\frac{H(k)}{1-\exp\left[j\frac{2\pi}{N}\left(k+\frac{1}{2}\right)\right]z^{-1}}
\end{aligned} \tag{2.105}$$

In this case the symmetry conditions for N odd are given by

$$|H(k)| = |H(N-1-k)| \qquad k=0,\ldots,\frac{N-1}{2}-1$$

$$\varphi(k) = \begin{cases} -\dfrac{2\pi}{N}\left(k+\dfrac{1}{2}\right)\dfrac{N-1}{2} & k=0,\ldots,\dfrac{N-1}{2}-1 \\ 0 & k=\dfrac{N-1}{2} \\ \dfrac{2\pi}{N}\left(N-k-\dfrac{1}{2}\right)\dfrac{N-1}{2} & k=\dfrac{N-1}{2},\ldots,N-1 \end{cases} \tag{2.106}$$

and the continuous frequency response is obtained, omitting the linear phase term, as

$$H(e^{j\omega}) = \frac{|H[(N-1)/2]|}{N}\frac{\sin\dfrac{\omega N}{2}}{\sin\dfrac{\omega}{2}} + \sum_{k=0}^{(N-3)/2}\frac{|H(k)|}{N}\left\{\frac{\sin\left\{N\left[\dfrac{\omega}{2}-\dfrac{\pi}{N}\left(k+\dfrac{1}{2}\right)\right]\right\}}{\sin\left[\dfrac{\omega}{2}-\dfrac{\pi}{N}\left(k+\dfrac{1}{2}\right)\right]} + \frac{\sin\left\{N\left[\dfrac{\omega}{2}+\dfrac{\pi}{N}\left(k+\dfrac{1}{2}\right)\right]\right\}}{\sin\left[\dfrac{\omega}{2}+\dfrac{\pi}{N}\left(k+\dfrac{1}{2}\right)\right]}\right\} \tag{2.107}$$

In the even case the symmetry conditions are given by

$$|H(k)| = |H(N-1-k)| \qquad k = 0, \ldots, \frac{N}{2} - 1$$

$$\varphi(k) = \begin{cases} -\dfrac{2\pi}{N}\left(k + \dfrac{1}{2}\right)\dfrac{N-1}{2} & k = 0, \ldots, \dfrac{N}{2} - 1 \\ \dfrac{2\pi}{N}\left(N - k - \dfrac{1}{2}\right)\dfrac{N-1}{2} & k = \dfrac{N}{2}, \ldots, \dfrac{N}{2} - 1 \end{cases} \tag{2.108}$$

with a corresponding continuous frequency response resulting, omitting the linear phase term, as

$$H(e^{j\omega}) = \sum_{k=0}^{(N/2)-1} \frac{|H(k)|}{N} \left\{ \frac{\sin\left\{N\left[\dfrac{\omega}{2} - \dfrac{\pi}{N}\left(k + \dfrac{1}{2}\right)\right]\right\}}{\sin\left[\dfrac{\omega}{2} - \dfrac{\pi}{N}\left(k + \dfrac{1}{2}\right)\right]} + \frac{\sin\left\{N\left[\dfrac{\omega}{2} + \dfrac{\pi}{N}\left(k + \dfrac{1}{2}\right)\right]\right\}}{\sin\left[\dfrac{\omega}{2} + \dfrac{\pi}{N}\left(k + \dfrac{1}{2}\right)\right]} \right\} \tag{2.109}$$

The approximation of a frequency response by means of functions of the form given by equations (2.97), (2.99), (2.107) and (2.109), is in general very poor as far as the inter-sample behaviour is concerned. Some improvements have been proposed to reduce the errors involved in the inter-sample behaviour.

The principle of these procedures can be easily appreciated with reference to the low-pass filter design. It is intuitively apparent that the problems in the approximation are caused by the discontinuity in the frequency response. In the low-pass case such a discontinuity is due to the very narrow transition band between the last sample in the passband taken to be equal to unity and the first sample in the stopband taken to be equal to zero. A discontinuity in the frequency domain implies a long impulse response and hence the error is caused by aliasing in the time domain due to the time limited nature of the impulse response. It is reasonable therefore to consider as in the window method that some reduction of the error can be obtained by increasing the width of the transition band (i.e. by making the discontinuity less abrupt). This can be achieved by allowing some frequency samples within the now lengthened transition band to assume values different from 0 or 1, thereby smoothing the transition between passband and stopband.

Now the problem becomes that of choosing these transition band samples to produce minimum error in the approximation of the passband and stopband.

One of the proposed methods relying on the above is based on a *steepest descent algorithm.*[14,33] A more efficient approach however is based on the formulation of this approximation problem as a *linear programming problem.*[12]

Linear programming theory deals with the problem of selecting between an infinite set of solutions, satisfying an undetermined set of linear conditions, the optimum solution that minimizes or maximizes an *objective function.* The objective function is taken to depend linearly on the variables of the problem.

Mathematically the problem can be formulated in two ways. The first formulation, known as the primal formulation, is as follows.

Find a vector $(x_1, x_2, \ldots, x_N)$ which minimizes the linear form

$$\sum_{i=1}^{N} c_i x_i \tag{2.110}$$

subject to the linear constraints

$$\begin{aligned} x_i &\geq 0 & i &= 1, \ldots, N \\ \sum_{i=1}^{N} a_{ij} x_i &\geq b_j & j &= 1, \ldots, M \qquad M < N \end{aligned} \tag{2.111}$$

This formulation of the problem assumes that the number of variables is greater than the number of linear constraints.

However a second formulation known as the dual formulation of the same linear programming problem can be expressed in the following way.

Find the vector $(y_1, \ldots, y_M)$, subject to the constraints

$$\sum_{j=1}^{M} a_{ij} y_j \leq c_j \qquad i = 1, \ldots, N \tag{2.112}$$

such that the linear functional

$$\sum_{j=1}^{M} b_j y_j \tag{2.113}$$

is maximized.

The importance of the formulation of our problem as a linear programming problem is that it is possible to show that the solution of the problem, if it exists, can be found in a finite number of steps, with a procedure guaranteed to converge. It is not possible here to deal with the many and different aspects of linear programming techniques, but due to their fundamental importance much research has been done and very efficient methods and software programs developed.

Let us now examine how the above digital filter design problem can be formulated as a linear programming problem using as an illustrative example the design of the linear phase low-pass filter.

In this example the approximation domain $(0, \pi)$ is divided in three regions as shown in Fig. 2.14. These regions are: (i) the passband, (1) where the magnitude of the frequency samples is fixed equal to unity; (ii) the stopband (3) where the magnitude of the samples is fixed equal to zero; (iii) the transition band (2), where the magnitude of the frequency samples is allowed to vary to minimize the approximation error. In Fig. 2.14 three samples T_1, T_2 and T_3 are chosen as variable.

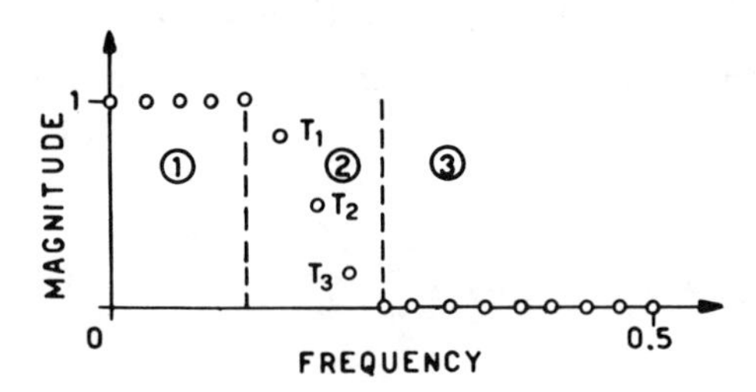

FIG. 2.14. Specifications of a low-pass frequency-sampling filters showing three variable coefficients (T_1, T_2, T_3).

The transfer functions of equations (2.97), (2.99), (2.107) and (2.109) can be written in the form

$$H(e^{j\omega}) = \sum_{k=0}^{M} |H(k)| S_k(e^{j\omega}) \tag{2.114}$$

and hence eq. (2.114) can be interpreted as a weighted sum of interpolation functions. This expression can be reformulated by evidentiating the variable frequency samples $T_k = |H(k)|$ in the form

$$H(e^{j\omega}) = B(e^{j\omega}) + \sum_{m=L_1}^{L_2} T_m S_m(e^{j\omega}) \tag{2.115}$$

where $B(e^{j\omega})$ represents the contribution of the fixed frequency samples to the frequency response.

A set of constraints involving the coefficients $\{T_m\}$ can then be obtained by writing the conditions necessary for the determination of the minimum value of the maximum error defined as

$$|W(e^{j\omega})[D(e^{j\omega}) - H(e^{j\omega})]| \tag{2.116}$$

where ω varies in the regions (1) and (3), $D(e^{j\omega})$ is the desired frequency response and $W(e^{j\omega})$ is an arbitrary positive function which allows us to specify different errors in different regions of the approximation intervals.

By indicating with δ the maximum error in the regions (1) and (3) and sampling the frequency response in a dense grid of frequencies, the following constraints can be obtained from the (2.116)

$$\begin{aligned} W(e^{j\omega_k})[D(e^{j\omega_k})-H(e^{j\omega_k})] &\le \delta \\ -W(e^{j\omega_k})[D(e^{j\omega_k})-H(e^{j\omega_k})] &\le \delta \end{aligned} \qquad k=0,\ldots,N_s \tag{2.117}$$

Upon using expression (2.115) we obtain

$$\begin{aligned} W(e^{j\omega_k})D(e^{j\omega_k})-W(e^{j\omega_k})B(e^{j\omega_k})-W(e^{j\omega_k})\sum_{m=L_1}^{L_2} T_m S_m(e^{j\omega_k}) &\le \delta \\ -W(e^{j\omega_k})D(e^{j\omega_k})+W(e^{j\omega_k})B(e^{j\omega_k})+W(e^{j\omega_k})\sum_{m=L_1}^{L_2} T_m S_m(e^{j\omega_k}) &\le \delta \end{aligned} \tag{2.118}$$

and hence the problem can be formulated in the following way.

Find the coefficients $\{T_m\}$ subject to the constraints

$$\begin{aligned} -W(e^{j\omega_k})\sum_{m=L_1}^{L_2} T_m S_m(e^{j\omega_k})-\delta &\le -W(e^{j\omega_k})D(e^{j\omega_k})+W(e^{j\omega_k})B(e^{j\omega_k}) \\ W(e^{j\omega_k})\sum_{m=L_1}^{L_2} T_m S_m(e^{j\omega_k})-\delta &\le W(e^{j\omega_k})D(e^{j\omega_k})-W(e^{j\omega_k})B(e^{j\omega_k}) \end{aligned} \tag{2.119}$$

so that $-\delta$ is maximized.

Obviously this last formulation is in the form of the dual problem of linear programming.

As an example of this design technique, in Fig. 2.15(a) and Fig. 2.15(b) are presented the results obtained by optimizing one and three coefficients in the transition band respectively.

IIH 2-D frequency sampling technique

The frequency sampling technique can also be applied to the design of 2-D FIR digital filters. The starting point of the procedure is the choice of a matrix of $N_1 \cdot N_2$ samples of the frequency response $\{H(k_1, k_2)\}$, $k_1 = 0, \ldots, N_1-1$, $k_2=0, \ldots, N_2-1$, from which the impulse response of the filter can be obtained using the 2-D inverse discrete Fourier transform (IDFT) from eq. (1.85).

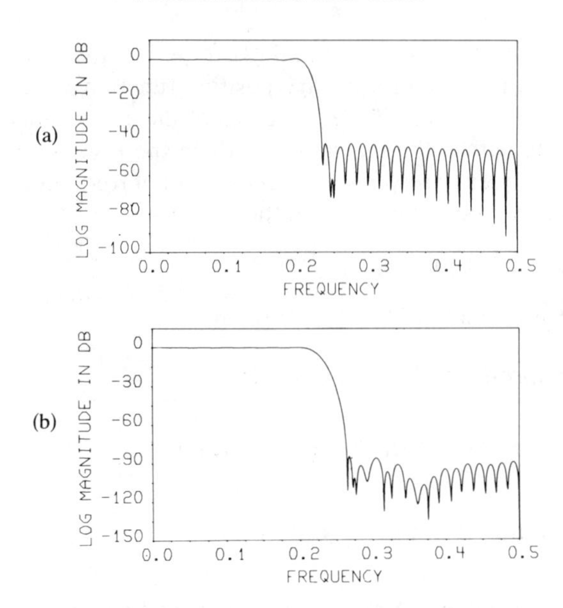

FIG. 2.15. Examples of frequency response of frequency-sampling low-pass digital filters: (a) $N=64$, 1 variable coefficient; (b) $N=64$, 3 variable coefficients.

The frequency response of the filter has the form of eq. (1.103) and on substituting for $h(n_1, n_2)$ it can be expressed as a function of the $\{H(k_1, k_2)\}$ as follows[23]

$$H(e^{j\omega_1}, e^{j\omega_2}) = \sum_{n_1=0}^{N_1-1}\sum_{n_2=0}^{N_2-1}\left[\frac{1}{N_1N_2}\sum_{k_1=0}^{N_1-1}\sum_{k_2=0}^{N_2-1} H(k_1,k_2)\exp\left[j2\pi\left(\frac{k_1n_1}{N_1}+\frac{k_2n_2}{N_2}\right)\right]\right] \times \exp\left[-j(n_1\omega_1+n_2\omega_2)\right] \tag{2.120}$$

which on summing over n_1 and n_2 results

$$H(e^{j\omega_1}, e^{j\omega_2}) = \frac{1}{N_1N_2}\sum_{k_1=0}^{N_1-1}\sum_{k_2=0}^{N_2-1} H(k_1, k_2) \times \frac{1-\exp(-jN_1\omega_1)}{1-\exp\left[+j\left(\frac{2\pi k_1}{N_1}-\omega_1\right)\right]}\,\frac{1-\exp(-jN_2\omega_2)}{1-\exp\left[+j\left(\frac{2\pi k_2}{N_2}-\omega_2\right)\right]}$$

$$= \sum_{k_1=0}^{N_1-1}\sum_{k_2=0}^{N_2-1} H(k_1, k_2)B(k_1, k_2, \omega_1, \omega_2) \tag{2.121}$$

where

$$B(k_1, k_2, \omega_1, \omega_2) = \frac{1}{N_1 N_2} \frac{1-\exp(-jN_1\omega_1)}{1-\exp\left[+j\left(\frac{2\pi k_1}{N_1}-\omega_1\right)\right]} \frac{1-\exp(-jN_2\omega_2)}{1-\exp\left[+j\left(\frac{2\pi k_2}{N_2}-\omega_2\right)\right]} \tag{2.122}$$

The above relation is the basis for the design of 2-D FIR digital filters using the frequency sampling technique as it expresses the interpolated frequency response as a linear combination of its frequency samples.

Let us now consider the symmetries in the frequency samples to yield a linear phase filter and a real impulse function.

By expressing the frequency samples in the form

$$H(k_1, k_2) = |H(k_1, k_2)|\, e^{j\varphi(k_1, k_2)} \tag{2.123}$$

the symmetry conditions are the following

$$|H(k_1, k_2)| = |H(k_1, N_2 - k_2)| = |H(N_1 - k_1, k_2)| = |H(N_1 - k_1, N_2 - k_2)|$$

$$\varphi(k_1, k_2) = \varphi(k_1) + \varphi(k_2) \tag{2.124}$$

with

$$\varphi(k_1) = \begin{cases} -\dfrac{2\pi}{N_1} k_1 \left(\dfrac{N_1 - 1}{2}\right) & k_1 = 0, \ldots, N_u \\ \dfrac{2\pi}{N_1}(N_1 - k_1)\left(\dfrac{N_1 - 1}{2}\right) & k_1 = N_u + 1, \ldots, N_1 - 1 \end{cases}$$

$$\varphi(k_2) = \begin{cases} -\dfrac{2\pi}{N_2} k_2 \left(\dfrac{N_2 - 1}{2}\right) & k_2 = 0, \ldots, N_v \\ \dfrac{2\pi}{N_2}(N_2 - k_2)\left(\dfrac{N_2 - 1}{2}\right) & k_2 = N_v + 1, \ldots, N_2 - 1 \end{cases} \tag{2.125}$$

and

$$N_u = \begin{cases} N_1/2 & N_1 \text{ even} \\ (N_1 - 1)/2 & N_1 \text{ odd} \end{cases}$$

$$N_v = \begin{cases} N_2/2 & N_2 \text{ even} \\ (N_2 - 1)/2 & N_2 \text{ odd} \end{cases} \tag{2.126}$$

with the additional constraints that if N_1 and N_2 are even, then

$$\varphi\left(k_1, \frac{N_2}{2}\right) = \varphi\left(\frac{N_1}{2}, k_2\right) = 0$$

$$\left|H\left(k_1, \frac{N_2}{2}\right)\right| = \left|H\left(\frac{N_1}{2}, k_2\right)\right| = 0 \tag{2.127}$$

By imposing these conditions on the above, the following expression can be obtained

$$H(e^{j\omega_1}, e^{j\omega_2}) = \exp\left(-j\left[\left(\frac{N_1-1}{2}\right)\omega_1 + \left(\frac{N_2-1}{2}\right)\omega_2\right]\right)$$

$$\times \frac{1}{N_1 N_2} \sum_{k_1=0}^{N_{12}} \sum_{k_2=0}^{N_{22}} |H_1(k_1, k_2)| \beta(\omega_1, k_1, N_1) \beta(\omega_2, k_2, N_2) \tag{2.128}$$

where

$$\begin{aligned} |H_1(0, 0)| &= |H(0, 0)|/4 \\ |H_1(0, k_2)| &= |H(0, k_2)|/2 \\ |H_1(k_1, 0)| &= |H(k_1, 0)|/2 \\ |H_1(k_1, k_2)| &= |H(k_1, k_2)| \end{aligned} \tag{2.129}$$

and

$$\beta(\omega, k, N) = \frac{\sin\left[\left(\frac{\omega}{2} - \frac{\pi k}{N}\right)N\right]}{\sin\left(\frac{\omega}{2} - \frac{\pi k}{N}\right)} + \frac{\sin\left[\left(\frac{\omega}{2} + \frac{\pi k}{N}\right)N\right]}{\sin\left(\frac{\omega}{2} + \frac{\pi k}{N}\right)} \tag{2.130}$$

with

$$\begin{aligned} N_{12} &= \begin{cases} N_1/2 - 1 & N_1 \text{ even} \\ (N_1 - 1)/2 & N_1 \text{ odd} \end{cases} \\ N_{22} &= \begin{cases} N_2/2 - 1 & N_2 \text{ even} \\ (N_2 - 1)/2 & N_2 \text{ odd} \end{cases} \end{aligned} \tag{2.131}$$

This expression produces a frequency response equal to the chosen frequency samples at the sampling points and an oscillatory behaviour between these samplint points. Hence it is necessary to use some techniques to reduce these passband and stopband oscillations.

As an example let us consider the problem of designing a low-pass circularly symmetric filter, with frequency samples defined on a matrix of N rows and N columns for $N = 16$.

The first quadrant as shown in Fig. 2.16, where the filter is observed, is defined when the radii R_1 and R_2 are fixed. The radius R_1 defines the passband whereas the radius R_2 defines the transition band. The samples in the region $(\omega_1^2 + \omega_2^2)^{1/2} \leq R_1$ are fixed equal to unity in magnitude whilst the samples in the region $R_1 < (\omega_1^2 + \omega_2^2)^{1/2} \leq R_2$ are allowed to vary freely

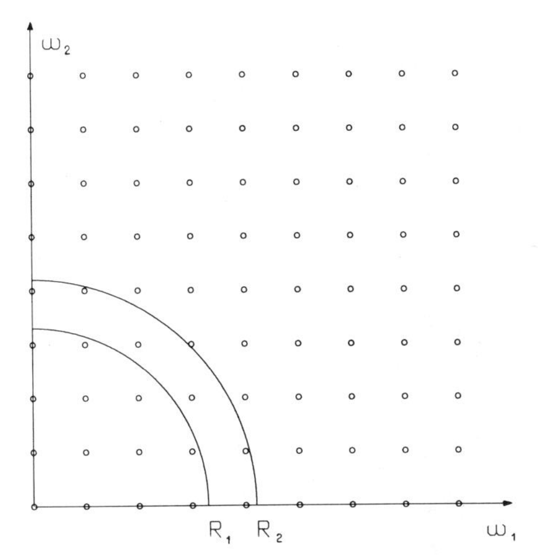

FIG. 2.16. Two-dimensional frequency sampling grid for a filter with $N_1 = N_2 = 16$.

so that a minimization of the error in the passband and/or in the stopband can be achieved.

The expression of the interpolated frequency response however can be written in the form

$$H(e^{j\omega_1}, e^{j\omega_2}) = H_f(e^{j\omega_1}, e^{j\omega_2}) + \sum_{k=1}^{M} T_k H_k(e^{j\omega_1}, e^{j\omega_2}) \tag{2.132}$$

where $H_f(e^{j\omega_1}, e^{j\omega_2})$ is the part corresponding to the fixed frequency samples. Hence in a manner similar to the 1-D case a linear programming system can be constructed, save that in this case a dense grid of points in 2-D must be chosen to produce the constraint equations. Thus the number of points and consequently the number of constraint equations may be very high and the procedure may be time consuming.

In Fig. 2.17(a) the filter with $R_1 = 3.3$, $R_2 = 4.2$, obtained without any optimization where the transition band samples are also fixed at 1, is shown. In Fig. 2.17(b) the result of the optimization with the linear programming technique is shown, where the samples between $R_1 = 3.3$ and $R_2 = 4.2$ are allowed to vary.[6]

III 1-D optimum filter design

The frequency sampling filters presented in the preceding sections are *optimum filters* in the case when designed by means of the linear programming techniques. The meaning of the word optimum in this case is

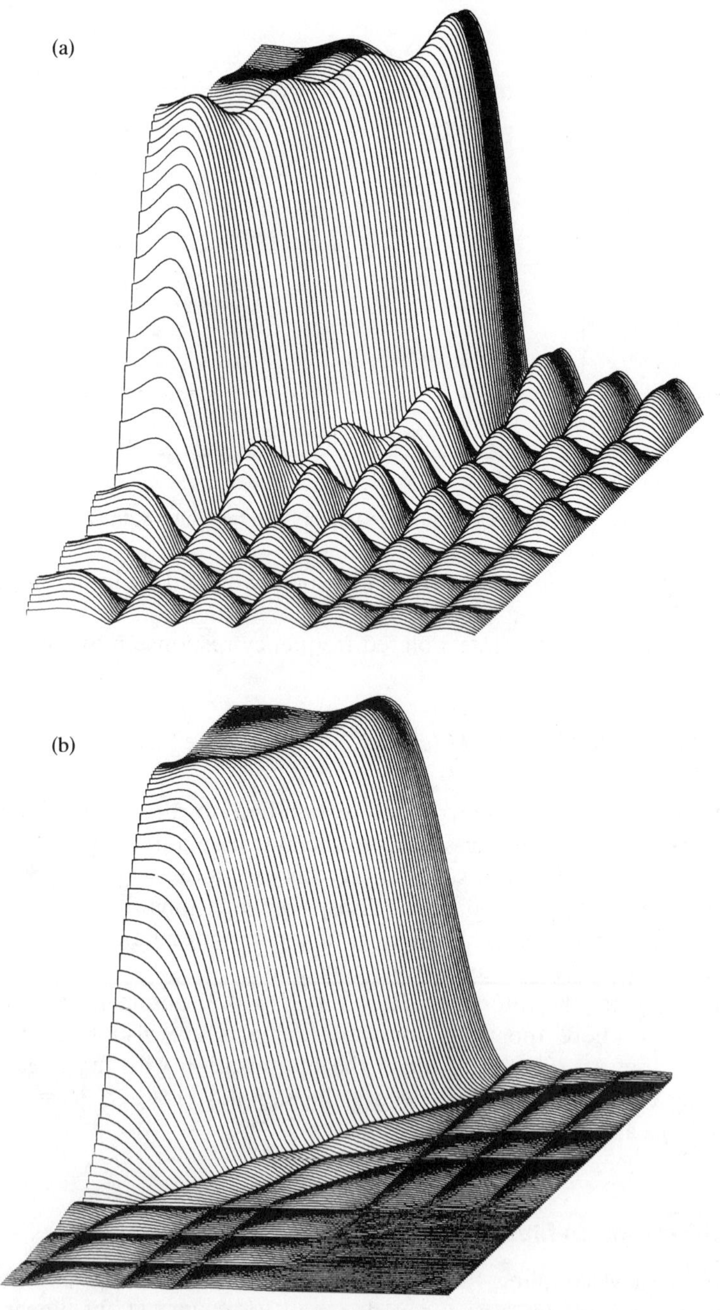

FIG. 2.17. Spatial frequency response of a 2-D frequency-sampling digital filter before the optimization (a) and after (b) with $N_1 = N_2 = 16$.

that they are those filters for which the maximum error is minimum in the passband and/or stopband, amongst all filters obtainable by varying n frequency samples in the transition band.

As in the frequency sampling case optimum filters must be defined very rigorously, firstly in so far as the variable parameters are concerned and secondly thc optimality criterion adopted for the approximation. Consequently several classes of optimum filters can be constructed, which are based on different choices of variable parameters and on the optimality criterion. As far as the parameters to be determined in the approximation are concerned, we shall consider those to be all the coefficients. That is to say all the coefficients of the filter are considered variable and are determined by the optimization procedure.

A first class of optimum filters which can be considered is the class of the filters which are optimum in the minimum $\|L_p\|$ norm sense where the definition of $\|L_p\|$ norm is given by the expression

$$\|L_p\| = \frac{1}{\omega_s}\int_0^{\omega_s} [D(e^{j\omega}) - H(e^{j\omega})]^p \, d\omega \tag{2.133}$$

and where $D(e^{j\omega})$ is the desired frequency response, $H(e^{j\omega})$ is the approximation function, the coefficients of which have to be obtained to yield minimum $\|L_p\|$.

The problem is in fact very simple for $p = 2$ where the norm reduces to a problem of minimum square error minimization and consequently after differentiation to the solution of a set of linear equations.

If p is different from 2, however, the problem is nonlinear and it becomes necessary to use a non linear optimization approach, which though possible it is in general less efficient than a linear method.

Another approach which has been considerably developed is based on the FIR design problem as a Chebyschev approximation problem corresponding to $p = \infty$ in eq. (2.133), but rather than considering eq. (2.133) in this case it is simpler to examine the problem afresh. Given an approximation function $H(e^{j\omega})$ and a desired function $D(e^{j\omega})$ defined on a compact set A of ω points (possibly the entire interval $0-\pi$ as any number of disjoint intervals) a weighted error functional in the approximation of $D(e^{j\omega})$ by $H(e^{j\omega})$ can be written in the following way

$$E(e^{j\omega}) = |W(e^{j\omega})[D(e^{j\omega}) - H(e^{j\omega})]| \tag{2.134}$$

where $W(e^{j\omega})$ is a positive weighting function, which allows us to define errors for different intervals. The Chebyshev optimality criterion consists in the minimization of the maximum error (maximum value of the (2.134)) in the set A. This is the general statement of the Chebyshev approximation which can be made more specific to our particular case, that is to the design of FIR digital filters of the four types considered in Section IIB.

It is convenient to begin by showing the four types of filters considered so far as having a transfer function to be used in the approximation as $H(e^{j\omega})$, in the unified form[31]

$$H(e^{j\omega}) = P(e^{j\omega})Q(e^{j\omega}) \tag{2.135}$$

where $Q(e^{j\omega})$ is a linear combination of cosine functions

$$Q(e^{j\omega}) = \sum_{n=0}^{N-1} a(n) \cos n\omega \tag{2.136}$$

and $P(e^{j\omega})$ is a suitable weighting function.

Ignoring the linear phase term, it can be seen that the type 1 filter transfer function of eq. (2.3) is directly expressible as a linear combination of cosine functions and hence in this case we have $P(e^{j\omega}) = 1$.

For the type 2 filters, the following expression can be written

$$\sum_{n=1}^{N} b(n) \sin n\omega = \sin \omega \sum_{n=0}^{N-1} \tilde{b}(n) \cos (n\omega) \tag{2.137}$$

where

$$\begin{aligned} b(1) &= \tilde{b}(0) - \tfrac{1}{2}\tilde{b}(2) \\ b(k) &= \tfrac{1}{2}[\tilde{b}(k-1) - \tilde{b}(k+1)], \qquad k = 2, \ldots, N-2 \\ b(N-1) &= \tfrac{1}{2}\tilde{b}(N-2) \\ b(N) &= \tfrac{1}{2}\tilde{b}(N-1) \end{aligned} \tag{2.138}$$

and obviously $P(e^{j\omega}) = \sin \omega$.

The corresponding relation for type 3 filters is written as

$$\sum_{n=1}^{N} c(n) \cos [\omega(n-\tfrac{1}{2})] = \cos (\omega/2) \sum_{n=0}^{N-1} \tilde{c}(n) \cos (n\omega) \tag{2.139}$$

where

$$\begin{aligned} c(1) &= \tilde{c}(0) + \tfrac{1}{2}\tilde{c}(1) \\ c(k) &= \tfrac{1}{2}[\tilde{c}(k-1) + \tilde{c}(k)], \qquad k = 2, \ldots, N-1 \\ c(N) &= \tfrac{1}{2}\tilde{c}(N-1) \end{aligned} \tag{2.140}$$

and therefore $P(e^{j\omega}) = \cos (\omega/2)$.

Finally for type 4 filters we have

$$\sum_{n=1}^{N} d(n) \sin [\omega(n-\tfrac{1}{2})] = \sin (\omega/2) \sum_{n=0}^{N-1} \tilde{d}(n) \cos n\omega \tag{2.141}$$

where

$$d(1)=\tilde{d}(0)-\tfrac{1}{2}d(1)$$
$$d(k)=\tfrac{1}{2}[\tilde{d}(k-1)-\tilde{d}(k)], \qquad k=2,\ldots,N-1 \tag{2.142}$$
$$d(N)=\tfrac{1}{2}\tilde{d}(N-1)$$

and hence $P(\mathrm{e}^{j\omega})=\sin(\omega/2)$.

The expression for the error involved in the approximation can now be written in the form

$$E(\mathrm{e}^{j\omega})=|W(\mathrm{e}^{j\omega})[D(\mathrm{e}^{j\omega})-P(\mathrm{e}^{j\omega})Q(\mathrm{e}^{j\omega})]|$$
$$=\left|W(\mathrm{e}^{j\omega})P(\mathrm{e}^{j\omega})\left[\frac{D(\mathrm{e}^{j\omega})}{P(\mathrm{e}^{j\omega})}-Q(\mathrm{e}^{j\omega})\right]\right| \tag{2.143}$$

which is valid everywhere except possibly at 0 and π.

Defining now

$$\hat{W}(\mathrm{e}^{j\omega})=W(\mathrm{e}^{j\omega})P(\mathrm{e}^{j\omega})$$
$$\hat{D}(\mathrm{e}^{j\omega})=\frac{D(\mathrm{e}^{j\omega})}{P(\mathrm{e}^{j\omega})} \tag{2.144}$$

we can write the expression of the error in the form

$$E(\mathrm{e}^{j\omega})=|\hat{W}(\mathrm{e}^{j\omega})[\hat{D}(\mathrm{e}^{j\omega})-Q(\mathrm{e}^{j\omega})]| \tag{2.145}$$

which can be used for any type of filter design.

The Chebyshev approximations have a very important property which is useful when seeking the solution of the minimization problem. This property is the so-called *alternation theorem* which in our case can be formulated in the following way.[9]

If $Q(\mathrm{e}^{j\omega})$ is a linear combination of N cosine functions as in eq. (2.136), then a necessary and sufficient condition for $Q(\mathrm{e}^{j\omega})$ be the unique, best weighted Chebyshev approximation of the continuous $\hat{D}(\mathrm{e}^{j\omega})$ on the set A is that the weighted error function $E(\mathrm{e}^{j\omega})$ to exhibit at least $(N+1)$ extremal frequencies on A. That is to say that there must be $(N+1)$ points on A, $\omega_0<\omega_1<\ldots<\omega_N$, such that $E(\mathrm{e}^{j\omega_i})=-E(\mathrm{e}^{j\omega_{i+1}})$, $i=0,\ldots,N-1$, and $|E(\mathrm{e}^{j\omega_i})|=\max_{\omega\in A}|E(\mathrm{e}^{j\omega})|$.

Several methods have been proposed for designing filters having such an equiripple behaviour in the error.[16,17,18,21,22] A most efficient method for which an efficient computer program is available is based on an iterative procedure known as the second Remez algorithm as follows.[31]

The alternation theorem states that the optimum solution must have at least $N+1$ extreme of the error. Let a set $\{\omega_k\}$ $k=0,\ldots,N$ of such extrema be in the approximation region so that we can write the following $N+1$ relations

$$\hat{W}(e^{j\omega_k})[\hat{D}(e^{j\omega_k})-Q(e^{j\omega_k})]=(-1)^k\delta \tag{2.146}$$

These relationships constitute a linear system of $N+1$ equations in $N+1$ variables, which variables are the N coefficients of the approximation function plus the unknown error. This system has a definite solution since the coefficient matrix is invertible as guaranteed by the fact that the functions of the basis satisfy the Haar condition.

Hence in principle the Remez algorithm takes the following form.

(1) $(N+1)$ ω_k are chosen and the system of eq. (2.146) is solved to produce the coefficients $a(n)$ and δ. In this way a cosine polynomial is obtained whose distance from the objective function is equal to δ in the ω_k points.

(2) The error $E(e^{j\omega})$ is compared on a dense grid and if $|E(e^{j\omega})|\leq\delta$ in the whole approximation region the solution obtained above is the optimum one. If at some points $|E(e^{j\omega})|>\delta$ then a new set of extremal frequencies is chosen by considering the $N+1$ where the error is maximum and of alternate sign.

It can be shown that with this procedure δ is forced to increase at every step and at the end it converges to the upper bound, which constitutes the solution of the problem. The starting set $\{\omega_k\}$ can be arbitrarily chosen. The procedure in this form however is not very efficient because at every step we have to solve a linear system.

Fortunately it is not really necessary to solve the system, but it is sufficient to evaluate δ through the formula

$$\delta=\frac{a_0\hat{D}(e^{j\omega_0})+a_1\hat{D}(e^{j\omega_1})+\ldots+a_N\hat{D}(e^{j\omega_N})}{\dfrac{a_0}{\hat{W}(e^{j\omega_0})}-\dfrac{a_1}{\hat{W}(e^{j\omega_1})}+\ldots+(-1)^N\dfrac{a_N}{\hat{W}(e^{j\omega_N})}} \tag{2.147}$$

where

$$a_k=\prod_{\substack{i=0\\ i\neq k}}^{N}\frac{1}{(\cos\omega_k-\cos\omega_i)} \tag{2.148}$$

Hence the values of $Q(e^{j\omega})$ at the N points $\omega_0,\ldots\omega_{N-1}$ can be directly computed in the form

$$Q(e^{j\omega_k})=\hat{D}(e^{j\omega_k})-(-1)^k\frac{\delta}{\hat{W}(e^{j\omega_k})}\qquad k=0,\ldots,N-1 \tag{2.149}$$

and an interpolation formula for computing $Q(e^{j\omega})$ and consequently $E(e^{j\omega})$ on a dense grid can be obtained by fitting a polynomial through the

N points using the Lagrange interpolation formula. Thus we have

$$Q(e^{j\omega}) = \frac{\sum_{k=0}^{N-1} \frac{\beta_k}{(\cos\omega - \cos\omega_k)} Q(e^{j\omega_k})}{\sum_{k=0}^{N-1} \frac{\beta_k}{(\cos\omega - \cos\omega_k)}} \tag{2.150}$$

where

$$\beta_k = \prod_{\substack{i=0 \\ i\neq k}}^{N-1} \frac{1}{\cos\omega_k - \cos\omega_i} \tag{2.151}$$

Three examples of filters obtained with the Remez exchange algorithm using the software program in McClellan *et al.*[28] are shown in Fig. 2.18. They are respectively a band-pass filter with $N = 255$, a full band differentiator with $N = 32$ and a full band Hilbert transformer with $N = 32$.

The Chebyshev or equiripple design method can also be formulated as a linear programming.[36] This formulation leads to an algorithm less efficient than the Remez algorithm, but it has the advantage of being more flexible in a sense which will be clarified in the following.

Starting from the (2.136) it is possible to write

$$-\delta \leq \hat{W}(e^{j\omega})\left[\hat{D}(e^{j\omega}) - \sum_{n=0}^{N-1} a(n)\cos n\omega\right] \leq \delta \tag{2.152}$$

and from this relation two sets of linear constraints can be obtained by sampling in a dense grid of points in the approximation interval

$$\begin{aligned} &\hat{W}(e^{j\omega_k})[\hat{D}(e^{j\omega_k}) - \sum_{n=0}^{N-1} a(n)\cos n\omega_k] \leq \delta \\ &-\hat{W}(e^{j\omega_k})\left[\hat{D}(e^{j\omega_k}) - \sum_{n=0}^{N} a(n)\cos n\omega\right] \leq \delta \end{aligned} \tag{2.153}$$

or equivalently

$$\begin{aligned} &-\hat{W}(e^{j\omega_k})\sum_{n=0}^{N-1} a(n)\cos n\omega_k - \delta \leq -\hat{W}(e^{j\omega})\hat{D}(e^{j\omega_k}) \\ &\hat{W}(e^{j\omega_k})\sum_{n=0}^{N-1} a(n)\cos n\omega_k - \delta \leq \hat{W}(e^{j\omega_k})\hat{D}(e^{j\omega_k}) \end{aligned} \tag{2.154}$$

where the coefficients $\{a(n)\}$ must be chosen to minimize the $-\delta$ value.

Thus the problem, as in the frequency sampling method, is presented as a linear programming problem, save that in this case the variables in the optimization are not some samples of the frequency response, but the

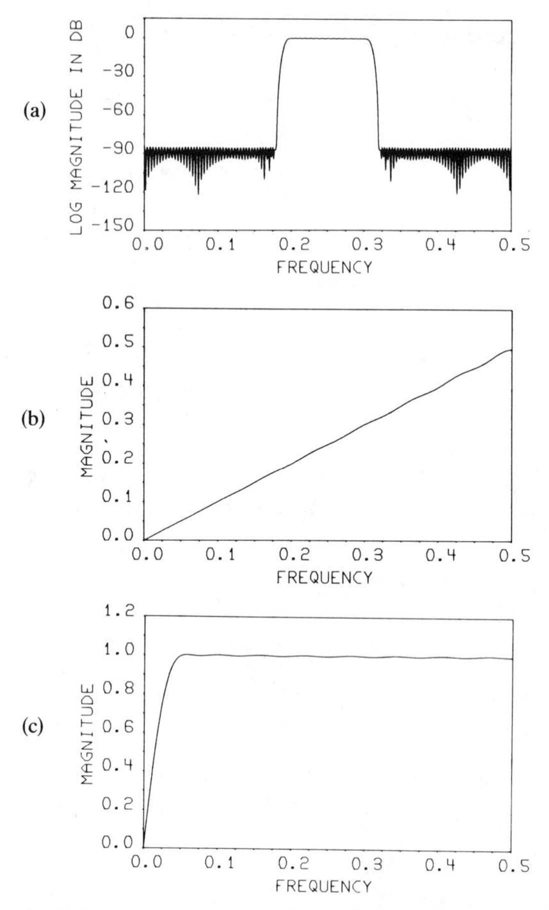

FIG. 2.18. Examples of frequency responses of optimum-type FIR digital filters: (a) band-pass filter ($N = 255$); (b) full-band differentiator ($N = 32$); (c) full-band Hilbert transformer ($N = 32$).

entire set of the coefficients of the impulse response or equivalently all the frequency samples of its DFT.

It must be observed that the solution obtained through the linear programming is exactly the same as the one obtained from the Remez exchange algorithm. In fact they are both optimization methods based on the same optimality criterion and the minimum is obtained by varying the same parameters.

Although the Remez exchange algorithm is more efficient in these problems, the importance of the linear programming approach is that it is more flexible and is directly generalizable to the 2-D filter design.[23] Its flexibility consists essentially in the fact that other linear constraints can be considered in addition to the (2.154). In the previous formulation constraints on the frequency response are only considered. However the linear programming approach allows us, for example, to introduce constraints on the coefficients $a(n)$ and in addition another linear combination on the coefficients, as, for example, the step response of the filter, which is given as

$$\varphi(n)=\sum_{m=0}^{n} h(m) \tag{2.155}$$

IIJ Analytical solutions to equiripple FIR filter design

It is possible to produce analytical solutions for a certain class of FIR filters, by using the equiripple properties of Chebyshev polynomials.

The general form of a Chebyshev polynomial is given by

$$T_n(x)=\cos(n \cosh^{-1} x) \tag{2.156}$$

A new function can be formed as follows

$$y(x)=1-\delta T_n(x) \tag{2.157}$$

which ripples about unity with a width $\pm\delta$. In this form the function is suitable for a filter response by transforming x in the form

$$x=x_0-(1+x_0)\cos^2\frac{\omega}{2} \tag{2.158}$$

The parameter x_0 can be chosen to produce a zero of transmission at π[10] or at some other frequency[19] or indeed to produce no zero of transmission at all.

IIK 2-D optimum filter design

The problem of designing 2-D optimum filters in the Chebyshev sense involves techniques where all the coefficients of the filter are allowed to vary in order to minimize the maximum error in the approximation region.

Unfortunately it is not possible to generalize directly multiple exchange algorithms as the Remez algorithm because the alternation problem is not directly generalizable in the 2-D case. However the problem is in principle tractable by means of the linear programming method, which can be applied in the same way as in the 1-D case since the transfer function of the filter is a linear combination of cosine and/or sine functions. Thus a set of constraints as the form given by eq. (2.154) can be obtained save that in this case the grid on which it is necessary to compute the value of the transfer function is a two-dimensional one, and that the variables are the elements of the impulse response matrix of the filter. This implies that the number of variables can be very high indeed and so will be the number of constraints. Consequently the problem becomes computationally intractable at least from direct use of linear programming packages without tailoring them to the constraints of this particular problem.

For this reason the maximum length of filters designed through a general and hence non-specifically tailored linear programming technique is limited as reported in Hu and Rabiner.[23] It may be possible, of course, to improve on this by a judicious re-examination of the linear programming formulation in a way that would cater for larger dimensionality systems.

IIL Transformations from 1-D filters to 2-D filters

An efficient technique has been proposed for the design of some classes of 2-D filters with values of N_1 and N_2 higher than the ones obtainable by means of the linear programming approach. This technique is based on frequency transformations of optimal 1-D filters to 2-D filters whereby in some cases optimality is preserved.[27]

For example if we consider the frequency response of a linear phase 2-D filter with N_1, N_2 odd, its real part has the form

$$H_R(e^{j\omega_1}, e^{j\omega_2}) = \sum_{n_1=0}^{(N_1-1)/2} \sum_{n_2=0}^{(N_2-1)/2} a(n_1, n_2) \cos n_1\omega_1 \cos n_2\omega_2 \tag{2.159}$$

On the other hand the real part of the frequency response of a 1-D filter with N odd can be written as

$$H(e^{j\omega}) = \sum_{n=0}^{(N-1)/2} a(n) \cos n\omega \tag{2.160}$$

If now we let $x = \cos \omega$, the following relationship holds

$$\cos n\omega = \cos n(\cos^{-1}(x)) = T_n(x) \tag{2.161}$$

where $T_n(x)$ is a Chebyshev polynomial of nth order. However

$$\sum_{n=0}^{(N-1)/2} a(n)T_n(x) = \sum_{n=0}^{(N-1)/2} \tilde{a}(n)x^n \tag{2.162}$$

and therefore eq. (2.160) can be written in the form

$$H(e^{j\omega}) = \sum_{n=0}^{(N-1)/2} \tilde{a}(n) \cos^n (\omega) \tag{2.163}$$

Hence with a transformation of variables of the form

$$\cos \omega = A \cos \omega_1 + B \cos \omega_2 + C \cos \omega_1 \cos \omega_2 + D \tag{2.164}$$

with A, B, C and D real, it is possible starting from eq. (2.163) to obtain an expression of the type

$$H(e^{j\omega_1}, e^{j\omega_2}) = \sum_{n_1=0}^{(N_1-1)/2} \sum_{n_2=0}^{(N_2-1)/2} \tilde{a}(n_1, n_2) \cos n_1\omega_1 \cos n_2\omega_2 \tag{2.165}$$

which can be easily transformed to the form

$$H(e^{j\omega_1}, e^{j\omega_2}) = \sum_{n_1=0}^{(N-1)/2} \sum_{n_2=0}^{(N-1)/2} a(n_1, n_2) \cos n_1\omega_1 \cos n_2\omega_2 \tag{2.166}$$

This form of course can be identified with eq. (2.156).

The form of mapping involved between the three variables can be obtained from eq. (2.164) by solving for ω_2

$$\omega_2 = \cos^{-1}\left[\frac{\cos \omega_1 - D - A \cos \omega_1}{B + C \cos \omega_1}\right] \tag{2.167}$$

and for every value of ω we have a line on the (ω_1, ω_2) plane which represents the locus of the points where $H(e^{j\omega_1}, e^{j\omega_2})$ assumes the same value as $H(e^{j\omega})$ in ω. The problem is now the choice of the coefficients of the transformation to obtain suitable contours in the (ω_1, ω_2) plane. Some techniques have been proposed to calculate the coefficients for arbitrary contours. Here we present the solution for a very important kind of filters, that is the circularly symmetric filters.

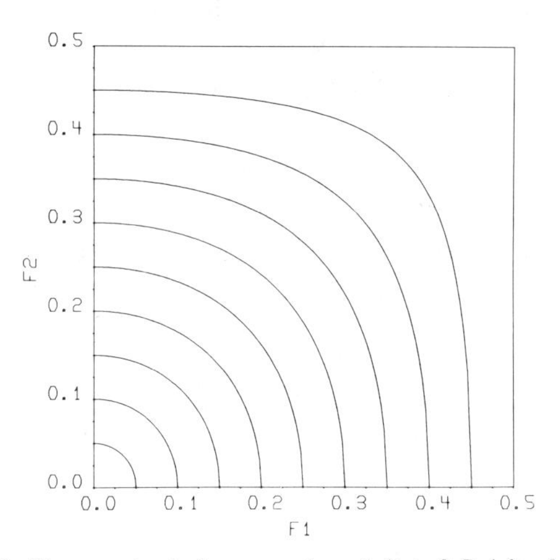

FIG. 2.19. The mapping in frequency from 1-D to 2-D (after McClellan).

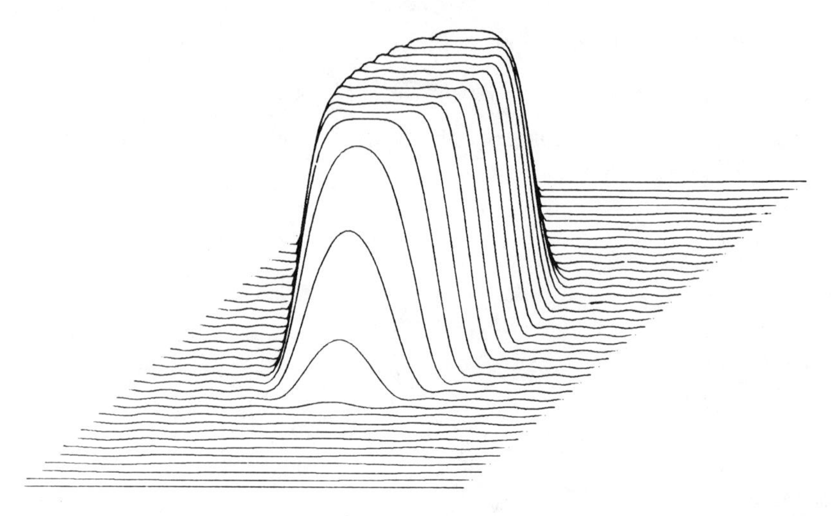

FIG. 2.20. Spatial frequency response of a digital filter designed by means of the McClellan transformation technique ($N_1 = N_2 = 27$).

Approximately circular contours can be obtained with the transformation (2.164) chosing the coefficients

$$A = B = C = -D = \tfrac{1}{2} \tag{2.168}$$

as shown in Fig. 2.19. It is quite apparent that the approximation to the circular form is good for small $r = (\omega_1^2 + \omega_2^2)^{\frac{1}{2}}$ and moreover that the error increases as the radius r increases. However for many filters of practical interest the approximation is very good and the design method is very efficient: a 1-D filter is required to be designed and then transformed to a 2-D filter directly in accordance with eq. (2.165). An example of spatial frequency response of a digital filter designed by means of this technique is shown in Fig. 2.20 ($N_1 = N_2 = 27$).

References

1. Blackman, R. B. and Tukey, J. W. (1958). "The Measurement of Power Spectra from the Point of View of Communications Engineering". Dover, New York.
2. Borchi, E., Cappellini, V. and Emiliani, P. L. (1975). A new class of FIR digital filters using a Weber-type weighting function. *Alta Frequenza* **44**, **8**, 469–70.
3. Borchi, E., Cappellini, V., and Emiliani, P. L. (1975). A new class of FIR digital filters with application to image processing. 1975 Proceedings: Midwest Symposium on Circuits and Systems, Montreal, Canada, 635–9.
4. Borchi, E., Cappellini, V., and Emiliani, P. L. (1975). A class of 1-D and 2-D FIR digital filters. 1975 Proceedings: Florence Conference on Digital Signal Processing, Firenze, Italy, 65–74.

5. Borchi, E., Cappellini, V. and Del Re, E. (1976). Minimum-bias windows defined through a quantum mechanical analogy. Report R7-76, Istituto di Elettronica dell'Università di Firenze, Italy.
6. Calzini, M., Cappellini, V. and Emiliani, P. L. (1975). Alcuni filtri numerici bidimensionali con risposta impulsiva finita. *Alta Frequenza* **44**, **12**, 747–53.
7. Cappellini, V. (1968). A class of digital filters with application to data compression. 1968 Proceedings: I.E.E.E. International Conference on Communications, Philadelphia, Pa., 788–93.
8. Cappellini, V. (1969). Design of some digital filters with application to spectral estimation and data compression. 1970 Proceedings: P.I.B. Symposium on Computer Processing in Communications, Polytechnic Institute of Brooklyn, Brooklyn, New York, **19**, 313–31.
9. Cheney, E. W. (1966). "Introduction to Approximation theory". McGraw-Hill, New York.
10. Constantinides, A. G. (1971). The design of linear phase nonrecursive lowpass digital filters having equiripple passbands. Symposium on Digital Filtering, Imperial College, London.
11. Gabor, D. (1946), Theory of communication, *I.E.E.J.* **3**, **93**, 429–41.
12. Gass, S. I. (1969). "Linear Programming". McGraw-Hill, New York.
13. Gold, B. and Rader, C. M. (1969). "Digital Processing of Signals". McGraw-Hill, New York.
14. Gold, B. and Jordan, K. L. (1969). A direct search procedure for designing finite duration impulse response filters. *I.E.E.E. Trans. Audio Electroacoustics* **AU-17**, **1**, 33–6.
15. Helms, H. D. (1968). Nonrecursive digital filters: design methods for achieving specifications on frequency response. *I.E.E.E. Trans. Audio Electroacoustics* **AU-16**, **3**, 336–42.
16. Helms, H. D. (1971). Digital filters with equiripple or minimax response. *I.E.E.E. Trans. Audio Electroacoustics* **AU-19**, **1**, 87–94.
17. Herrmann, O. (1970). Design of nonrecursive digital filters with linear phase. *Electronics Letters* **6**, **11**, 328–9.
18. Herrmann, O. and Schüssler, H. W. (1970). Design of nonrecursive digital filters with minimum phase. *Electronics Letters* **6**, **11**, 329–30.
19. Herrmann, O., Rabiner, L. R. and Chan, D. S. K. (1973). Practical design rules for optimum finite impulse response low-pass digital filters. *Bell System Techn. J.* **52**, **6**, 769–99.
20. Hilberg, W. and Rothe, P. G. (1971). The general uncertainty relation for real signals in communication theory. *Information and Control* **18**, 103–25.
21. Hofstetter, E., Oppenheim, A. V. and Siegel, J. (1971). A new technique for the design of nonrecursive digital filters. Proceedings: 5th Annual Princeton Conference on Information Sciences and Systems, 64–72.
22. Hofstetter, E., Oppenheim, A. V. and Siegel, J. (1971). On optimum nonrecursive digital filters. 1971 Proceedings: 5th Allerton Conference on Circuit and System theory, 789–98.
23. Hu, J. V. and Rabiner, L. R. (1972). Design techniques for two-dimensional digital filters. *I.E.E.E. Trans. Audio Electroacoustics* **AU-20**, **4**, 249–57.
24. Huang, T. S. (1972). Two-dimensional windows. *I.E.E.E. Trans. Audio Electroacoustics* **AU-20**, **1**, 88–9.
25. Kaiser, J. F. (1966). Digital filters. *In* "System Analysis by Digital Computer" (F. F. Kuo and J. F. Kaiser, eds), pp. 218–77. Wiley, New York.

26. Kay, J. and Silverman, R. A. (1957). On the uncertainty relation for real signals. *Information and Control* **1**, 64–75.
27. McClellan, J. H. (1973). The design of two dimensional digital filters by transformations. Proceedings: 7th Annual Princeton Conference on Information Sciences and Systems, 247–51.
28. McClellan, J. H., Parks, T. W. and Rabiner, L. R. (1973). A computer program for designing optimum FIR linear phase digital filters. *I.E.E.E. Trans. Audio Electroacoustics* **AU-21**, **6**, 506–26.
29. Palmer, D. F. (1969). Bias criteria for the selection of spectral windows. *I.E.E.E. Trans. Information Theory*, **IT-15**, **5**, 613–15.
30. Papoulis, A. (1973). Minimum-bias windows for high-resolution spectral estimates. *I.E.E.E. Trans. Information Theory* **IT-19**, **1**, 9–12.
31. Parks, T. W. and McClellan, J. H. (1972). Chebyshev approximation for nonrecursive digital filters with linear phase. *I.E.E.E. Trans. Circuit Theory* **CT-19**, 184–94.
32. Pollack, H. O. and Landau, H. J. (1961). Prolate spheroidal wave functions, Fourier analysis and uncertainty, Part 2. *Bell System Techn. J.* **40**, **1**, 65–84.
33. Rabiner, L. R., Gold, B. and McGonegal, C. A. (1970). An approach to the approximation problem for nonrecursive digital filters. *I.E.E.E. Trans. Audio Electroacoustics* **18**, **2**, 83–106.
34. Rabiner, L. R. and Shafer, R. W. (1971). Recursive and nonrecursive realizations of digital filters designed by frequency sampling techniques. *I.E.E.E. Trans. Audio Electroacoustics* **19**, **3**, 200-207.
35. Rabiner, L. R. and Shafer, R. W. (1972). Correction to recursive and nonrecursive realizations of digital filters designed by frequency sampling techniques. *I.E.E.E. Trans. Audio Electroacoustics* **AU-20, 1,** 104–5.
36. Rabiner, L. R. (1972). Linear program design of finite impulse response (FIR) digital filters. *I.E.E.E. Trans. Audio Electroacoustics* **AU-20**, **4**, 280-8.
37. Rossi, C. (1967). Window functions for non-recursive digital filters. *Electronics Letters* **3**, **12**, 559–61.
38. Slepian, D. and Pollack, H. O. (1961). Prolate spheroidal wave functions, Fourier analysis and uncertainty. Part 1. *Bell System Techn. J.* **40**, **1**, 43–64.

Chapter 3

Design Methods of Infinite Impulse Response Digital Filters

IIIA Introduction

In this chapter we examine some design methods of IIR 1-D and 2-D digital filters.

The contents are by no means meant to be exhaustive in nature. We include those aspects of design of the above kind of filters which we consider to be both useful and representative of the available methods. Thus in the 1-D case for IIR filter design we put emphasis on the bilinear transformation as applied to the derivation of either a digital filter transfer function or indeed a digital-filter structure from analogue filters. Direct design methods are also studied in some detail and their attributes examined. Low sensitivity structures based on the Wave Digital Filter concepts of Fettweis and the Linear Transformation Digital Filter concept of Constantinides are also examined and examples given.

The 2-D case is similarly treated where available methods for the design of IIR 2-D digital filters are given with emphasis on the more useful ones from the practical point of view.

Software for the application of the design methods contained in this chapter are given in Chapter 6.

IIIB Basic principles of transformation from continuous time filters to digital filters

The existence of analogue continuous time filters has made the design of digital filters possible from many points of view. Early attempts and successes were concentrated on transferring the continuous time filter transfer function into a digital filter transfer function by making use of some importance correspondences between the two. The direct or matched z-transform,[39] the impulse invariance technique[36] and the bilinear transform are representative of such methods. We outline here the direct z-transform method and examine the bilinear z-transform only. The impulse invariance technique has been shown to be very unsatisfactory[36] in that it relies on keeping the impulse response of the digital filter equal to the sampled version of the impulse response of the analogous continuous time filter and hence it does not have a direct filter frequency interpretation (such unwanted effects as aliasing, etc. take place here).

IIIB1 The direct z-transform method

In this case the idea is to transfer the poles and zeros of a given analogue filter transfer function $H(s)$ into a corresponding set of poles and zeros for a digital filter transfer function $G(z)$ in accordance with the following.

Let a numerator or denominator factor of $H(s)$ be $(s+\alpha_i)$. Then the corresponding factor of $G(z)$ will be $1-z^{-1}\,\mathrm{e}^{-\alpha_i T}$. Thus we have the mapping

$$(s+\alpha_i)\to(1-z^{-1}\,\mathrm{e}^{-\alpha_i T}) \tag{3.1}$$

through which all poles and zeros are transformed to the digital domain. It has been shown by Golden[36] that in certain cases a *guard filter* is sometimes required of the form $(1+z^{-1})^N$ where N is suitably chosen. In fact this guard filter is really part of the whole procedure with N having a specific value given by the order of the zero of $H(s)$ at infinity. Thus the factor $(1+z^{-1})^N$ represents, in accordance with the mapping above, the multiple zero which $H(s)$ may have at infinity and which zero is naturally transformed to the edge of the band at π/T in the digital filter transfer function $G(z)$.

Although this procedure is straightforward it does not yield very satisfactory results particularly for wide-band filters and an alternative and most powerful technique has been developed based on the bilinear transformation.[35]

IIIB2 The bilinear transformation technique

The bilinear z-transformation is a direct replacement in $H(s)$ of the complex frequency variable s by

$$s\to k\frac{1-z^{-1}}{1+z^{-1}} \tag{3.2}$$

where k is a real positive constant chosen as follows. The frequency axis Ω in the analogue case from eq. (3.2) transforms to the unit circle $z^{-1} = \exp(-j\omega T)$ so that

$$\Omega = k \tan \frac{\omega T}{2} \tag{3.3}$$

and hence a critical frequency Ω_c in the analogue filter response will correspond to a critical frequency ω_c in the digital filter response where

$$\Omega_c = k \tan \frac{\omega_c T}{2} \tag{3.4}$$

Thus knowing the two critical frequencies the constant k can be calculated from eq. (3.3). The bilinear transformation in its form given by eqs (3.2)–(3.4) facilitates the transfer of analogue continuous time filters of a specific kind (i.e., L.P., H.P., etc.) to be transformed to the same kind (i.e. L.P., H.P., etc.) digital filters.

Such a transformation is possible for piece-wise constant frequency responses only and not for example for integrators or other filters having responses which are not piece-wise constant.

Stability of the digital filter obtained by such means is guaranteed if the original continuous filter prototype is stable because stable regions in one domain are mapped into stable regions in the other domain through the transformation of eq. (3.2).

It is perhaps instructive to examine the bilinear transformation from the graphical point of view which facilitates some understanding of its mechanism of operation.

In Fig. 3.1 we show a continuous filter response which is to be transformed to the digital domain by replacing the Ω-axis in this figure by the relationship of eq. (3.4). This relationship is shown under the continuous response in Fig. 3.1. The replacement mentioned above is achieved merely by projecting onto the transformation relationship the corresponding points which are then reconstructed to the left of the figure as shown. Hence it is quite apparent that the entire analogue filter response from $-\infty$ to $+\infty$ is compressed between the limits $-\pi/T$ to π/T in the digital domain and periodically repeated thereafter.

IIIC Transformations of passive ladder filters

In the above sub-section we considered the problem of transforming a transfer function $H(s)$ in the analogue filter domain to a transfer function $G(z)$ in the digital filter domain. The question of realization of $G(z)$

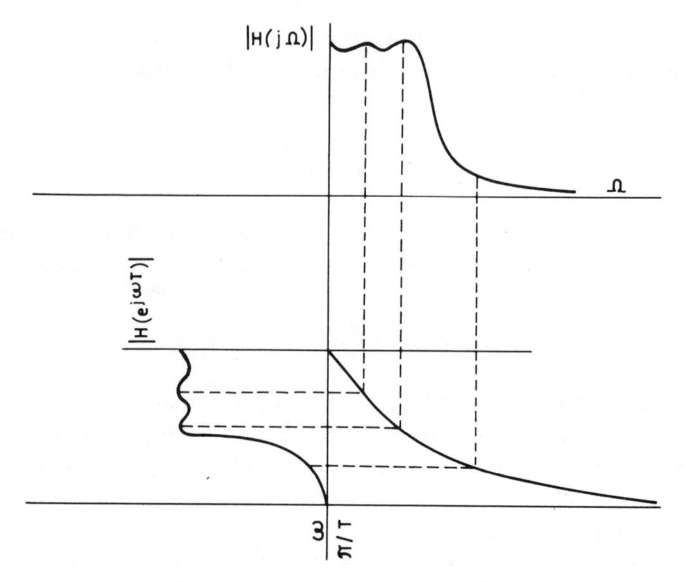

FIG. 3.1. Graphical representation of the bilinear transform.

however was not considered since by having a specific $G(z)$ many and different structures may be employed for its realization.[35,44]

However a question arises, which is of paramount importance, as to what structure should be used so as to have low sensitivity to coefficient variations.[38] It is well known in analogue filter design that for lossless doubly terminated ladder filters such as the example of Fig. 3.2 the sensitivity in

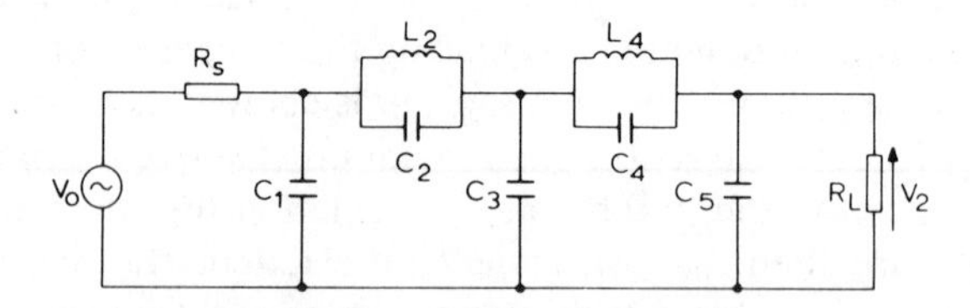

FIG. 3.2. A doubly terminated lossless ladder filter.

the passband to variations in the reactive components of the ladder is low and indeed at points of maximum power transfer it is zero. By the term sensitivity here we mean the differential sensitivity which is usually represented and used in its relative form

$$S_x^{|H(s)|} = \frac{x}{|H(s)|} \frac{\partial |H(s)|}{\partial x} \tag{3.5}$$

Fettweis first suggested the use of one-port reflection coefficients for the transfer of the analogue filter properties from the analogue domain to the digital filter domain and produced the *wave digital filter* structure.[31,32,33,53]

The reason why reflection coefficients should be used is that direct signal flow graph representation of the ladder filter in terms of its V–I relationship leads to structures in the digital filter domain using the bilinear transformation which are unrealizable in that they contain loops without delay.[31,18]

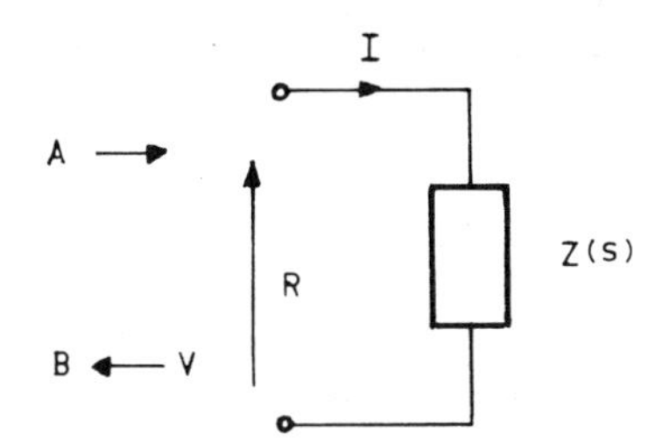

FIG. 3.3. One-port description of a typical ladder component.

Basically a one-port impedance Z as in Fig. 3.3 can be described by its reflection coefficient as

$$B = A\frac{Z-R}{Z+R} \tag{3.6}$$

where R is an arbitrary reference resistance. The numerator of eq. (3.6), if delay-free loops are to be avoided, must contain an overall factor of z^{-1}, i.e. it must vanish for $z^{-1}=0$ or in view of eq. (3.2) it must vanish for $s=k$. In the normalized analogue case this is equivalent to saying that it must vanish at $s=1$. Hence we must have

$$Z(1)-R=0 \tag{3.7}$$

Thus by treating all impedances of a ladder as one-ports these can be suitably transformed where eq. (3.7) will have to be applied with different values of R to all the ladder impedances. The interconnection of these one-ports however requires an impedance level adjustment which in the Fettweis nomenclature corresponds to an *adaptor*. For example if we take the series situation of Fig. 3.4 with $R_1 \ldots R_N$ as shown, the interconnection means that

$$i_1 = i_2 = \ldots i_N \tag{3.8}$$

and

$$V_1 + V_2 + \ldots + V_N = 0 \tag{3.9}$$

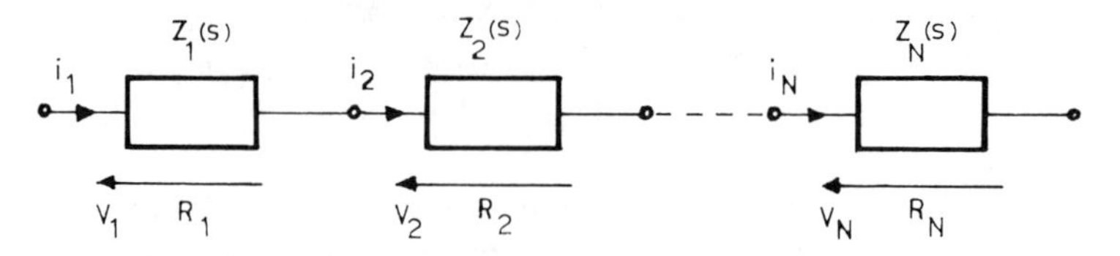

FIG. 3.4. Series connection of impedances.

in addition

$$\begin{aligned} A_r &= V_r + R_r \cdot I_r \\ B_r &= V_r - R_r \cdot I_r \end{aligned} \qquad r = 1, \ldots, N \tag{3.10}$$

and hence

$$B_r = A_r - \alpha_r \cdot A_0, \tag{3.11}$$

where

$$\begin{aligned} \alpha_r &= 2R_r/R \\ A_0 &= A_1 + A_2 + \cdots + A_N \\ R &= R_1 + R_2 + \cdots + R_N \end{aligned} \tag{3.12}$$

and

$$\alpha_1 + \alpha_2 + \ldots + \alpha_N = 2 \tag{3.13}$$

The digital structure realizing these conditions is known as a *series adaptor*[31] and it is represented symbolically as in Fig. 3.5.

Similar structures can be derived for the parallel connection of one-ports.

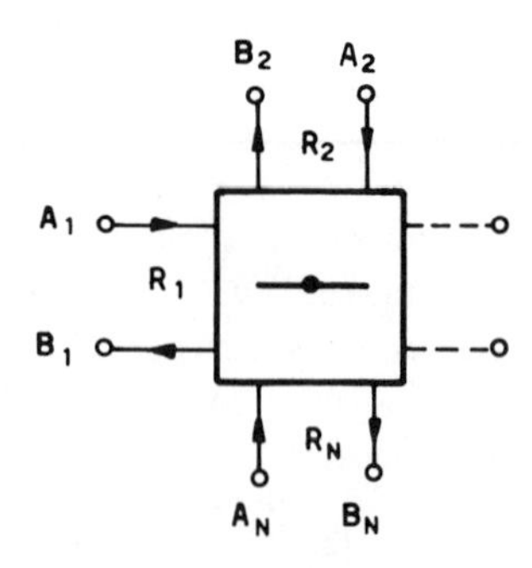

FIG. 3.5. A series adaptor.

A two-port series-parallel adaptor is shown in Fig. 3.6 which facilitates the interconnection of two ports of different normalization resistances.

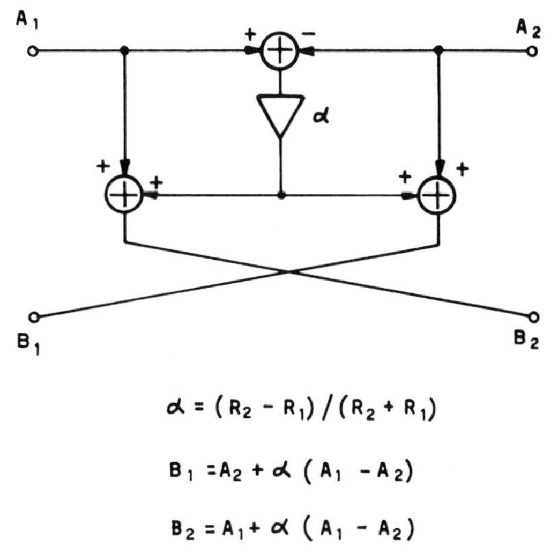

FIG. 3.6. A series-parallel adaptor.

A simple example of a series tuned circuit equivalent is shown in Fig. 3.7. A parallel tuned circuit and its digital equivalent structure is shown in

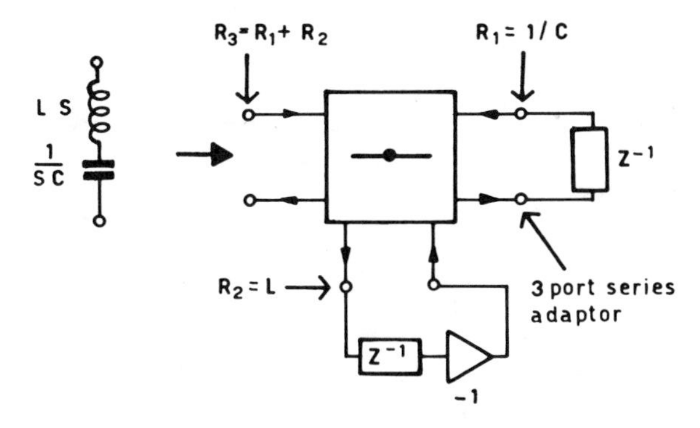

FIG. 3.7. Series tuned circuit and its digital equivalent.

Fig. 3.8. The above theory has also been applied to derive the filter of Fig. 3.9 where the original passive filter is also shown. Note that the digital realization structure requires adders, multipliers and delays. The number of delays, of course, is consistent with the number of storage elements of the original passive filter. Fettweis has produced some solutions where it would be possible in the case of capacitive loops or inductive cut-sets to reduce the overall number of delays in the digital realization.[32]

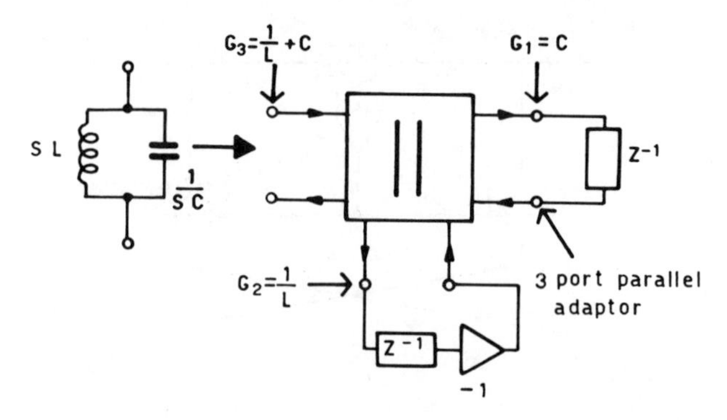

FIG. 3.8. Parallel tuned circuit and its digital equivalent.

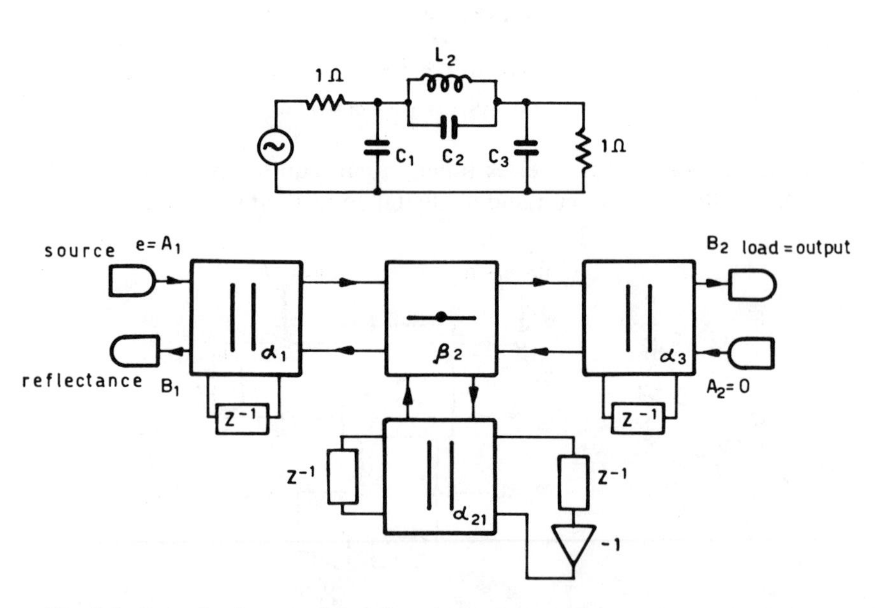

FIG. 3.9. Example of a wave digital filter obtained by the Fettweis one-port approach.

A more direct approach for the wave digital filters is to treat the ladder elements of the passive ladder as two-ports in which case to realize a ladder we need floating impedances and shunt admittances.

In the general two-port case of Fig. 3.10 as examined by Constantinides we have[18,19,40,41]

$$\begin{bmatrix} A_i \\ B_i \end{bmatrix} = \begin{bmatrix} 1 & R_i \\ 1 & -R_i \end{bmatrix} \begin{bmatrix} V_i \\ I_i \end{bmatrix} \tag{3.14}$$

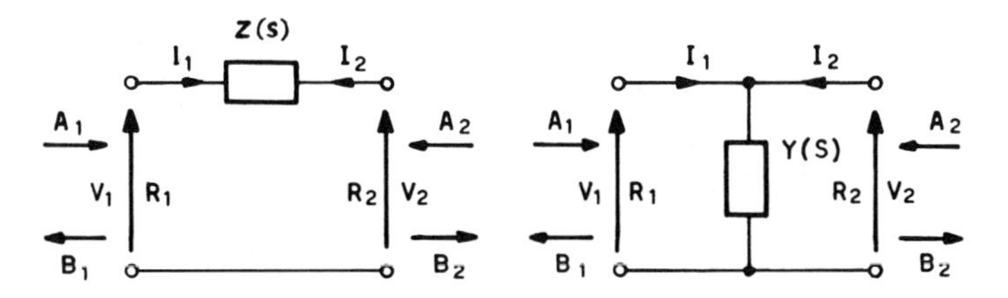

FIG. 3.10. Two-port representation of ladder elements.

for port $i=1$ or 2. In addition for any two-port we have†

$$\begin{bmatrix} V_1 \\ I_1 \end{bmatrix} = \begin{bmatrix} A & B \\ C & D \end{bmatrix} \begin{bmatrix} V_2 \\ I_2 \end{bmatrix} \tag{3.15}$$

and hence we can write

$$\begin{bmatrix} A_1 \\ B_1 \end{bmatrix} = \begin{bmatrix} 1 & R_1 \\ 1 & -R_1 \end{bmatrix} \begin{bmatrix} A & B \\ C & D \end{bmatrix} \begin{bmatrix} 1 & +R_2 \\ 1 & -R_2 \end{bmatrix}^{-1} \begin{bmatrix} A_2 \\ B_2 \end{bmatrix} \tag{3.16}$$

Rearranging eq. (3.16) we have

$$\begin{bmatrix} B_1 \\ B_2 \end{bmatrix} = \begin{bmatrix} S_{11} & S_{12} \\ S_{21} & S_{22} \end{bmatrix} \begin{bmatrix} A_1 \\ A_2 \end{bmatrix} \tag{3.17}$$

which is the two-port scattering-parameter form based on voltage waves. For the two cases of a floating impedance and a shunt admittance we have the scattering parameters as shown in Table 3.1.

Interconnection of a series arm impedance Z with a shunt arm admittance Y requires as indicated in Fig. 3.11 that the reflected wave from one port at the interconnection be equal to the incident wave at the other adjacent port of interconnection. With the two-port normalization resistances equal at these adjacent ports we shall have, therefore,

$$\begin{aligned} A_1^{(2)} &= B_2^{(1)} \\ A_2^{(1)} &= B_1^{(2)} \end{aligned} \tag{3.18}$$

Such an interconnection, however, produces delay-free loops unless there exists a delay in one of the branches indicated in Fig. 3.11. Consequently from Table 3.1 we have

$$\left.\begin{aligned} \text{either} \quad & R_1 - R_2 + Z(1) = 0 \\ \text{or} \quad & R_2 - R_1 + Z(1) = 0 \end{aligned}\right\} \text{series } Z(s) \tag{3.19}$$

$$\left.\begin{aligned} \text{and either} \quad & G_1 - G_2 - Y(1) = 0 \\ \text{or} \quad & G_2 - G_1 - Y(1) = 0 \end{aligned}\right\} \text{shunt } Y(s) \tag{3.20}$$

† Note that I_1 and I_2 both flow into the network and thus eq. (3.15) has B and D of opposite sign to the normal chain matrix form.

$S_{11}=\dfrac{R_2-R_1+Z(s)}{R_2+R_1+Z(s)}$	$\dfrac{G_1-G_2-Y(s)}{G_1+G_2+Y(s)}$
$S_{12}=\dfrac{2R_1}{R_2+R_1+Z(s)}$	$\dfrac{2G_2}{G_1+G_2+Y(s)}$
$S_{21}=\dfrac{2R_2}{R_2+R_1+Z(s)}$	$\dfrac{2G_1}{G_1+G_2+Y(s)}$
$S_{22}=\dfrac{R_1-R_2+Z(s)}{R_2+R_1+Z(s)}$	$\dfrac{G_2-G_1-Y(s)}{G_1+G_2+Y(s)}$
$S_{11}+S_{12}=1$	$S_{21}-S_{11}=1$
$S_{21}+S_{22}=1$	$S_{12}-S_{22}=1$

TABLE 3.1.

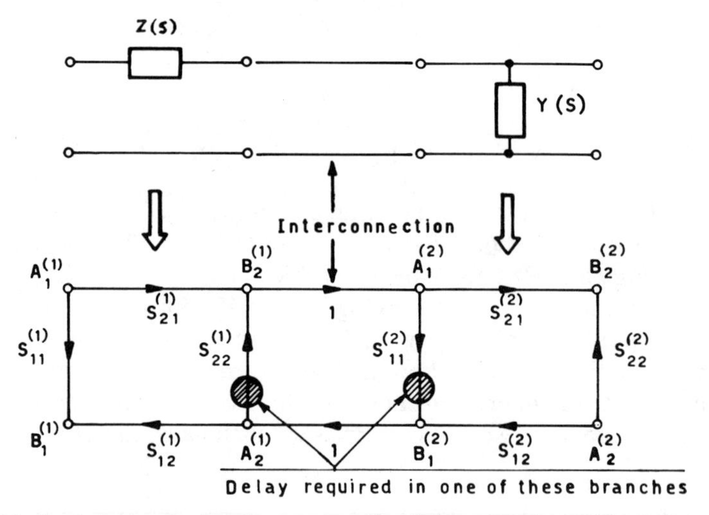

FIG. 3.11. Equivalent wave digital representation of the connection of Fig. 3.10 based on two-port scattering parameters.

Extension of the above two-port approach and consequent generalizations which involve an inherently large number of choices for the realization structures is possible by using the Constantinides approach through linear transformation.[21,22]

The approach we use here is to treat each element (either series or shunt) of a ladder network independently as a two-port network on which we perform the linear transformations. This approach has been employed successfully for the voltage scattering parameter based transformation for the design of digital filters above† and indeed of active filters[58,20] indicating the versatility of the approach.

In the generalization of the above[5] we have a two-port network for which at port 1 we have a voltage V_1 and a current I_1 flowing into the network, and at port 2 we have a voltage V_2 and a current I_2 flowing into the nctwork again. These variables are related through a set of equations, and for our case we shall use the modified chain matrix description as the most convenient. The usual chain matrix description requires I_1 to flow into the network and I_2 out of the network. However, this is inconvenient in our approach and we modify the signs of B and D to allow I_2 to flow into the network. Thus, we have as before

$$\begin{bmatrix} V_1 \\ I_1 \end{bmatrix} = \begin{bmatrix} A & B \\ C & D \end{bmatrix} \begin{bmatrix} V_2 \\ I_2 \end{bmatrix} \tag{3.21}$$

with B and D appropriately modified.

It is restated here that direct use of V–I relationships of the type expressed by eq. (3.21) for the transfer of an analogue element (i.e. resistor, capacitor, inductor or a source) into the digital domain produces unrealizable digital-filter structures.[31,18,19] The use of scattering parameters as originally described by Fettweis[31,32] and also in the two-port form as described by Constantinides[18] and Lawson and Constantinides[19] removed this constraint of unrealizability.

The scattering parameter description, however, is a special case of a more general approach based on linear transformation of variables from which we develop new design methods which we refer to as the invariant voltage ratio method (i.v.r.) and the modified transfer admittance method (m.t.a.). To achieve this we define a new set of variables for the two ports of the network as follows:[21]

$$\begin{bmatrix} X_1 \\ Y_1 \end{bmatrix} = P \begin{bmatrix} V_1 \\ I_1 \end{bmatrix} \quad \text{and} \quad \begin{bmatrix} X_2 \\ Y_2 \end{bmatrix} = Q \begin{bmatrix} V_2 \\ I_2 \end{bmatrix} \tag{3.22}$$

where P and Q are two 2×2 nonsingular matrices. It will be observed from eq. (3.22) that when

$$P = \begin{bmatrix} 1 & R_1 \\ 1 & -R_1 \end{bmatrix} \quad \text{and} \quad Q = \begin{bmatrix} 1 & R_2 \\ 1 & -R_2 \end{bmatrix} \tag{3.23}$$

we have the voltage-based scattering parameter description where $X_i = A_i$, the incident wave, and $Y_i = B_i$, the reflected wave, at port $i = 1, 2$.

† It should be emphasized here that the Fettweis structures and the structures obtained for the port approach are the same except in the case of tuned circuit where the second method has fewer adders.

In view of eq. (3.21) we can express a linear relationship between the set of transformed variables given by

$$\begin{bmatrix} X_1 \\ Y_1 \end{bmatrix} = PTQ^{-1} \begin{bmatrix} X_2 \\ Y_2 \end{bmatrix} \tag{3.24}$$

where

$$T = \begin{bmatrix} A & B \\ C & D \end{bmatrix} \tag{3.25}$$

In general P and Q must be chosen so that:

(i) the application of the transformation to the overall ladder network is equivalent to applying the same transformation to the individual constituent elements of the ladder and cascading the resulting digital structures to obtain the entire digital filter. This implies that there should be no possibility of generating delay-free loops between adjacent digital structures equivalent to ladder elements or at least there should be some freedom of choice in the form of a free parameter, so that, when it assumes a specific value, delay-free loops disappear. This is the realizability condition mentioned earlier and studied by several authors.[31,19]

(ii) The resulting digital-filter structure must have a transfer function which is precisely the transfer function of the original ladder bilinearly transformed. We can allow a modification of the original transfer function in the form of a real constant factor.

(iii) No coefficients in the digital-filter structure should have infinite values as in such a case we would have a structure which cannot be constructed. (The structure will be unrealizable but the unrealizability in this case is, of course, not the same as that discussed in (i) above.)

Our motivation is to design digital-filter structures in the most general sense allowing for example the use of voltages as one set of variables so that the structure will have a transfer function which is precisely the voltage transfer ratio of the original ladder. A slight modification can allow the output variable at port 2 to be the current rather than the voltage and in this case we have in the transfer function a scalar constant factor (dependent on the load resistance). Transformations that fulfil our aims are given below.

In the invariant voltage ratio (i.v.r.) method we have

$$P = \begin{bmatrix} 1 & 0 \\ 1 & -R_1 \end{bmatrix} \quad \text{and} \quad Q = \begin{bmatrix} 1 & R_2 \\ 1 & 0 \end{bmatrix} \tag{3.26}$$

It will be observed from conditions given by eq. (3.26) in conjunction with eq. (3.22) that $X_1 = V_1$ and $Y_2 = V_2$. If we consider X_i as the stimuli and Y_i

as the responses we shall have a transfer function in terms of the transformed variables which is precisely the voltage transfer ratio of the original analogue network.

The stimulus–response relationship mentioned above requires a reformulation of eq. (3.24) whereby the responses (Y_i) are expressed in terms of the stimuli (X_i). In this case we can write by simple algebraic manipulation

$$\begin{bmatrix} Y_1 \\ Y_2 \end{bmatrix} = \begin{bmatrix} \sigma_{11} & \sigma_{12} \\ \sigma_{21} & \sigma_{22} \end{bmatrix} \begin{bmatrix} X_1 \\ X_2 \end{bmatrix} \tag{3.27}$$

where for the i.v.r. method

$$\begin{aligned} \sigma_{11} &= (A + CR_1 - BG_2 + DR_1G_2)/(A - BG_2) \\ \sigma_{12} &= R_1G_2/(A - BG_2) \\ \sigma_{21} &= 1/(A - BG_2) \\ \sigma_{22} &= BG_2/(A - BG_2) \end{aligned} \tag{3.28}$$

The quantities A, B, C, D are the elements of the matrix of eq. (3.21) and as we are interested in ladder elements that are passive and reciprocal we shall have $AD - BC = -1$. (The negative sign comes from taking I_2 flowing into the network.)

The derivation of the digital equivalent structures to the ladder elements can now be put into effect. We note that we have to deal with ladder elements which may be located in (a) the series arm position and (b) the shunt arm position. These two cases will correspond to matrix of eq. (3.21) being

$$\text{(a)} \quad \begin{bmatrix} 1 & -Z \\ 0 & -1 \end{bmatrix} \qquad \text{(b)} \quad \begin{bmatrix} 1 & 0 \\ Y & -1 \end{bmatrix} \tag{3.29}$$

We must also consider the interconnection of these structures. Interconnection may be viewed once again from eq. (3.21) as having a matrix given by

$$T = \begin{bmatrix} 1 & 0 \\ 0 & -1 \end{bmatrix} \tag{3.30}$$

These three matrices are sufficient to determine the form of the parameters of eq. (3.28) from which the digital-filter structure is immediately evident. Thus we have for the series arm Z

$$\begin{aligned} \sigma_{11} &= (R_2 - R_1 + Z)/(R_2 + Z) \\ \sigma_{12} &= R_1/(R_2 + Z) \\ \sigma_{21} &= R_2/(R_2 + Z) \\ \sigma_{22} &= Z/(R_2 + Z) \end{aligned} \tag{3.31}$$

for the shunt arm Y

$$\begin{aligned}\sigma_{11} &= (G_1 - G_2 - Y)/G_1 \\ \sigma_{12} &= G_2/G_1 \\ \sigma_{21} &= 1 \\ \sigma_{22} &= 0\end{aligned} \tag{3.32}$$

and for the interconnection

$$\begin{aligned}\sigma_{11} &= 1 - R_1/R_2 \\ \sigma_{12} &= R_1/R_2 \\ \sigma_{21} &= 1 \\ \sigma_{22} &= 0\end{aligned} \tag{3.33}$$

Thus, if adjacent elements have $R_1 = R_2$, the interconnection reduces to

$$\begin{aligned}\sigma_{11} &= 0 = \sigma_{22} \\ \sigma_{12} &= 1 = \sigma_{21}\end{aligned} \tag{3.34}$$

(This is precisely the same state of affairs that exists with scattering parameters where the incident wave at one port is the reflected wave from the other port when the two ports have the same normalization resistance.)

We shall simplify the problem further by considering Z and Y as being a single inductance, as by so doing we can modify any equivalent digital-filter structure by making use of the Constantinides digital frequency transformations[10,11,16] (see Section IIID) to derive structures for capacitive Z and Y or, indeed, when they happen to be tuned circuits of the series or parallel kind.

We have for $Z = sL$ and $Y = 1/sL$ the following relationship for s replaced by the bilinear transformation

$$s \to (1 - z^{-1})/(1 + z^{-1}) \tag{3.35}$$

in a manner similar to the one described in the literature.[31,53,18,19,41,5,40] For the series arm we have

$$\begin{aligned}Z &= L\frac{1 - z^{-1}}{1 + z^{-1}} \\ \sigma_{11} &= \frac{(R_2 - R_1 + L) + z^{-1}(R_2 - R_1 - L)}{(R_2 + L) + z^{-1}(R_2 - L)} \\ \sigma_{12} &= \frac{R_1(1 + z^{-1})}{(R_2 + L) + z^{-1}(R_2 - L)}\end{aligned} \tag{3.36}$$

$$\sigma_{21} = \frac{R_2(1+z^{-1})}{(R_2+L)+z^{-1}(R_2-L)}$$

$$\sigma_{22} = \frac{L(1-z^{-1})}{(R_2+L)+z^{-1}(R_2-L)}$$

and for the shunt arm

$$Y = \frac{1}{L}\frac{1+z^{-1}}{1-z^{-1}}$$

$$\begin{aligned} \sigma_{11} &= \frac{[(G_1-G_2)L-1]-z^{-1}[(G_1-G_2)L+1]}{G_1L(1-z^{-1})} \\ \sigma_{12} &= G_2/G_1 \\ \sigma_{21} &= 1 \\ \sigma_{22} &= 0 \end{aligned} \tag{3.37}$$

In a manner similar to the two-port wave approach it is argued that avoidance of delay-free loops between adjacent structures is achieved when the constant term in the numerators of either σ_{11} or σ_{22} above is set equal to zero. It will be observed from the above expressions that this is not possible for σ_{22} but it is possible for σ_{11}. Hence we have $R_1 = R_2 + L$ for the shunt arm. (The bilinear transformation is not a dimensionless quantity. As it replaces the complex frequency s it must be construed as having a unity factor associated with it, measured in rad/s. Hence such expressions as $R_1 = R_2 + L$ are in fact $R_1(\Omega) = R_2(\Omega) + L(H) \cdot \text{rad/s}$ which are in fact dimensionally correct.) The corresponding structures for these elements are shown in Fig. 3.12(a) and Fig. 3.12(b) respectively.

In the modified transfer admittance (m.t.a.) method we have the transformation matrices

$$P = \begin{bmatrix} 0 & 1 \\ G_1 & -1 \end{bmatrix} \quad \text{and} \quad Q = \begin{bmatrix} G_2 & 1 \\ 0 & -1 \end{bmatrix} \tag{3.38}$$

which produce the σ-parameters as defined in eq. (3.27)

$$\begin{aligned} \sigma_{11} &= (AG_1R_2 - CR_2 - BG_1 + D)/(CR_2 - D) \\ \sigma_{12} &= G_1R_2/(CR_2 - D) \\ \sigma_{21} &= 1/(CR_2 - D) \\ \sigma_{22} &= -CR_2/(CR_2 - D) \end{aligned} \tag{3.39}$$

and again for passive reciprocal networks in which we are interested, we

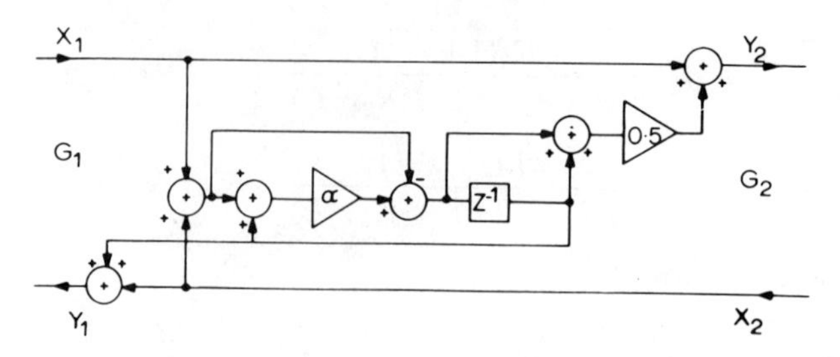

FIG. 3.12(a). Digital equivalent of a series inductance under i.v.r. formulation. $\alpha = (R_2 - L)/(R_2 + L)$, $R_1 = R_2 + L$.

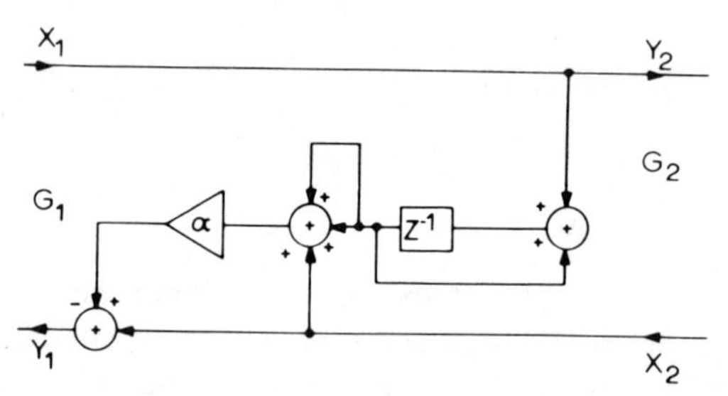

FIG. 3.12(b). Digital equivalent of a shunt inductance under i.v.r. formulation. $\alpha = 1/G_1L$, $G_1 = G_2 + 1/L$.

also have $AD - BC = -1$. In a manner parallel to above, we have for the series arm Z

$$\begin{aligned}
\sigma_{11} &= (R_2 - R_1 + Z)/R_1 \\
\sigma_{12} &= R_2/R_1 \\
\sigma_{21} &= 1 \\
\sigma_{22} &= 0
\end{aligned} \tag{3.40}$$

and for the shunt arm Y

$$\begin{aligned}
\sigma_{11} &= (G_1 - G_2 - Y)/(G_2 + Y) \\
\sigma_{12} &= G_1/(G_2 + Y) \\
\sigma_{21} &= G_2/(G_2 + Y) \\
\sigma_{22} &= -Y/(G_2 + Y)
\end{aligned} \tag{3.41}$$

For the same reasons as given above we consider $Z = sL$ and $Y = 1/sL$ for which we obtain the realizable situation after the relevant delay-free loop constraint is imposed, obtaining for the series arm Z

$$
\begin{aligned}
Z &= L\frac{1-z^{-1}}{1+z^{-1}} \\
\sigma_{11} &= \frac{2L}{R_1}\,\frac{z^{-1}}{1+z^{-1}} \\
\sigma_{12} &= 1+\frac{L}{R_1} \\
\sigma_{21} &= 1 \\
R_2 &= R_1+L
\end{aligned}
\tag{3.42}
$$

and for the shunt arm Y

$$
\begin{aligned}
Y &= \frac{1}{L}\,\frac{1+z^{-1}}{1-z^{-1}} \\
\sigma_{11} &= \frac{-2z^{-1}}{(1+LG_2)+(1-LG_2)z^{-1}} \\
\sigma_{12} &= \frac{(1+LG_2)(1-z^{-1})}{(1+LG_2)+(1-LG_2)z^{-1}} \\
\sigma_{21} &= \frac{LG_2(1-z^{-1})}{(1+LG_2)+(1-LG_2)z^{-1}} \\
\sigma_{22} &= -\frac{1+z^{-1}}{(1+LG_2)+(1-LG_2)z^{-1}} \\
(G_1-G_2)L &= 1
\end{aligned}
\tag{3.43}
$$

The corresponding structures are shown in Fig. 3.13(a) and Fig. 3.13(b) respectively.

The frequency transformations[10,11,16] may be usefully employed to derive digital-filter structures for other elements of the ladder. In the frequency transformation problem we are interested in transforming a given filter (e.g. low-pass) into some other filter having different passbands and stopbands.[12,15] In the analogue case the complex variable s, for example, is replaced by $s/(s^2+\omega_o^2)$ for a low-pass to band-stop transformation. Equivalently, we can say an inductor is replaced by a parallel tuned circuit. The same state of affairs is applicable to the digital-filter structures but the variable to be replaced is z^{-1}. From these transformations we have chosen a convenient form that will achieve our aim. These are given in Table 3.2.

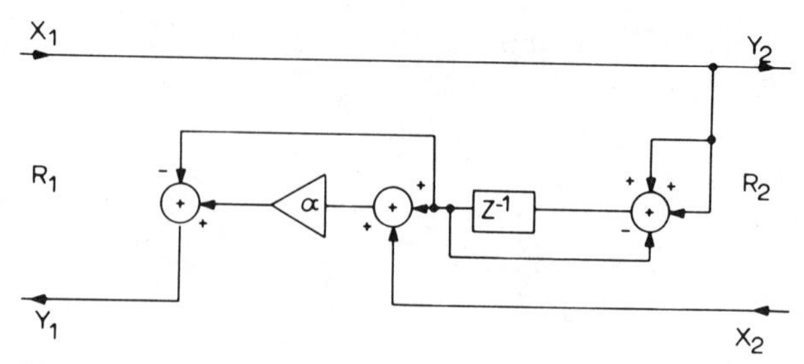

FIG. 3.13(a). Digital equivalent of a series inductance under m.t.a. formulation. $\alpha = R_2/R_1$, $R_1 = R_2 + L$.

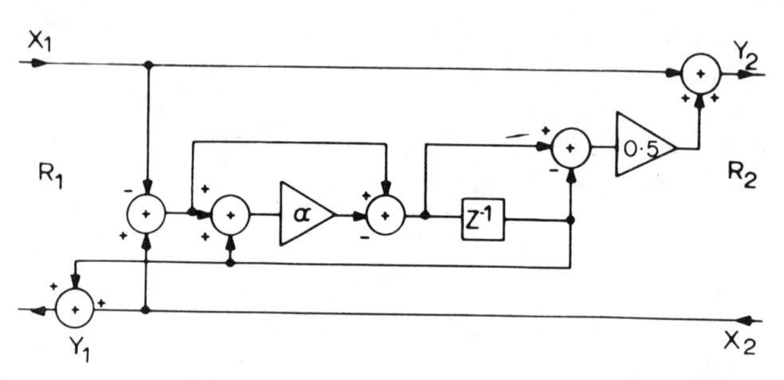

FIG. 3.13(b). Digital equivalent of a shunt inductance under m.t.a. formulation. $\alpha = (G_2L-1)/(G_2L+1)$, $G_1 = G_2 + 1/L$.

	Given	
	L (inductor)	$\longrightarrow z^{-1}$
C (capacitor)	replace L by $1/C$	replace z^{-1} by $-z^{-1}$
L_s, C_s (series)	$L_s + \dfrac{1}{C_s}$	$-z^{-1}\dfrac{z^{-1}+\beta}{1+\beta z^{-1}}$ $\beta = \dfrac{(1-L_sC_s)}{(1+L_sC_s)}$
L_p, C_p (parallel)	$\dfrac{1}{C_p + 1/L_p}$	$z^{-1}\dfrac{z^{-1}+\beta}{1+\beta z^{-1}}$ $\beta = \dfrac{(1-L_pC_p)}{(1+L_pC_p)}$

TABLE 3.2.

In Fig. 3.14 we show digital-filter structures that correspond to a parallel tuned circuit and which can be used to replace z^{-1}, *in situ*, in the structures of Figs 3.12 and 3.13.

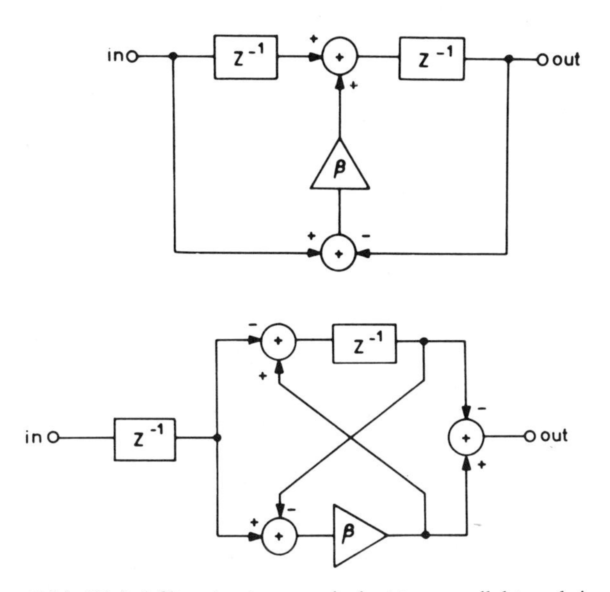

FIG. 3.14. Digital-filter structures equivalent to a parallel tuned circuit.

So far we have examined the linear transformations as applied to two-port ladder elements. The two one-port elements of the source and load are examined below.

In Fig. 3.15(a) we show the resistive source needed for a doubly terminated filter for which we have

$$V_0 = [1 \quad R_s] \begin{bmatrix} V \\ I \end{bmatrix} \tag{3.44}$$

and for the i.v.r. case

$$\begin{bmatrix} X \\ Y \end{bmatrix} = \begin{bmatrix} 1 & 0 \\ 1 & -R \end{bmatrix} \begin{bmatrix} V \\ I \end{bmatrix} \tag{3.45}$$

whereas for the m.t.a. case

$$\begin{bmatrix} X \\ Y \end{bmatrix} = \begin{bmatrix} 0 & 1 \\ G & -1 \end{bmatrix} \begin{bmatrix} V \\ I \end{bmatrix} \tag{3.46}$$

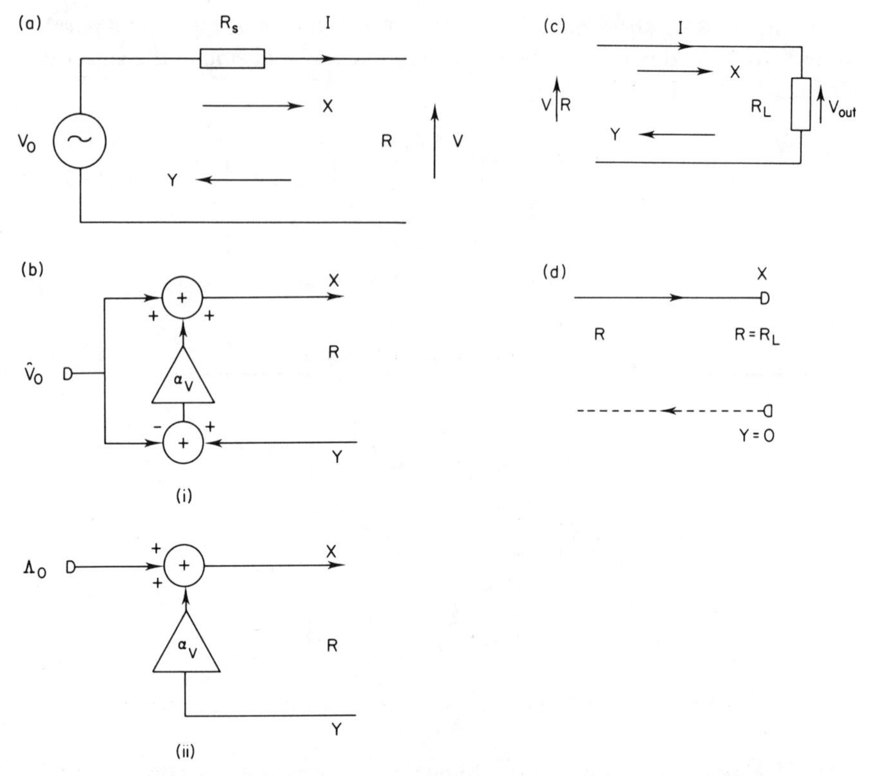

FIG. 3.15(a). Resistive voltage source. (b) Digital equivalents of resistive voltage sources: (i) i.v.r. formulation $\alpha_v = R_s/(R+R_s)$. (ii) m.t.a. formulation $\Lambda_0 = V_0/(R+R_s)$ $\alpha_v = -R/(R+R_s)$. (c) Resistive load. (d) Digital equivalent of resistive load under i.v.r. and m.t.a. formulations.

Hence by direct contribution of eq. (3.44) with eqs (3.45) and (3.46), we obtain the necessary sources for the digital-filter structures given by

$$
\begin{aligned}
X &= \alpha Y + (1-\alpha)V_0 \\
\alpha &= R_s/(R+R_s)
\end{aligned}
\tag{3.47}
$$

for the i.v.r. source and by

$$
\begin{aligned}
X &= \alpha Y + \Lambda_0 \\
\alpha &= -R/(R+R_s) \\
\Lambda_0 &= V_0/(R+R_s)
\end{aligned}
\tag{3.48}
$$

for the m.t.a. source.

The structures for these sources are shown in Fig. 3.15(b).

The resistive load of Fig. 3.15(c) may be transformed in the same way as above. It is observed that

$$V_2 = R_L I_2 \tag{3.49}$$

which in conjunction with eq. (3.45) and eq. (3.46) yields

$$\begin{aligned} Y &= \beta X \\ \beta &= (R_L - R)/R_L \end{aligned} \tag{3.50}$$

for the i.v.r. resistive load and

$$\begin{aligned} Y &= \beta X \\ \beta &= (R_L - R)/R \end{aligned} \tag{3.51}$$

for the m.t.a. resistive load.

The corresponding structures are shown in Fig. 3.15(d) for $R = R_L$.

A fifth order filter has been chosen to illustrate the application of the proposed methods. The analogue filter is shown in Fig. 3.16.

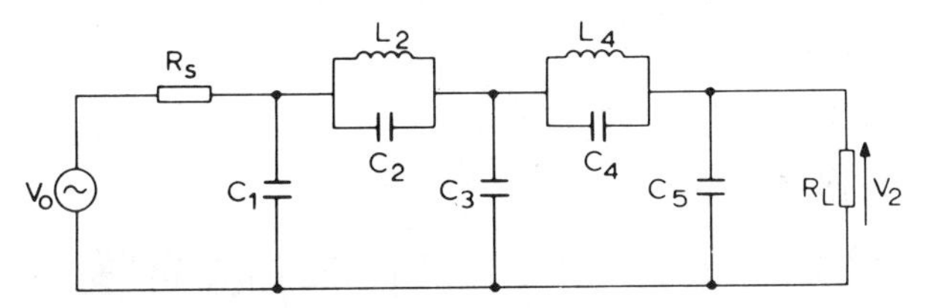

FIG. 3.16. Example of a fifth-order analogue filter. $R_s = R_L = 1$; $C_1 = 1.14359$; $C_2 = 0.20169$; $L_2 = 1.15336$; $C_3 = 1.66967$; $C_4 = 0.59903$; $L_4 = 0.83376$; $C_5 = 0.87504$.

Figures 3.17(a) and 3.17(b) show, respectively, the i.v.r. and m.t.a. structures where the equivalent ladder structures are shown in block form. Each block is obtained from the structures of Figs 3.12 and 3.13 according to the approach adopted for the transformation.

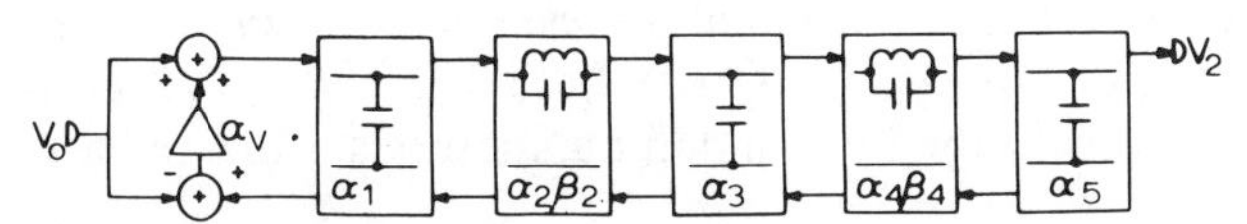

FIG. 3.17(a). I.v.r. digital filter realization of fifth-order analogue filter. $\alpha_v = 0.655165$; $\alpha_1 = 0.601911$; $\alpha_2 = -0.415417$; $\beta_2 = 0.622559$; $\alpha_3 = 0.645250$; $\alpha_4 = -0.020859$; $\beta_4 = 0.333825$; $\alpha_5 = 0.466678$.

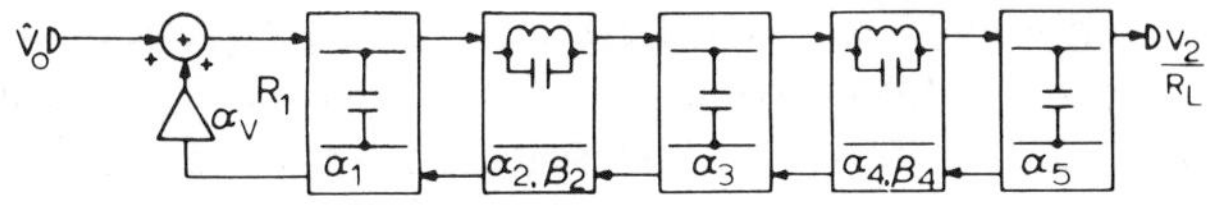

FIG. 3.17(b). M.t.a. digital filter realization of fifth-order analogue filter. $\alpha_v = -0.344835$; $\alpha_1 = -0.203821$; $R_1 = 0.526334$; $\alpha_2 = 0.292292$; $\beta_2 = 0.622559$; $\alpha_3 = -0.290499$; $\alpha_4 = 0.489571$; $\beta_4 = 0.333825$; $\alpha_5 = 0.066644$.

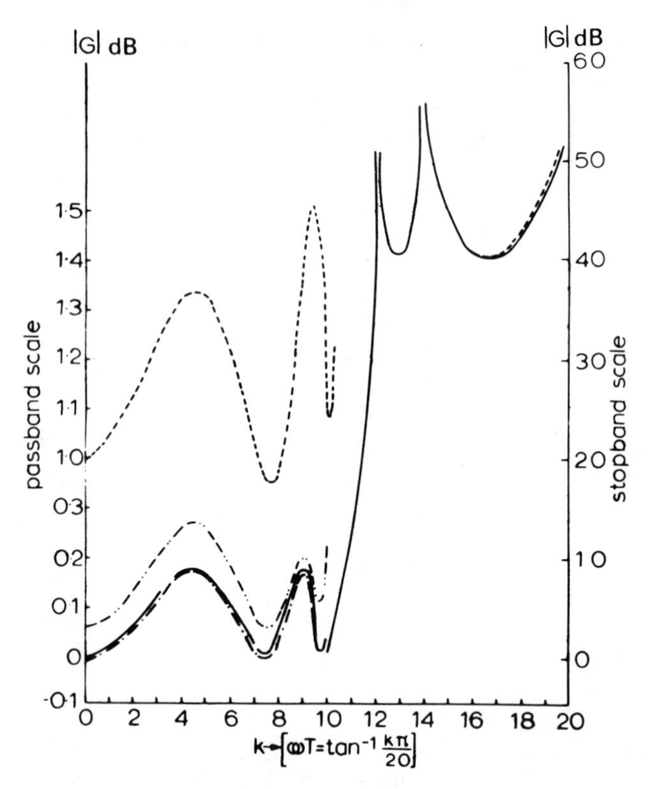

FIG. 3.18. Attenuation response i.v.r. ——— 6 d.p.; —·—· 3 d.p.; ——— ideal; ——·· 2 d.p.; - - - - - - - - 1 d.p.; d.p. = decimal places.

Figure 3.18 shows the attenuation characteristics of the i.v.r. filter when the coefficient is expressed to one decimal place, two places, three and six places, and more than six decimal places. Similarly for the m.t.a. structure in Fig. 3.19. It will be observed from these figures that the sensitivity of the structure derived by the i.v.r. and m.t.a. methods is low indeed.

IIID 1-D frequency transformations

The problem of transforming a given digital filter of some specific critical frequency to another filter of different critical frequency is basically a frequency transformation problem. This is achieved by mapping the complex variable z^{-1} through appropriate functions in accordance with *Constantinides' frequency transformations.*[10,11,16]

It is usually assumed that one has a low-pass filter of a definite cutoff frequency, say β rad/s, from which other low-pass, high-pass, band-pass or

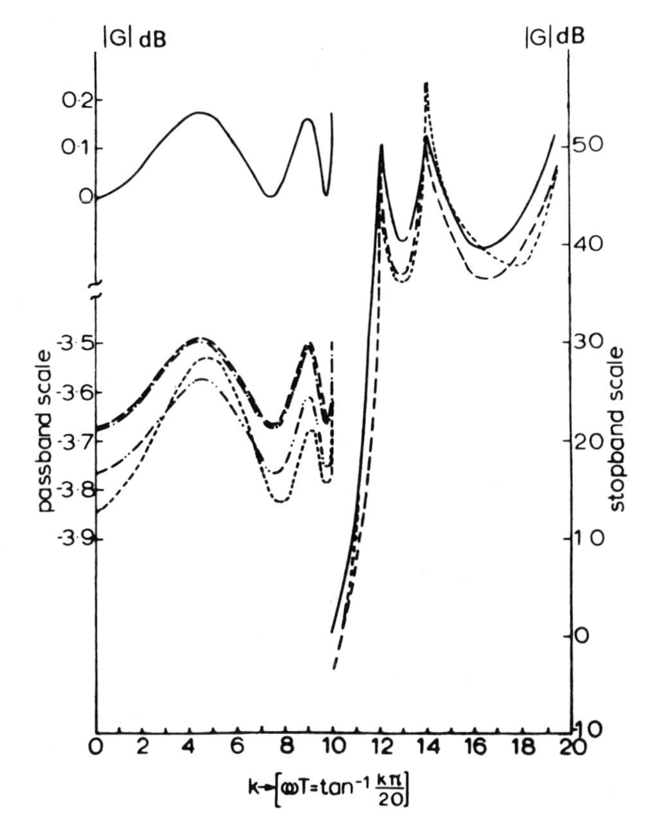

FIG. 3.19. Attenuation response m.t.a. ——— 6 d.p.; —·—· 3 d.p.; ——— ideal; ——·· 2 d.p.; -------- 1 d.p.; d.p. = decimal places.

band-stop filters are required to be derived. The low-pass prototype is usually normalized in frequency with respect to a quarter of the sampling frequency. That is to say that a low-pass digital filter is said to be *normalized* when

$$\beta = \frac{1}{4}\left(\frac{2\pi}{T}\right) \tag{3.52}$$

This value of β is very convenient in that it also corresponds to the cutoff frequency of a low-pass digital filter resulting from a bilinearly transformed analogue filter in accordance with eq. (3.2) with $k = 1$.

The transformations required for the above kind of filters are given in Table 3.3.

It is interesting to note that there are some special cases of these transformations which are particularly simple and have useful properties.

Filter type	Transformation	Associated design formulas
Low-pass	$\dfrac{z^{-1}-\alpha}{1-\alpha z^{-1}}$	$\alpha=\dfrac{\sin\left(\dfrac{\beta-\omega_c}{2}\right)T}{\sin\left(\dfrac{\beta+\omega_c}{2}\right)T}$
High-pass	$-\dfrac{z^{-1}+\alpha}{1+\alpha z^{-1}}$	$\alpha=-\dfrac{\cos\left(\dfrac{\beta-\omega_c}{2}\right)T}{\cos\left(\dfrac{\beta+\omega_c}{2}\right)T}$
Band-pass	$-\dfrac{z^{-2}-\dfrac{2\alpha k}{k+1}z^{-1}+\dfrac{k-1}{k+1}}{\dfrac{k-1}{k+1}z^{-2}-\dfrac{2\alpha k}{k+1}z^{-1}+1}$	$\alpha=\cos\omega_0 T=\dfrac{\cos\left(\dfrac{\omega_2+\omega_1}{2}\right)T}{\cos\left(\dfrac{\omega_2-\omega_1}{2}\right)T}$ $k=\cot\left(\dfrac{\omega_2-\omega_1}{2}\right)T\tan\dfrac{\beta T}{2}$
Band-elimination	$\dfrac{z^{-2}-\dfrac{2\alpha}{1+k}z^{-1}+\dfrac{1-k}{1+k}}{\dfrac{1-k}{1+k}z^{-2}-\dfrac{2\alpha}{1+k}z^{-1}+1}$	$\alpha=\dfrac{\cos\left(\dfrac{\omega_2+\omega_1}{2}\right)T}{\cos\left(\dfrac{\omega_2-\omega_1}{2}\right)T}=\cos\omega_0 T$ $k=\tan\left(\dfrac{\omega_2-\omega_1}{2}\right)T\tan\dfrac{\beta T}{2}$

TABLE 3.3. Frequency transformations from a low-pass digital filter prototype of cutoff frequency β.

For example in the low-pass to high-pass case one can replace z^{-1} by $-z^{-1}$ to yield a high-pass filter of cutoff frequency

$$\omega_c=\frac{\pi}{T}-\beta \tag{3.53}$$

It is particularly simple to implement in that it implies more reversal of signs of z^{-1} in a realization structure of the odd powers of z^{-1} only.

In the band-pass case an arithmetically symmetric filter can be obtained when $\alpha=0$ and $k=1$ for which situation we have

$$\begin{aligned}\omega_1&=\frac{1}{2}\left(\frac{\pi}{T}-\beta\right)\\ \omega_2&=\frac{1}{2}\left(\frac{\pi}{T}+\beta\right)\end{aligned} \tag{3.54}$$

and $\omega_0=\dfrac{1}{2}\pi/T$.

When $\alpha \neq 0$ but with $k = 1$ we have a band-pass filter of bandwidth

$$\omega_2 - \omega_1 = \beta \tag{3.55}$$

and of centre frequency controllable by the parameter α. This is useful in variable centre frequency constant bandwidth applications. Finally for the band-pass case when $k \neq 1$ and $\alpha = 0$ the centre frequency of the band is treated as a quarter of the sampling frequency but the bandwidth is controllable by the parameter. This case is useful in many situations when variable bandwidth band-pass filtering is required at a constant centre frequency, e.g. in data compressions applications.

Similar situations exist for the band-stop case but these will not be repeated here as the principles are identical to the above.

It is of particular interest in the application of the frequency transformations to the Fettweis structures for wave digital filters or to the Constantinides structures for the linearly transformed filters to have the structural form of the transformations which can be used. The transformations of Table 3.3 are directly applicable to such cases and some structural forms are shown in Fig. 3.14. The transformations appropriate for structural replacement are as shown in Table 3.2. The results of application of the frequency transformations are shown in the example of Fig. 3.20.

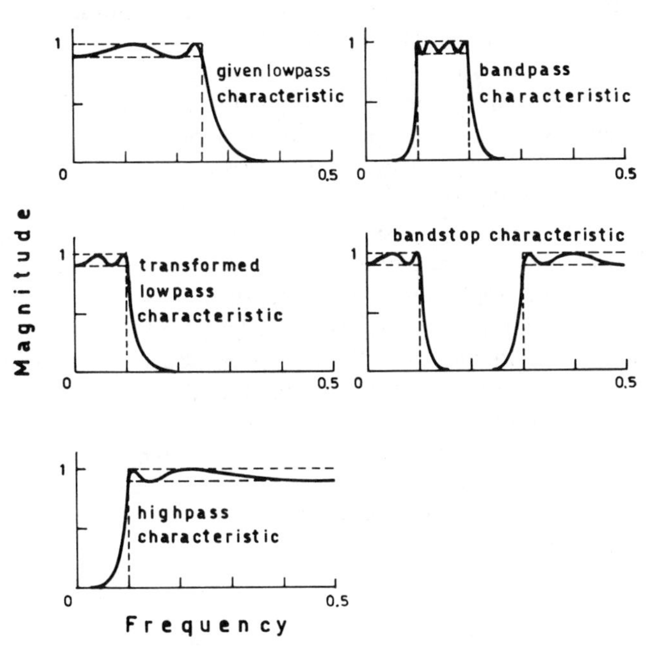

FIG. 3.20. Frequency transformations.

IIIE Analytic techniques for IIR digital filter design

The square of the amplitude characteristic of a IIR digital filter can be determined from the transfer function by setting $z^{-1} = \exp(-j\omega)$. (The sampling period T is assumed to be unity.) Within the passband of the filter the characteristic is required to be as close to 1 as possible whereas within the stopband it should be as near to zero as possible. On this basis different types of filters can be determined analytically from the kind of approximation employed in the two bands. For example Butterworth filters are easily derivable by setting the square of the amplitude characteristic equal to

$$|H(e^{j\omega})|^2 = \frac{1}{1+\left[\dfrac{f(\omega)}{f(\omega_c)}\right]^{2n}} \tag{3.56}$$

where $f(\omega)$ can be $\tan \omega/2$ as in references[51,8,9] for low-pass filters or indeed $\sin \omega/2$. The frequency ω_c is the required cut-off frequency of the digital filter.

From eq. (3.56) it is straightforward to determine the location of the pole and zero positions on the z-plane by substituting the appropriate expression for $f(\omega)$ in terms of z^{-1} (e.g. $\tan \omega/2 = -j(1-z^{-1})/(1+z^{-1})$) and hence the poles in the stable region and the zero for minimum phase can be chosen to construct the transfer function.

In addition to the usual Butterworth response filters, Chebyshev and elliptic kind of filter have been designed by this means.[8,9,44,52]

The optimality of the equiripple approximation makes the Chebyshev filter response desirable for the monotonic stopband case, whereas the elliptic response is desirable for the case when both the passband and stopband are equiripple approximations to different constants respectively.

It is desirable sometimes, however, to have equiripple passbands and arbitrarily defined stopbands in that the stringency of the elliptic equiripple stopband may result to an over-designed filter. In such a case we can employ the following procedure as proposed by Constantinides[13,14] and subsequently used by Deckzy for the derivation of some filter responses.[25,26] The procedure relies on making use of the following transformation which in concept has a correspondence with an equivalent transformation used in passive filter design.[45]

Let $W = \xi + j\eta$, $\xi = \text{Re } W$, $\eta = \text{Im } W$, and $z = e^{sT}$, $s = \sigma + j\omega$. Then the transformation

$$W^2 = \frac{z^{-2} - 2\cos\omega_b \cdot z^{-1} + 1}{z^{-2} - 2\cos\omega_a \cdot z^{-1} + 1} \tag{3.57}$$

maps the arc $(|\omega_a|, |\omega_b|)$, $|\omega_b|>|\omega_a|$, of the unit circle C: $|z^{-1}|=1$ of the z^{-1} plane onto the entire imaginary axis η of the W plane, such that, if

$$|\arg z_1^{-1}| \leq |\arg z_2^{-1}|$$

then

$$|\eta_1| \geq |\eta_2|$$

where $|\omega_a|<|\arg z_1^{-1}|$, $|\arg z_2^{-1}|<\omega_b$ and $|\eta_1|$ and $|\eta_2|$ correspond to the points z_1^{-1} and z_2^{-1}, respectively.

Some properties of the transformation (3.57) can be noted. Consider the real axis ξ of the W plane, so that,

$$\xi^2 = \frac{z^{-2} - 2 \cos \omega_b \cdot z^{-1} + 1}{z^{-2} - 2 \cos \omega_a \cdot z^{-1} + 1}$$

and hence,

$$z^{-2} - 2\frac{(\xi^2 \cos \omega_a - \cos \omega_b)}{\xi^2 - 1} \cdot z^{-1} + 1 = 0$$

Let

$$\psi = \frac{\xi^2 \cos \omega_a - \cos \omega_b}{\xi^2 - 1} \tag{3.58}$$

such that

$$z^{-2} - 2\psi z^{-1} + 1 = 0$$

Therefore

$$z^{-1} = \psi \pm \sqrt{\psi^2 - 1} \tag{3.59}$$

If z^{-1} of (3.59) is to lie on the unit circle, then $|z^{-1}|=1$, and hence from (3.59)

$$\psi^2 < 1 \qquad \text{for } |z^{-1}|=1 \tag{3.60}$$

Condition (3.60), however, when it is applied to (3.58) imposes some constraints on the range of values that ξ can assume, to ensure that $|z^{-1}|=1$.

That is,

$$\psi = \frac{\xi^2 \cos \omega_a - \cos \omega_b}{\xi^2 - 1} < 1$$

and

$$\psi = \frac{\xi^2 \cos \omega_a - \cos \omega_b}{\xi^2 - 1} > -1$$

Thus for $\psi = \pm 1$ we have

$$\frac{\xi^2 \cos \omega_a - \cos \omega_b}{\xi^2 - 1} = \pm 1$$

i.e.

$$\xi = \pm \frac{\sin\left(\frac{\omega_b}{2}\right)}{\sin\left(\frac{\omega_b}{2}\right)} = \pm\alpha \qquad \text{for } \psi = +1$$

and

$$\xi = \pm \frac{\cos\left(\frac{\omega_b}{2}\right)}{\cos\left(\frac{\omega_a}{2}\right)} = \pm\beta \qquad \text{for } \psi = -1$$

Since $|\omega_a| < |\omega_b|$, then

$$\alpha > 1 \quad \text{and} \quad \beta < 1$$

and for condition (3.60) to prevail, it follows that ξ must lie within the intervals

$$-\beta < \xi < \beta, \quad \alpha < \xi < \infty, \qquad -\infty < \xi < +\infty$$

or

$$|\xi| < \beta \quad \text{and} \quad \alpha < |\xi| < \infty \tag{3.61}$$

If ξ lies outside the above intervals, that is in the regions

$$\beta < |\xi| < \alpha \tag{3.62}$$

then condition (3.60) does not hold, and therefore, for any value of ξ in this region there is no value of z^{-1} on the unit circle and the converse is also true.

However, for $\beta < |\xi| < \alpha$, then $\psi^2 > 1$ and therefore z^{-1} of (3.59) becomes real and not equal to unity. That is

$$z^{-1} = \psi + \sqrt{\psi^2 - 1}$$

and

$$z^{-1} = \psi - \sqrt{\psi^2 - 1}$$

These two values of z^{-1} are seen to be reciprocal and therefore for any value of ξ in the region given by (3.62), there correspond two values of z^{-1}

on the real axis of the z^{-1} plane, one of which lies inside the unit circle and the other one outside the unit circle.

While confining the discussion to the W plane, consider the function

$$F(W)=\prod_{i=1}^{n}\left[\frac{W+\alpha_i}{\alpha_i-W}\right] \tag{3.63}$$

where all α_i are real and constant.

When W is completely imaginary then (3.63) can be written

$$|F(W)|=\left|\prod_{i=1}^{n}\left[\frac{(W+\alpha_i)}{(\alpha_i-W)}\right]\right|=1 \tag{3.64}$$

Thus, in view of (3.63) and (3.64), for imaginary values of W, the function $F(W)$ takes the form,

$$F(W)=\mathrm{e}^{jf(W)} \tag{3.65}$$

As a result of (3.65) the function

$$P^2(W)=|1\pm F(W)|^2 \tag{3.66}$$

when W is imaginary, becomes

$$P^2(W)=2[1\pm\cos f(W)] \tag{3.67}$$

This function represents an equiripple variation in amplitude within the extreme values of 0 and 4.

In view of this result, let us define the square of the amplitude characteristic of a digital filter as follows,

$$|G(\omega)|^2=\frac{1}{1\pm\frac{\varepsilon^2}{4}P^2(W)} \tag{3.68}$$

so that, when W is completely imaginary $|G(\omega)|^2$ is equiripple in nature.

Let the arc given by $-\omega_a\le\omega\le\omega_b$ represent the passband of the digital filter, which is transformed to the imaginary axis of the W plane. Hence the passband of the amplitude characteristic given by (3.68) becomes equiripple.

In the foregoing argument we have succeeded in producing an equiripple passband, but no mention was made of the shape of the stopband. Thus, the stopband can be shaped at will by choosing appropriate values of α_i in (3.63).

Let us now consider the case of a low-pass digital filter.[17] If ω_a of the transformation (3.57) is made equal to zero then the passband stretches from 0 to $|\omega_b|$. Thus, the point ω_b becomes the cutoff point ω_c, where ω_c is the cutoff frequency, i.e.

$$W^2 = \frac{z^{-2} - 2\cos\omega_c z^{-1} + 1}{z^{-2} - 2z^{-1} + 1} \tag{3.69}$$

The passband $|\omega| < \omega_c$ is transformed to the entire imaginary axis of the W plane, whereas the stopband is transformed inside the interval $\pm\cos\omega_c/2$ of the real axis, as it can be seen from (3.58) in conjunction with (3.60) by setting $\omega_a = 0$.

Thus, the range of values that α_i of (3.63) can take, for real transmission zeros, is given by

$$-\cos\left(\frac{\omega_c}{2}\right) \le \alpha_i \le \cos\left(\frac{\omega_c}{2}\right) \tag{3.70}$$

Now let α_i lie in the above range and consider (3.63), which can be written in the form,

$$P^2(W) = \left[\prod_i \left(\frac{W+\alpha_i}{W-\alpha_i}\right)^{\frac{1}{2}} + \prod_i \left(\frac{W-\alpha_i}{W+\alpha_i}\right)^{\frac{1}{2}}\right]^2$$

When W is on the imaginary axis then any factor of the first product can be written in the form

$$\frac{W+\alpha_i}{W-\alpha_i} = e^{2j\theta_i} \qquad \theta_i = \tan^{-1}\frac{W}{j\alpha_i} \tag{3.71}$$

Hence, $P^2(W)$ becomes

$$\begin{aligned} P^2(W) &= \left[\exp\left(j\sum_i \theta_i\right) + \exp\left(-j\sum_i \theta_i\right)\right]^2 \\ P(W) &= 2\cos\left(\sum_i \theta_i\right) \end{aligned} \tag{3.72}$$

Now the low-pass transformation of (3.69) when z^{-1} is on the unit circle, takes the form

$$W^2 = \frac{\cos\omega - \cos\omega_c}{\cos\omega - 1} \tag{3.73}$$

Let

$$\cos\omega = \frac{1-t^2}{1+t^2}$$

where

$$t=\tan\frac{\omega}{2}$$

such that (3.73) becomes

$$W^2=\left(\frac{1+\cos\omega_c}{2}\right)\cdot\frac{t^2-t_c^2}{t^2}$$

where $t_c=\tan\omega_c/2$.

Now let

$$x=\frac{t}{t_c},\qquad \gamma=\cos\frac{\omega_c}{2}$$

so that

$$W^2=\gamma^2\left(\frac{x^2-1}{x^2}\right) \tag{3.74}$$

From (3.71) and (3.74) we have

$$\begin{aligned}\theta_i&=\tan^{-1}\frac{W}{j\alpha_i}\\&=\tan^{-1}\left(\frac{\gamma\sqrt{1-x^2}}{\alpha_i x}\right)\end{aligned}$$

or

$$\tan\theta_i=\frac{\gamma\sqrt{1-x^2}}{\alpha_i x}$$

Therefore

$$\cos\theta_i=\frac{\alpha_i x}{\sqrt{\gamma^2-(\gamma^2-\alpha_i^2)x^2}}$$

or

$$\theta_i=\cos^{-1}\frac{\alpha_i x}{\sqrt{\gamma^2-(\gamma^2-\alpha_i^2)x^2}} \tag{3.75}$$

In view of (3.75), (3.72) becomes

$$P(W)=2\cos\left(\sum_i\cos^{-1}\frac{\alpha_i x}{\sqrt{\gamma^2-(\gamma^2-\alpha_i^2)x^2}}\right) \tag{3.76}$$

and hence the squared amplitude characteristic becomes

$$|G(\omega)|^2 = \frac{1}{1+\varepsilon^2 \cos^2\left(\sum_i \cos^{-1} \frac{\alpha_i x}{\sqrt{\gamma^2-(\gamma^2-\alpha_i^2)x^2}}\right)} \tag{3.77}$$

The above form of the square of the amplitude characteristic is the most general form for low-pass digital filters having equiripple passbands and arbitrary stopbands. Because of this arbitrary nature of the stopband, it is possible to impose the necessary constraints on it to ensure that it behaves in a prescribed manner.

For Chebyshev low-pass filters the stopband is monotonic, whereas for elliptic filters the stopband is equiripple. It will be shown presently that these two forms of amplitude characteristics may be obtained from (3.77).

Consider the case of (3.76) when all α_i are made equal to γ. This means that there is no transmission zero of the amplitude characteristics in the stopband (except at the Nyquist frequency) and thus the monotonic transition region extends over the entire stopband. Therefore, the stopband becomes monotonically decreasing.

In (3.76) setting $\alpha_i = \gamma$ results in the following function

$$P(W) = 2\cos\left(\sum_i \cos^{-1} x\right)$$

or

$$P(W) = 2\cos(n\cos^{-1} x)$$

and hence (3.77) becomes

$$|G(\omega)|^2 = \frac{1}{1+\varepsilon^2\cos^2(n\cos^{-1} x)}$$

which is the square of a Chebyshev-type amplitude characteristic.[8,51,52]

Returning to (3.76) and setting $n = 2$, we obtain

$$P(W) = 2\cos\left(2\cos^{-1}\frac{\alpha x}{\sqrt{\gamma^2-(\gamma^2-\alpha^2)x^2}}\right)$$

let

$$\cos^{-1}\frac{\alpha x}{\sqrt{\gamma^2-(\gamma^2-\alpha^2)x^2}} = \theta$$

so that

$$P(W) = 2[2\cos^2\theta - 1]$$

and hence

$$\frac{P(W)}{2}=2\left[\frac{ax}{\sqrt{\gamma^2-(\gamma^2-\alpha^2)x^2}}\right]^2-1,$$

i.e.

$$\frac{P(W)}{2}=-\frac{\gamma^2+\alpha^2}{\gamma^2-\alpha^2}\cdot\frac{x^2-\left(\dfrac{\gamma^2}{\gamma^2+\alpha^2}\right)}{x^2-\left(\dfrac{\gamma^2}{\gamma^2-\alpha^2}\right)} \tag{3.78}$$

Equation (3.78) is a second order approximant which, when written using the symbolic notation of the elliptic low-pass digital filters,[9] is seen to be a second order equiripple approximant. Thus, if

$$\frac{\gamma^2+\alpha^2}{\gamma^2-\alpha^2}=C \tag{3.79}$$

then

$$\frac{\gamma^2}{\gamma^2+\alpha^2}=\frac{C+1}{2C}$$

and

$$\frac{\gamma^2}{\gamma^2-\alpha^2}=\frac{C+1}{2}$$

which are identical to the conditions given in Constantinides.[9]

From (3.79) it can be seen that as α approaches γ, C approaches infinity and hence the stopband ripple tends to zero, which is in accordance with the previous result on Chebyshev filters.

In the general case of elliptic filters perhaps it may be helpful to treat the problem as follows. Let $P^2(W)$ be given by

$$P^2(W)=\left[1\pm\prod_i\left(\frac{W-\alpha_i}{W+\alpha_i}\right)\right]^2 \tag{3.80}$$

and also let $\prod_i (W+\alpha_i)^2=M+N$, where M is the even part and N the odd part of the product. Thus,

$$P^2(W)=\left[1\pm\frac{M-N}{M+N}\right]^2$$

i.e.

$$P^2(W)=\left[\frac{2M}{M+N}\right]^2 \tag{3.81}$$

or

$$P^2(W)=\left[\frac{2N}{M+N}\right] \tag{3.82}$$

Equation (3.81) corresponds to (3.80) with a positive sign, whereas (3.82) corresponds to (3.80) with a negative sign.

From (3.81) and (3.82) we have

$$P^2(W)=\frac{4}{\left[1+\frac{N}{M}\right]^2} \tag{3.83}$$

or

$$P^2(W)=\frac{4}{\left[1+\frac{M}{N}\right]^2} \tag{3.84}$$

Since the above two forms are similar, let us consider one of them, e.g. (3.83), and let us define

$$e^2(W)\triangleq\frac{4}{P^2(W)}=\left[1+\frac{N}{M}\right]^2$$

Generalizing the constraints of the second order, we may say that $P^2(W)$ must show equiripple behaviour varying between $+\infty$ and some constant $C>4$ in the range $0<|W|<\cos(\omega_c T/2)=W_2$ and $P^2(W_2)=\infty$. This means that $e^2(W)$ varies between 0 and $4/C$ with $e^2(W_2)=0$.

Thus we have to find a $P^2(W)$ of the form given by (3.80) such that $\min P^2(W)$ is maximum on $0<|W|<W_2$, which is identical to saying that $\max[1+N/M]$ is minimum on $0<|W|<W_2$. Hence the problem is reduced to the standard Chebyshev problem, which is to find a real rational function $-N/M$ (or $-M/N$), N odd M even, in W, which is the best approximation in the Chebyshev sense to unity in the range $0<|W|<W_2=\cos\omega_2 T/2$, and the solution of this problem is well known.

Note that eq. (3.77) represents amplitude characteristics of low-pass digital filters having equiripple passbands and arbitrary defined stopbands. If zeros of transmission are required in the stopband then the parameters α_i must be chosen so that

$$|\alpha_i|\leq\gamma \tag{3.85}$$

The conventional Chebyshev response, however, requires the stopband to be monotonically decreasing up to the Nyquist frequency.[8] This requirement puts a different constraint on α_i from that given by eq. (3.85).

Let all α_i be identical to α. Then, dropping all constant factors, $P(W, \alpha_i)$ of eq. (3.76) becomes

$$P(W, \alpha) = \cos\left[n \cos^{-1} \frac{\alpha x}{\sqrt{\gamma^2 - (\gamma^2 - \alpha^2)x^2}}\right] \tag{3.86}$$

which is an nth-order Chebyshev polynomial in X, where

$$X = \frac{\alpha x}{\sqrt{\gamma^2 - (\gamma^2 - \alpha^2)x^2}} \tag{3.87}$$

Since the stopband is required in this case to be monotonically decreasing we must have

$$\frac{\mathrm{d}|G|^2}{\mathrm{d}\omega} \leq 0 \qquad \forall \omega_c \leq \omega \leq \frac{\Omega_s}{2} \tag{3.88}$$

in the baseband. But

$$\frac{\mathrm{d}|G|^2}{\mathrm{d}\omega} = -\left\{\frac{1}{1+\varepsilon^2 P^2(\omega)}\right\}^2 2P(\omega)\frac{\mathrm{d}P(\omega)}{\mathrm{d}\omega}\varepsilon^2$$

hence

$$P(\omega)\frac{\mathrm{d}P(\omega)}{\mathrm{d}\omega} > 0$$

and, since $\mathrm{d}\omega/\mathrm{d}x > 0$, $\forall \omega$ we obtain

$$P(\omega)\frac{\mathrm{d}P(\omega)}{\mathrm{d}x} \geq 0 \tag{3.89}$$

From eqs (3.86) and (3.89) we obtain

$$P(\omega)\frac{\mathrm{d}P(\omega)}{\mathrm{d}x} = -\cos Y \sin Y \frac{\mathrm{d}Y}{\mathrm{d}x} \geq 0 \tag{3.90}$$

where Y is given by $Y = n \cos^{-1} X$, and X is given by eq. (3.87). It follows from eq. (3.90) that†

$$|\alpha| \geq \gamma \tag{3.91}$$

Thus if α satisfies expression (3.91) the stopbands of the digital filters of eq. (3.57) will be monotonic.

† The function $P(W, \alpha)$ of eq. (3.90) can be written in two forms: (i) $P(W, \alpha) = \frac{1}{2}\{[X + j\sqrt{(1-X^2)}]^n + [X + j\sqrt{(1-X^2)}]^{-n}\}$ for $X < 1$, (ii) $P(W, \alpha) = \frac{1}{2}\{[X - \sqrt{(X^2-1)}]^n + [X - \sqrt{(X^2-1)}]^{-n}\}$ for $X > 1$ in the stopband $x > 1$ and hence $X > 1$, and to derive the condition given by eq. (3.91) we must use the second expression for $P(W, \alpha)$.

Since there is theoretically an infinity of choices for α satisfying eq. (3.91), it follows that we can have an infinity of low-pass digital filters having equiripple passbands and monotonic stopbands. From this family of filters, however, we choose an optimum one on the basis of maximum sharpness of the amplitude characteristic at the cutoff frequency.

At cutoff frequency we have $x = 1$, and hence the slope is given by

$$\text{slope} = -\frac{\varepsilon^2}{(1+\varepsilon^2)^{\frac{3}{2}}} T \frac{n^2}{2\alpha^2} \cot \frac{\omega_c}{2} \tag{3.92}$$

with

$$|\alpha| \geq \cos(\omega_c/2)$$

Thus the absolute value of the slope is a maximum when α is a minimum. But minimum α corresponds to the case when $|\alpha| = \gamma$, in which case eq. (3.92) becomes

$$(\text{slope})_\gamma = -\left(\frac{n^2}{\sin \omega_c}\right) \frac{\varepsilon^2}{(1+\varepsilon^2)^{\frac{3}{2}}} \tag{3.93}$$

Equation (3.93) gives the maximum possible slope of a low-pass digital filter for a given order n and ripple factor ε. The case of $|\alpha| = \gamma$ corresponds to a $P(W, \alpha) = P(\omega)$ function given by

$$P(\omega) = \cos(n \cos^{-1} x) \tag{3.94}$$

which is immediately recognized as the Chebyshev function given elsewhere.[8]

An interesting case of eq. (3.86) corresponds to that when $\alpha = 1$, when

$$P(\omega) = \cos\left\{n \cos^{-1} \frac{\sin(\omega/2)}{\sin(\omega_c/2)}\right\} \tag{3.94$'$}$$

which is the sine formulation of the conventional Chebyshev problem.[46] In view of the optimum property of eq. (3.94) it follows that eq. (3.94)′ will give a less sharp filter at cutoff. If we take the condition of maximum sharpness at cutoff as a necessary condition, the solution of eq. (3.94) is unique.

IIIF Design techniques of 2-D IIR digital filters

IIIF1 Shanks design technique

The design technique proposed by Shanks[54] consists of mapping (1-D) into (2-D) filters with arbitrary directivity in a (2-D) frequency response plane.

These filters are called rotated filters because they are obtained by rotating (1-D) filters.

Suppose a (1-D) continuous filter whose impulse response is real, is given in its factored form:

$$H_1(s) = H_0\left[\prod_{i=1}^{m}(s-q_i)\Big/\prod_{i=1}^{n}(s-p_i)\right] \tag{3.95}$$

where H_0 is a scalar gain constant. The zero locations q_i and the pole locations p_i may be complex, in which case their conjugates are also present in the corresponding product. The cutoff angular frequency for this filter is assumed to be unity.

The filter given in eq. (3.95) can also be viewed as a (2-D) filter that varies in one dimension only and could be written as follows:

$$H_2(s_1, s_2) = H_1(s_2) = H_0\left[\prod_{i=1}^{m}(s_2-q_i)\Big/\prod_{i=1}^{n}(s_2-p_i)\right] \tag{3.96}$$

Rotating clockwise the (s_1, s_2) axis through an angle β by means of the transformation

$$s_1 = s_1' \cos\beta + s_2' \sin\beta \tag{3.97a}$$

$$s_2 = -s_1' \sin\beta + s_2' \cos\beta \tag{3.97b}$$

will result in a filter whose frequency response is rotated by an angle $-\beta$ with respect to the frequency response of (3.96)

$$H_2(s_1', s_2') = H_0 \frac{\prod_{i=1}^{m}[(s_2' \cos\beta - s_1' \sin\beta) - q_i]}{\prod_{i=1}^{n}[(s_2' \cos\beta - s_1' \sin\beta) - p_i]} \tag{3.98}$$

$H_2(s_1', s_2')$ describes a continuous (2-D) filter in the new co-ordinate system of s_1' and s_2'. The corresponding digital version of the above filter is obtained through the application of the (2-D) bilinear z-transform defined by the following two equations:

$$s_1' = \frac{1-z_1}{1+z_1} \tag{3.99a}$$

$$s_2' = \frac{1-z_2}{1+z_2} \tag{3.99b}$$

In the above, it is assumed that the sample interval T is the same in both directions. Substituting eq. (3.99) into (3.98) will result in:

$$H(z_1, z_2) = A \prod_{i=1}^{M} \frac{a_{11}^i + a_{21}^i z_1 + a_{12}^i z_2 + a_{22}^i z_1 z_2}{b_{11}^i + b_{21}^i z_1 + b_{12}^i z_2 + b_{22}^i z_1 z_2} \tag{3.100}$$

where:

$$A = H_0$$

$$M = \max(m, n)$$

$$\left.\begin{aligned} a_{11}^i &= \cos\beta - \sin\beta - \tfrac{1}{2}Tq_i \\ a_{21}^i &= \cos\beta + \sin\beta - \tfrac{1}{2}Tq_i \\ a_{12}^i &= -\cos\beta - \sin\beta - \tfrac{1}{2}Tq_i \\ a_{22}^i &= -\cos\beta + \sin\beta - \tfrac{1}{2}Tq_i \end{aligned}\right\} \quad \text{for } 1 \le i \le m \tag{3.101}$$

$$a_{11}^i = a_{21}^i = a_{12}^i = a_{22}^i = 1, \qquad \text{for } m < i \le M$$

$$\left.\begin{aligned} b_{11}^i &= \cos\beta - \sin\beta - \tfrac{1}{2}Tp_i \\ b_{21}^i &= \cos\beta + \sin\beta - \tfrac{1}{2}Tp_i \end{aligned}\right\} \quad \text{for } 1 \le i \le n$$

$$\left.\begin{aligned} b_{12}^i &= -\cos\beta - \sin\beta - \tfrac{1}{2}Tp_i \\ b_{22}^i &= -\cos\beta + \sin\beta - \tfrac{1}{2}Tp_i \end{aligned}\right\} \quad \text{for } 1 \le i < n$$

$$b_{11}^i = b_{21}^i = b_{12}^i = b_{22}^i = 1 \qquad \text{for } n < i \le M$$

Unfortunately this technique as it stands does not guarantee the stability of the designed filter and the approach suffers from warping effects of the bilinear transformation on the frequency response. A modification to this technique has been introduced by Costa and Venetsanopoulos.[23]

IIIF2 Costa and Venetsanopoulos design technique

In this method it is shown that the rotated filters can be used in designing circularly symmetric (2-D) recursive filters.[23] A number of rotated filters whose angles of rotation are uniformly distributed over 180° results in a filter having a magnitude response which approximates a circularly symmetric cutoff boundary by a polygon. This polygon has an even number of sides because each rotated filter contributes two sides of the polygon.

It has been proved that stability of the designed filter is ensured if the following two conditions hold:

(i) $270° \le \beta \le 360°$, where β is the angle of rotation, and

(ii) $C_i < 0$ for $i = 1, 2, \ldots, M$, where $C_i = \text{Re}\, p_i$ and p_i represents the location of a pole.

This has been derived through the knowledge of the stability constraints on the coefficients of the denominator $B(z_1, z_2)$ of a second order (2-D) recursive digital filter of the form of eq. (3.100).

IIIF3 A transformation technique for (2-D) recursive filters

This technique is based on the use of a two-variable reactance function as a transformation applied to a 1-D low-pass filter.[1]

A two-variable reactance function can be written[47,57]

$$f(s_1, s_2) = \frac{a_1 s_1 + a_2 s_2}{1 + b s_1 s_2} \tag{3.102}$$

The method makes use of the second-order two-variable reactance function, eq. (3.102) as a transformation applied to a 1-D low-pass filter of the form

$$H(s) = \sum_{i=1}^{N} \frac{A_i}{s + B_i} \tag{3.103}$$

to realize a first-quadrant, 2-D recursive filter. Second-, third-, and fourth-quadrant filters are obtained by the same transformation but replacing s_2 by $-s_2$ for second, s_1 by $-s_1$ and s_2 by $-s_2$ for the third, and s_1 by $-s_1$ for the fourth quadrant filter. These four one-quadrant sections are cascaded and the bilinear transformation is used in order to obtain the digital version of the filter.

The cutoff boundary of the filter depends on the choice of the cutoff frequency of the 1-D filter, eq. (3.103) and the coefficients of the two-variable reactance function eq. (3.102). The resulting filter is a zero-phase recursive filter having symmetry with respect to the ω_1 and ω_2 axis. Since the z-transform of the filter has already been decomposed into four one-quadrant filters, the decomposition technique of Pistor[49] is avoided and in addition the resulting filter is stable.

To illustrate the proposed design technique several low-pass (2-D) recursive digital filters are constructed from third-order Butterworth and Chebyshev filters as low-pass prototypes. Figure 3.21 represents the amplitude characteristic of a (2-D) recursive digital filter obtained from a third-order Butterworth low-pass prototype with the a_1 and a_2 coefficients of the numerator of the two-variable reactance function set to unity. Figure 3.22 is the amplitude spectrum of a (2-D) recursive digital filter obtained from a third-order Chebyshev filter with a_1 and a_2 set to unity and $b = 0.1$.

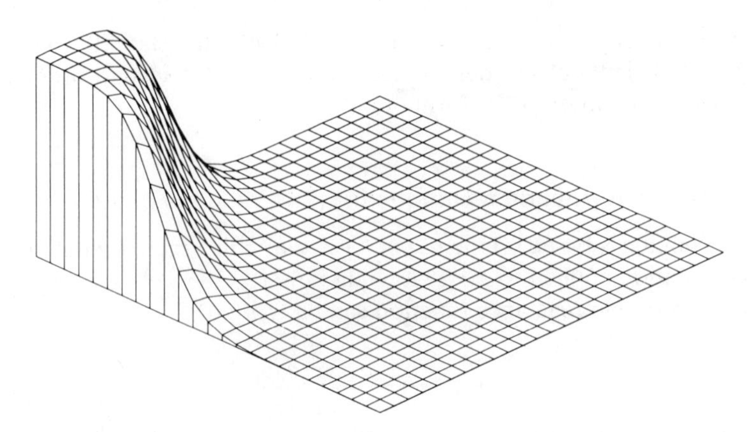

FIG. 3.21. Amplitude characteristic of a 2-D recursive filter designed by means of a 2-D reactance transformation from a third-order 1-D Butterworth filter.

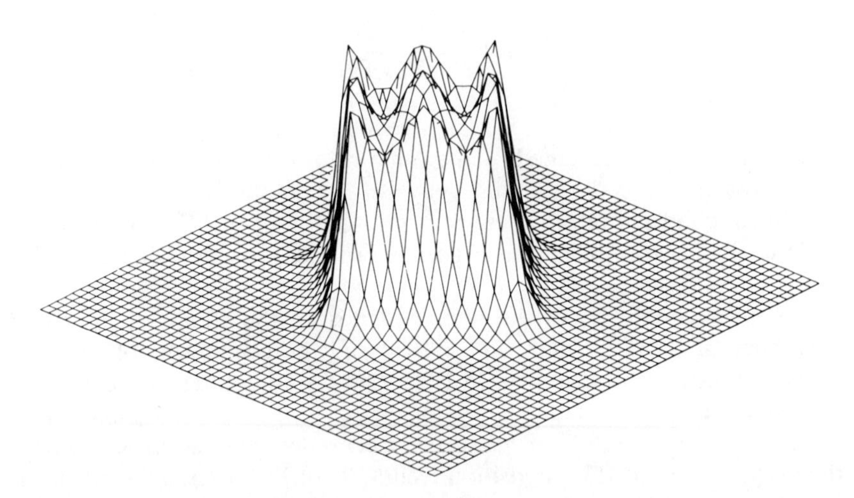

FIG. 3.22. Amplitude characteristic of a 2-D recursive filter designed by means of a 2-D reactance transformation from a third-order 1-D Chebyshev filter.

IIIF4 1-D to 2-D mapping of the squared magnitude function

In Chapter 1 it was shown that starting from a first-quadrant filter with transfer function $H(z_1, z_2)$ and impulse response $h_1(n_1, n_2)$ it is possible to define the corresponding second-, third- and fourth-quadrant filters by

$$h_1(n_1, n_2) = h_2(n_1, -n_2) = h_3(-n_1, -n_2) = h_4(-n_1, n_2) \qquad (3.104)$$

with transfer functions[3]

$$H(z_1, z_2) = H_2(z_1, z_2^{-1}) = H_3(z_1^{-1}, z_2^{-1}) = H_4(z_1^{-1}, z_2) \qquad (3.105)$$

The cascade of these four filters is a zero-phase digital filter whose frequency response can be expressed as

$$G(\omega_1, \omega_2) = \frac{\sum_{m=0}^{N_s} \sum_{n=0}^{N_s} p(m, n) \cos(m\omega_1) \cos(n\omega_2)}{\sum_{k=0}^{K_s} \sum_{l=0}^{K_s} q(k, l) \cos(k\omega_1) \cos(l\omega_2)} \qquad (3.106)$$

where the $p(m, n)$ and $q(k, l)$ can be expressed through the convolution of the coefficients of the four filters. Thus the problem of designing zero-phase digital filters can be divided in two parts:[3] (a) evaluation of the coefficients $p(m, n)$ and $q(k, l)$ of the overall transfer function to satisfy the requirements of the problem and (b) factorization of the transfer function obtained in the four component filters, which recurse in different directions.

To avoid designing in two dimensions,[3] which takes a great deal of computation time, the first step of the procedure can be performed by transformation of the squared magnitude frequency response of a 1-D digital filter which has the form

$$|H(\omega)|^2 = G(\omega) = \frac{\sum_{n=0}^{N_s} p(n) \cos(n\omega)}{\sum_{k=0}^{K_s} q(k) \cos(k\omega)} \qquad (3.107)$$

$G(\omega)$ can be transformed in a 2-D function of the form of eq. (3.106) by the mapping introduced by McClellan for nonrecursive digital filters†[43]

$$\cos\omega = A\cos(\omega_1) + B\cos(\omega_2) + C\cos(\omega_1)\cos(\omega_2) + D \qquad (3.108)$$

The form of this mapping can be controlled by a suitable choice of the coefficients A, B, C, D; by selecting $A = B = C = -D = \frac{1}{2}$, the mapping gives approximately circular symmetry frequency responses.

Having found the coefficients $p(m, n)$ and $q(k, l)$, it is now necessary to determine the matrices $a(m, n)$ and $b(k, l)$ to be used in the recursive evaluation of the output. Due to the fact that $G(\omega_1, \omega_2)$ is zero-phase and positive, the method introduced by Pistor (based on the cepstrum properties), can be used for the factorization of the matrix $q(k, l)$ in four matrices defined on different quadrants. The numerator can be factorized

† It is important to point out that the method can be applied to any form of the 1-D filter transfer function (direct, cascade, parallel). In the cascade and parallel cases, every section can be transformed and stabilized independently.

in the same way or it can be realized as a linear phase nonrecursive digital filter in cascade with the recursive ones.

However the decomposition procedure is approximate, due to the truncation which is performed. When the error in the approximation of the denominator is large, the numerator of the filter, considered as a nonrecursive filter in cascade with the recursive ones, can be used to correct the errors introduced by truncating the series. To this end, it can be observed that the function to be approximated through the numerator does not present discontinuities and is generally smooth; a simple optimization method can be used based on the minimization of the weighted squared error.

In Figs 3.23 and 3.24 examples of this technique are presented.[3] In Fig. 3.23 the transfer function is shown, obtained by starting from the squared

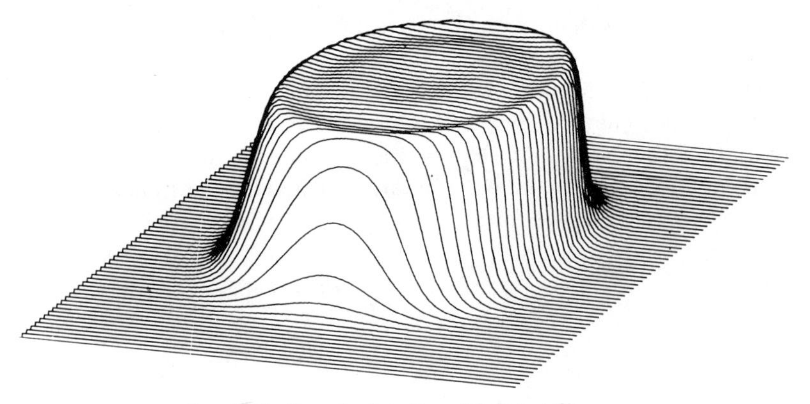

FIG. 3.23. Transfer function of 2-D test filter.

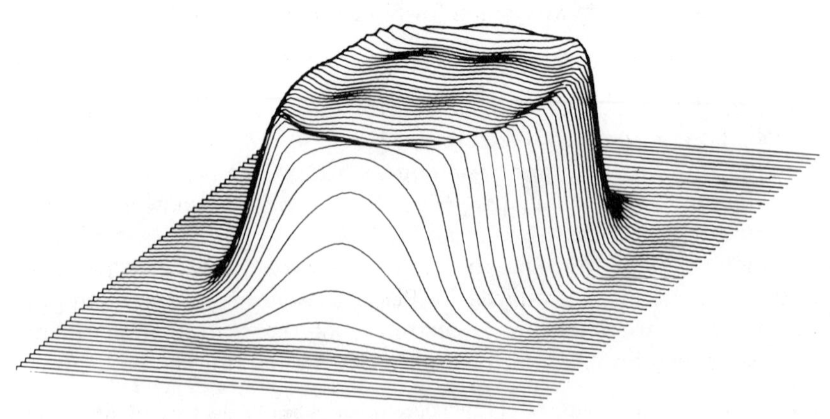

FIG. 3.24. Frequency response of 2-D recursive digital filter designed by means of 1-D to 2-D mapping of the squared magnitude function.

magnitude frequency response of a fourth-order Chebyshev filter with a 2% ripple, a normalized frequency of $f_c = 0.25$ and a frequency at -20 dB of $f_{-20} = 0.35$.

Figure 3.24 shows the stabilized filter frequency response having the denominator decomposed in four filters with $K_s = 4$ and being the numerator, with $N_s = 6$, used to correct part of the errors introduced in the truncation of the denominator. The maximum inband oscillation is 0.048 and the maximum outband oscillation is 0.018.

IIIG Algorithmic methods for the design of 1-D IIR filters

The transformation techniques described in the previous paragraphs are very useful where standard forms of filters as low-pass, band-pass, stop-band filters have to be designed. If arbitrary frequency or impulse responses have to be synthetized some algorithmic design procedures must in general be used.

In the following paragraphs some of these methods will be summarized trying to point out the important features with reference to the following points: (i) approximation criterion, (ii) convergence of the algorithm, (iii) computational efficiency. In general these design procedures can be divided into two classes, with reference to the domain of approximation. The first class deals with the procedures to meet some requirements in the frequency domain, while the second starts from requirements on the impulse response of the filter.

IIIG1 Frequency domain techniques

As shown in Chapter 1 it is impossible to have real-time IIR filters with linear phase. In general the frequency response of these filters will be a complex function defined by a magnitude and a phase or group delay and hence different specification for approximation frequency are possible which are the following three cases: (i) approximation of magnitude or squared magnitude of the frequency response, (ii) approximation of group delay response, (iii) approximation of both responses, that is to say of the magnitude and group delay.

For a filter realized in the cascade form the transfer function can be written as

$$H(z) = k_0 \prod_{i=1}^{N} H_i(z) = k_0 \prod_{i=1}^{N} \frac{z^2 + a_{1i}z + a_{2i}}{z^2 + b_{1i}z + b_{2i}} \tag{3.109}$$

This is the form commonly used in the approximation procedures. Let us now discuss the form of its magnitude, group delay on squared magnitude as a function of the coefficients and as a function of the polar co-ordinates of poles and zeros.[25]

The magnitude function can be obtained as a function of the coefficients starting from the (3.109) in the following form

$$|H(e^{j\omega})| = k_0 \prod_{i=1}^{N} \left| \frac{(1+a_{2i})\cos\omega + a_{1i} + j(1-a_{2i})\sin\omega}{(1+b_{2i})\cos\omega + b_{1i} + j(1-b_{2i})\sin\omega} \right| \tag{3.110}$$

whilst for the group delay, using the general result of Deczky,[24] the following expression can be written

$$\tau(e^{j\omega}) = \sum_{i=1}^{N} \left[\mathrm{Re}\left(\frac{2\cos\omega + b_{1i} + j2\sin\omega}{(1+b_{2i})\cos\omega + b_{1i} + j(1-b_{2i})\sin\omega} \right) - \mathrm{Re}\left(\frac{2\cos\omega + a_{1i} + j2\sin\omega}{(1+a_{2i})\cos\omega + a_{1i} + j(1-a_{2i})\sin\omega} \right) \right] \tag{3.111}$$

A simpler form can be obtained using as variables the polar co-ordinates of poles (r_{pi}, φ_{pi}) and zeros (r_{oi}, φ_{oi}) of the filter. In this case we have

$$|H(e^{j\omega})| = k_0 \prod_{i=1}^{N} \frac{\{1-2r_{oi}\cos(\omega-\varphi_{oi}) + r_{oi}^2\}^{\frac{1}{2}}\{1-2r_{oi}\cos(\omega+\varphi_{oi}) + r_{oi}\}^{\frac{1}{2}}}{\{1-2r_{pi}\cos(\omega-\varphi_{pi}) + r_{pi}^2\}^{\frac{1}{2}}\{1-2r_{pi}\cos(\omega+\varphi_{pi}) + r_{pi}\}^{\frac{1}{2}}} \tag{3.112}$$

and

$$\tau(e^{j\omega}) = \sum_{i=1}^{N} \left[\frac{1-r_{pi}\cos(\omega-\varphi_{pi})}{1-2r_{pi}\cos(\omega-\varphi_{pi}) + r_{pi}^2} + \frac{1-r_{pi}\cos(\omega+\varphi_{pi})}{1-2r_{pi}\cos(\omega+\varphi_{pi}) + r_{pi}^2} - \frac{1-r_{oi}\cos(\omega-\varphi_{oi})}{1-2r_{oi}\cos(\omega-\varphi_{oi}) + r_{oi}^2} - \frac{1-r_{oi}\cos(\omega+\varphi_{oi})}{1-2r_{oi}\cos(\omega+\varphi_{oi}) + r_{oi}^2} \right] \tag{3.113}$$

Finally the squared magnitude of the filter can be expressed in the form

$$|H(e^{j\omega})|^2 = k_0 \prod_{i=1}^{N} \frac{1 + c_{1i}\cos\omega + c_{2i}\cos 2\omega}{1 + d_{1i}\cos\omega + d_{2i}\cos 2\omega} = \prod_{i=1}^{N} \frac{N_i(\omega)}{D_i(\omega)} \tag{3.114}$$

where the coefficients $c_{ij} d_{ij}$ can be obtained from the nonperiodic autocorrelation of the sequences $(1, a_{1i}, a_{2i})$ and $(1, b_{1i}, b_{2i})$ respectively.

Two different but, in the limit, related error criteria have been used in the IIR filter approximation problems; the minimum p-error criterion and the minimax criterion. Using these error criteria the approximation problem can be given in the following general form.

Given a real-valued function $f(\omega)$ defined on a compact set Ω and an approximation function $g(\omega, c)$ which depends on a vector c of parameters $c \equiv (c_0, c_1 \ldots c_N)$, the vector c^* must be determined so that

(a) for the minimum p-error criterion we have

$$\left\{\int_\Omega w(\omega)|g(\omega, c^*)-f(\omega)|^p \, d\omega\right\}^{1/p} \le \left\{\int_\Omega w(\omega)|g(\omega, c)-f(\omega)|^p \, d\omega\right\}^{1/p} \tag{3.115}$$

(b) for the minimax error criterion we have

$$\max_{\omega\in\Omega} w(\omega)|g(\omega, c^*)-f(\omega)| \le \max_{\omega\in\Omega} w(\omega)|g(\omega, c)-f(\omega)| \tag{3.116}$$

for all c, where $w(\omega)$ is an optional positive function. Criterion (a) tends to criterion (b) in the limit as p tends to ∞. In (3.115), (3.116) $f(\omega)$ can be any one of the functions considered before, that is the squared magnitude, the magnitude, the group delay or any combination of these functions. A final remark about the stability problem is now necessary. If we are approximating the squared magnitude (3.114) it is not possible to control the stability of the filter, in the sense that the filter is definitely unstable as shown in Chapter 1. Hence the approximation procedure must be performed without any control of the stability and the result obtained is then factorized in its minimum phase and maximum phase parts from which the stable part in chosen.

If we are approximating the magnitude and/or the group delay it is possible to control the stability of the filter. Using as variables the polar co-ordinates of zeros and poles the stability can be controlled by modifying the approximation routines to limit these values to less than 1 for the radii of the poles. Stability control can also be introduced in procedures based on the coefficients of the filter as optimization variables. In this case, however, the procedure is more involved in that at every step of the approximation procedure a test of the stability of the filter has to be performed, and after the test is carried out the step in the variation of the coefficients has to be reduced if the filter has been found to be unstable.

IIIG2 Approximation methods based on the minimum p-error criterion

A first approach to the design of IIR filters with arbitrary magnitude has been proposed by Steiglitz,[55] using a minimum p-error criterion with $p = 2$ (minimum squared error) and a nonlinear optimization technique.

The function to be minimized is the magnitude function specified on a discrete set of frequencies and is given by

$$E(c)=\sum_{i=1}^{M}(|H(e^{j\omega_i})|-H_i)^2 \tag{3.117}$$

where the H_i are the desired values at the frequencies ω_i, $i=1,\ldots,M$ and the vector c is chosen as the vector of the coefficients of the second order sections and K_o

$$c\equiv(a_{1i}, a_{2i}, b_{1i}, b_{2i}, K_o), \qquad i=1\ldots N \tag{3.118}$$

The error functional (3.117) is a nonlinear function of the coefficients. So a nonlinear technique, as for example the Fletcher and Powell algorithms[34] must be used in the minimization. The use of this algorithm is very convenient because a programmed version is available in the standard IBM libraries[56] and only the gradient of the function c is necessary in the optimization procedure.

In the original paper the form of the gradient is presented together with a description of the organization of the program. Moreover a FORTRAN code of the program is available.[48]

We observe that in this procedure the approximation function is the magnitude of the filter, using as variable the coefficients of the z-transfer function. This means that the stability is not guaranteed as a result of the approximation procedure. So the obtained filter has to be transformed in its minimum phase version using the pole inversion procedure (Chapter 1, Section IQ). Sometimes the filter so obtained can be again optimized and a new lower minimum of $E(c)$ can be obtained. So the entire procedure consists in the error minimization, the stability test and a new optimization if a pole inversion occurs.

An extension of this method is due to Deczky.[25,27] The main differences in the approximation procedure are the following: (i) extension to the $p>2$ case, (ii) extension to the approximation of group delay, and magnitude plus group delay, (iii) use of the weighted norm of the error, (iv) use of the co-ordinates of poles and zeros of the second order sections as optimization parameters, which makes easier the control of the stability of the filter.

The distance function or error criterion in this case assumes the form

$$E_{2p}(c)=\sum_{k=1}^{M} w(\omega_k)|g(\omega_k, c)-f(\omega_k)|^{2p} \tag{3.119}$$

where the function $g(\omega, c)$ can be the magnitude of the filter, the group delay or as some combination of these two functions.

This technique can be used to design all pass sections of the form

$$\frac{a_{2i}z^2 + a_{1i}z + 1}{z^2 + a_{1i}z + a_{2i}} \tag{3.120}$$

having a prescribed group delay; this is of course very important in phase equalization problems.

Some modifications to the standard IBM routine for the Fletcher and Powell algorithm have to be made to facilitate a stable filter derivation as the output of the optimization procedure. This can be done by constraining the magnitude of the polar position of the poles r_{pi} to be less than 1.

IIIG3 Design methods based on the minimax error of polynomials

An equiripple approximation to an arbitrary function by means of a rational function can be obtained using a minimum p-error approximation method, when $p \to \infty$. Those methods which we do not describe in detail, because they are conceptually similar to those presented above, have been studied intensively recently.[6,2] Several problems have been considered with respect to the determination of the best form of the error function for large p as for example the choice of the starting point in the approximation, the check of optimality of the solution, the choice of the grid on which the errors are computed. In addition other algorithms than the Fletcher Powell case have been considered for the nonlinear optimization.

Another general proposed method is based on the use of the second Remez algorithm for the approximation of arbitrary functions by means of rational approximation functions. The Remez algorithm is a multiple exchange algorithm similar to the one presented in the FIR case and it is based on the extension of the alternation theorem to the rational approximations.

This theorem states that there is unique rational approximation which minimizes the weighted maximum error in the set X. Moreover if this unique rational function is given by

$$R_{MN}(\omega, c) = \frac{\sum_{j=0}^{M} a(j)\omega^j}{\sum_{j=0}^{N} b(j)\omega^j} = \frac{N(\omega)}{D(\omega)} \tag{3.121}$$

where $N(\omega)/D(\omega)$ is irreducible, then the number of points at which the error

$$E(\omega) = w(\omega)|R_{MN}(\omega, c) - f(x)| \tag{3.122}$$

assumes the maximum value r^*_{MN} with alternate signs is not less than $L+2$, where $L=M+N$.

Based on this theorem the second Remez algorithm can be summarized in the following form.

(1) Choice of an initial set of $L+1$ points $\{\omega_i\}$ ordered so that

$$\omega_i^{(1)} < \omega_{i+1}^{(1)}, \qquad i=1,\ldots,L+1$$

(2) Solution of the nonlinear system

$$R_{MN}(\omega_i^{(1)}) = f(\omega_i^{(1)}) - (-1)^i \frac{E}{w(\omega_i^{(1)})}, \qquad i=1,\ldots L+2 \qquad (3.123)$$

of $L+2$ equations in $L+2$ unknowns ($L+1$ coefficient and the error value E).

(3) Replacement of each extremum point with the nearest extremum of the same sign and return to step (2).

The algorithm has the same form as the polynomial approximation method presented in Chapter 2, but for the fact that the solution of a nonlinear system substitutes the solution of a linear system in the FIR case.

However a fundamental difference is that whilst in the FIR case the convergence of the algorithm to the equiripple solution is guaranteed irrespective of the starting point in the approximation, in the rational case the convergence is guaranteed only if the starting point in the approximation is "sufficiently" near to the optimum, in the minimax sense, so particular care has to be taken in the choice of the starting point of the algorithm.

Deczky[25,26,27] applied this algorithm to the magnitude or group delay approximation of an IIR digital filter in the cascade form, using as parameters in the approximation the co-ordinates of the zeros and poles. In particular he pointed out that the expression of the group delay is not simply a ratio of polynomials, but nonlinear constraints exist between the coefficients of the polynomials and these constraints can lead to situations where the best Chebyshev approximation may not be equiripple. Thus some modifications must be made to the procedure to obtain the optimum solution in the minimax sense. Details of these modifications can be found together with a discussion on the starting point in the approximation. Further details can be found in specialized texts, for example, Cheney.[7]

Other approaches have been proposed for equiripple approximations which are limited to design squared magnitude functions of filters in direct form, because in this case the numerator and denominator can be written as linear combinations of the coefficients.

A first method is based on the differential correction algorithm.[29,30] Our problem is to approximate a function $f(\omega)$ by means a rational approximation $g_k(\omega, c) = N_k(\omega)/D_k(\omega)$ where k represents the kth pass in the iteration procedure. After the computation of the Chebyshev norm of error

$$\Delta_k = \max_{\omega \varepsilon \Omega} w(\omega)|g_k(\omega, c) - f(\omega)| \tag{3.124}$$

where $w(\omega)$ is a positive weighting function and Ω can be the union of disjoint intervals, the functional δ_k can be defined in the following way $(w(\omega) = 1)$

$$\delta_k = \max_{\omega \varepsilon \Omega} \left\{ \frac{|f(\omega)D(\omega) - N(\omega)| - \Delta_k D(\omega)}{D_k(\omega)} \right\} \tag{3.125}$$

where $g(\omega, c) = N(\omega)/D(\omega)$, and $N(\omega)$ and $D(\omega)$ have the forms

$$\begin{aligned} N(\omega) &= c(0) + \sum_{n=1}^{N} 2c(n) \cos n\omega \\ D(\omega) &= d(0) + \sum_{n=1}^{M} 2d(n) \cos n\omega \end{aligned} \tag{3.126}$$

and c is the vector of coefficients $\{c(n)\}$, $\{d(n)\}$. Varying the vector c the functional δ_k is minimized, with the constraints $|d(n)| \leq 1$ and the values of coefficients thus obtained are used to form the $(k+1)$th rational approximation $g_{k+1}(\omega, c) = N_{k+1}(\omega)/D_{k+1}(\omega)$, which becomes the starting function for another iteration of the algorithm.

It is possible to prove that this algorithm converges to the minimax approximation and the approximation is reached when at some iteration $\delta_k \geq 0$. Moreover only the speed of convergence depends on the initial conditions where the convergence is not affected by this choice. A reasonable set of initial conditions could be $D_0(\omega) = 1$ and for Δ_0 a guessed value of the optimum norm.

The problem is now how to find the minimum value of δ_k. This can be obtained by means of linear programming. In fact the relation (3.125) can be written in the form

$$\delta_k \geq \frac{|f(\omega)D(\omega) - N(\omega)| - \Delta_k D(\omega)}{D_k(\omega)} \tag{3.127}$$

which can be rewritten in the form

$$\begin{aligned} f(\omega)D(\omega) - N(\omega) + \Delta_k D(\omega) + \delta_k D_k(\omega) &\geq 0 \\ N(\omega) - f(\omega)D(\omega) + \Delta_k D(\omega) + \delta_k D_k(\omega) &\geq 0 \end{aligned} \tag{3.128}$$

These relations are linear in the variables $\{c(n)\}$, $\{d(n)\}$, and can be solved by means of linear programming, where the functional to be minimized is only δ_k.

The denominator $D(\omega)$ generated by this algorithm is strictly greater than zero, on the contrary if the numerator is not constant, constraints have to be included to present the numerator from becoming less than zero or else some normalization must be introduced to correct for the oscillations around the zero value in the regions where the desired function is zero.

The problem of the approximation of a squared magnitude function using the minimax error criteria can also be solved using linear programming techniques as follows.[50]

Find the coefficients $\{c(n)\}$ and $\{d(n)\}$ such that

$$-\varepsilon(\omega) \leq \frac{N(\omega)}{D(\omega)} - F(\omega) \leq \varepsilon(\omega) \tag{3.129}$$

where $\varepsilon(\omega)$ is a tolerance function which allows us to define the acceptable error on different regions of the frequency axis. From (3.129) it is possible to obtain a set of linear inequalities in the form

$$\begin{aligned} N(\omega) - D(\omega)[F(\omega) + \varepsilon(\omega)] &\leq 0 \\ -N(\omega) + D(\omega)[F(\omega) - \varepsilon(\omega)] &\leq 0 \end{aligned} \tag{3.130}$$

with the additional constraints

$$\begin{aligned} -N(\omega) &\leq 0 \\ -D(\omega) &\leq 0 \end{aligned} \tag{3.131}$$

To solve the preceding problem as linear programming a new auxiliary variable δ is subtracted from the left side of each constraint equations, obtaining the set of relations

$$\begin{aligned} N(\omega) - D(\omega)[F(\omega) + \varepsilon(\omega)] - \delta &\leq 0 \\ -N(\omega) + D(\omega)[F(\omega) - \varepsilon(\omega)] - \delta &\leq 0 \\ -N(\omega) - \delta &\leq 0 \\ -D(\omega) - \delta &\leq 0 \end{aligned} \tag{3.132}$$

and a linear programming problem is formulated as the minimization of δ under the constraints (3.132). If the minimum value of δ is equal to 0, then a solution to the linear problem exists and the coefficients can be obtained through the linear programming technique. If $\delta_{\min}$ is greater then zero then it is not possible to find a solution to the approximation problems. Thus either $f(\omega)$ or $\varepsilon(\omega)$ must be modified to obtain a solution.

Using this method an iterative procedure can be constructed, when, for example, the tolerance function $\varepsilon(\omega)$ for a given number of sections and a given $f(\omega)$ is tightened to obtain the best minimax approximation to $f(\omega)$ with a rational function of fixed order.

IIIG4 Time domain techniques

In Sections IIIG2 and IIIG3 some design methods for filters with arbitrary frequency responses were considered. Let us now briefly consider methods for designing filters the impulse responses of which are required to be approximations in some sense to an arbitrary sampled impulse response.[4]

The impulse response of an IIR digital filter is related to its rational transfer function, that is to the coefficients of its finite difference equation, by the relationship

$$\sum_{n=0}^{\infty} h(n)z^{-n} = \frac{\sum_{k=0}^{M} a(k)z^{-k}}{\sum_{k=0}^{N} b(k)z^{-k}} \tag{3.133}$$

where the sequence $\{h(n)\}$, $n=0,\ldots,\infty$ is the impulse response of the filter and $\{a(k)\}$ and $\{b(k)\}$ are the coefficients of the filter. The time domain approximation problem consists in the choice of the coefficients $\{a(k)\}$ and $\{b(k)\}$, so that $h(n)$ satisfy the relations

$$h(n)=d(n) \qquad n=1,\ldots,L \tag{3.134}$$

where $\{d(n)\}$ is the desired impulse response.

If $L=M+N$ the problem can be solved directly because the constraints (3.134) reduce to a system of linear equations. The output of the filter (3.133) to the $\delta(n)$ sequence, that is the impulse response of the filter can be written for $a(0)=b(0)=1$ as

$$h(n)=b(1)y(n-1)+b(2)y(n-2)+\ldots+b(N)y(n-N)+a(n) \quad 1<n\leq M \tag{3.135}$$

$$h(n)=b(1)y(n-1)+b(2)y(n-2)+\ldots+b(N)y(n-N) \qquad n>M \tag{3.136}$$

In this case it is possible to solve the system (3.136) and to find the coefficients $b(k)$ which produce an $h(n)=d(n)$ for $n=M+1\ldots,M+N$, and then using this set of coefficients the system (3.134) can be solved to obtain the $a(k)$ for which $h(n)=d(n)$ for $n=1,\ldots M$.

This technique, known as the Padé approximation technique, can be used also to approximate frequency responses using as $d(n)$ samples the first few terms of the Fourier series of the function $H(e^{j\omega})$. Unfortunately it is not possible to guarantee the quality of this approximations nor can we guarantee in the frequency domain case stability of the resulting function. However this approach can be used to obtain a good starting point for more sophisticated design techniques as the second Remez algorithm or the minimum p-error algorithms.

If $L>M+N$ the problems expressed by the relations of eq. (3.134) can be solved using both nonlinear and linear approximation techniques. In the first case the minimum squared error functional

$$E(c)=\sum_{k=1}^{L} w(k)[d(k)-h(k)]^2 \tag{3.137}$$

can be minimized using the coefficients $\{a(k)\}$ and $\{b(k)\}$ as variables in the minimization, in the second case linear programming techniques can be used to minimize the maximum difference between the $\{h(n)\}$ and $\{d(n)\}$.

IIIH Recursive 2-D filters with arbitrary frequency response

A minimum p-error technique can be used also to design 2-D recursive filters. In this case the problem can be formulated in the following way.[42]

Given a desired 2-D magnitude function $|H_d(e^{j\omega_1}, e^{j\omega_2})|$ we wish to design a recursive 2-D digital filter whose transfer function has a magnitude that is an approximation of $|H_d(e^{j\omega_1}, e^{j\omega_2})|$ according to the minimum p-error criterion. If we consider $N\times M$ samples of the $|H_d(e^{j\omega_1}, e^{j\omega_2})|$ function, H_{ij} $i=1,\ldots,N$, $j=1,\ldots,M$ it is possible to construct the error functional

$$E(c)=\sum_{i=1}^{N}\sum_{j=1}^{M} [|H(e^{j\omega_{1i}}, e^{j\omega_{2i}})|-H_{ij}]^p \tag{3.138}$$

where $|H(e^{j\omega_1}, e^{j\omega_2})|$ is the magnitude of the frequency response of the approximation filter $H(z_1^{-1}, z_2^{-1})$. E is a function of the coefficients of $H(z_1^{-1}, z_2^{-1})$ and it can be minimized by means of a nonlinear minimization method.

Due to the fact that stability and testing for stability are difficult problems in the 2-D IIR design,[37,54] the form of $H(z_1^{-1}, z_2^{-1})$ can be so chosen to facilitate the control of stability in the minimization procedure. This can be obtained using in the approximation a transfer function expressed as the cascade of first and second order sections of the form

$$H(z_1^{-1}, z_2^{-1})=A\prod_{e=1}^{L}\frac{\sum_{i=0}^{M_1}\sum_{j=0}^{M_2} a_{ij}^{(e)} z_1^{-i} z_2^{-j}}{1+\sum_{i=0}^{N_1}\sum_{j=0}^{N_2} b_{ij}^{(e)} z_1^{-i} z_2^{-j}} \quad (i+j\neq 0) \tag{3.139}$$

In this case it is possible to test the stability of the filters using simple existing relationships. We remind the reader that for a second order section with a denominator expressed in the form $D(z_1, z_2) = 1 + b_{21}z_1 + b_{12}z_2 + b_{22}z_1z_2$, the stability conditions are $|1/b_{12}|$ 1, $|(1 + b_{12})/(b_{21} + b_{22})| > 1$, $|(1 + b_{12})/(b_{21} - b_{22})| > 1$. If any section becomes unstable the incremental step in the values of coefficients in the minimization procedure can then be decreased.

A second algorithm that can be used for 2-D IIR filter design is the differential correction algorithm, which we described in the preceding section. This method can be used to design filters with arbitrary squared magnitude,[30] with the consequence that it is not possible to consider the stability problems during the approximation procedure. Hence the approximation must be followed by the factorization of the obtained squared magnitude function in stable recursive filters, that, eventually, can be used in cascade as parallel combinations to get a zero phase filter if factorization is at all possible.

Using this algorithm it is possible to obtain an equiripple approximation to the desired squared magnitude function, that is an optimum approximation according to the minimax criterion. Unfortunately the factorization procedure destroys this optimally because the necessary truncation of the obtained infinite coefficient filters introduce errors which are different in different zones of the ω_1, ω_2 plane.

References

1. Ahmadi, M., Constantinides, A. G. and King, R. A. (1976). Design technique for a class of stable two-dimensional recursive digital filters. Proceedings: I.E.E.E. International Conference on Acoustics, Speech and Signal Processing, Philadelphia, Pa.
2. Bandler, J. W. and Bardakjian, B. J. (1973). Least *p*th optimization of recursive digital filters. *I.E.E.E. Trans. Audio Electroacoustics* **AU-21**, **5**, 460–70.
3. Bernabo, M., Cappellini, V. and Emiliani, P. L. (1976). Design of 2-dimensional recursive digital filters. *Electronics Letters* **12, 11,** 288–9.
4. Brophy, F. and Salazar, A. C. (1974). Recursive digital filter synthesis in the time domain. *I.E.E.E. Trans. Acoustics, Speech and Signal Processing* **ASSP-22**, **1**, 45–55.
5. Bukowski, J. V. (1975). A comparative study of two linear transformations for deriving digital filters from LC ladder networks. D. I. C. Dissertation, Imperial College of Science and Technology, London.
6. Charalambous, C. (1975). Minimax optimization of recursive digital filters using recent minimax results. *I.E.E.E. Trans. Acoustics, Speech, Signal Processing* **ASSP-23**, **4**, 333–45.
7. Cheney, E. W. (1966). "Introduction to Approximation Theory". McGraw-Hill, New York.

8. Constantinides, A. G. (1967). Synthesis of Chebyshev digital filters. *Electronics Letters* **3**, **3**, 124–6.
9. Constantinides, A. G. (1967). Elliptic digital filters. *Electronics Letters* **3**, **6**, 255–6.
10. Constantinides, A. G. (1967). Frequency transformations for digital filters. *Electronics Letters* **3**, **11**, 487–9.
11. Constantinides, A. G. (1968). Frequency transformations for digital filters. *Electronics Letters* **4**, **7**, 115–16.
12. Constantinides, A. G. (1968). Comment on recursive bandpass digital filter. *I.E.E.E. Proceedings* **56**, **9**, 1604–5.
13. Constantinides, A. G. (1968). Digital filters with equiripple passband. First I.E.E.E. International Symposium on Circuit Theory.
14. Constantinides, A. G. (1969). Digital filters with equiripple passband. *I.E.E.E. Trans. Circuit Theory* **CT-16**, 4, 535–8.
15. Constantinides, A. G. (1969). Design of bandpass digital filters. *I.E.E.E. Proceedings* **57**, **6**, 1229–31.
16. Constantinides, A. G. (1970). Spectral transformations for digital filters. *I.E.E. Proceedings* **117**, **8**, 1585–90.
17. Constantinides, A. G. (1970). Family of equiripple lowpass digital filters. *Electronics Letters* **6**, **11**, 351–3.
18. Constantinides, A. G. (1974). Alternative approach to the design of wave digital filters. *Electronics Letters* **10**, **5**, 59–60.
19. Constantinides, A. G. and Lawson, S. S. (1974). The design and analysis of wave digital filters that imitate the behaviour of lossless transmission line filters. Proceedings: Colloque Problèmes de sensibilité en synthèse des circuits, Brussels, 79–96.
20. Constantinides, A. G. and Haritantis, G. (1975). Wave active filters. *Electronics Letters* **11**, **12**, 254–6.
21. Constantinides, A. G. (1976). Design of digital filters from LC ladder networks. *I.E.E. Proceedings* **123**, **12**, 1307–12.
22. Constantinides, A. G. (1976). Digital filters derivable from analogue networks. International Conference on Information Theory, Patras, Greece.
23. Costa, J. M. and Venetsanopoulos, A. N. (1974). Design of circularly symmetric two-dimensional recursive filters. *I.E.E.E. Trans. Acoustics, Speech and Signal Processing* **ASSP-22**, **6**, 432–43.
24. Deczky, A. G. (1969). General expression for the group delay of digital filters. *Electronics Letters* **5**, **25**, 663–5.
25. Deczky, A. G. (1972). Synthesis of recursive digital filters using the minimum *p*-error criterion. *I.E.E.E. Trans. Audio Electroacoustics* **AU-20**, **4**, 257–63.
26. Deczky, A. G. (1973). Computer aided synthesis of digital filters in the frequency domain. Technical Dissertation, n. 4980, ETH, Zurich, Switzerland.
27. Deczky, A. G. (1974). Equiripple and minimax (Chebyshev) approximation for recursive digital filters. *I.E.E.E. Trans. Acoustics, Speech and Signal Processing* **ASSP-22**, **2**, 98–111.
28. Dubois, E. and Blostein, M. L. (1975). A circuit analysis method for the design of recursive two-dimensional digital filters. 1975 Proceedings: I.E.E.E. International Symposium on Circuits and Systems, Boston.
29. Dudgeon, D. E. (1974). Recursive filter design using differential correction. *I.E.E.E. Trans. Acoustics, Speech and Signal Processing* **ASSP-22**, **6**, 443–8.

30. Dudgeon, D. E. (1975). Two-dimensional recursive filter design using differential correction. *I.E.E.E. Trans. Acoustics, Speech and Signal Processing* **ASSP-23**, **3**, 264–7.
31. Fettweis, A. (1972). Digital filter structures related to classical filter networks. *Arch. Electron. Uebertragungstech.* **25**, 78–89.
32. Fettweis, A. (1974). Wave digital filters with reduced number of delays. *J. Circuit Theory and Applications* **2**, 319–30.
33. Fettweis, A. and Meerkötter, K. (1974). Suppression of parasitic oscillations in wave digital filters. *I.E.E.E. Proceedings* **ISCAS**, 682–6.
34. Fletcher, R. and Powell, M. J. D. (1963). A rapidly convergent descent method for minimization. *Comput. J.* **6**, **2**, 163–8.
35. Golden, R. M. and Kaiser, J. F. (1964). Design of wideband sampled-data filters. *Bell System Techn. J.* **43**, **2**, 1533–46.
36. Golden, R. M. (1968). Digital filter synthesis by sampled data transformations. *I.E.E.E. Trans. Audio Electroacoustics* **AU-16**, **3**, 321–9.
37. Huang, T. S. (1972). Stability of two-dimensional recursive filters. *I.E.E.E. Trans. Audio Electroacoustics* **AU-20**, **2**, 158–63.
38. Jackson, L. B. (1969). An analysis of limit cycles due to multiplication rounding in recursive digital filters. Proceedings: 7th Annual Allerton Conference on Circuit and Systems Theory, 69–79.
39. Kelly, L. C. and Holmes, J. N. (1964). Computer processing of signals with particular reference to simulation of electric filter networks. Joint Speech Research Unit, Ruislip, Middlesex, England, Res. Report JU14-1.
40. Lawson, S. S. and Constantinides, A. G. (1975). A method for deriving digital filter structures from classical filter networks. *I.E.E.E. Proceedings* **ISCAS**, 170–73.
41. Lawson, S. S. (1975). Digital filter structures from classical analogue networks. Ph. D. Thesis, University of London.
42. Maria, G. A. and Fahmy, M. M. (1974). An Lp technique for two-dimensional digital recursive filters. *I.E.E.E. Trans. Acoustics, Speech and Signal Processing* **ASSP-22**, **1**, 15–21.
43. McClellan, J. H. (1973). The design of two-dimensional digital filters by transformations. Proceedings: 7th Annual Princeton Conference on Information Sciences and Systems, 247–51.
44. Oppenheim, A. V. and Shafer, R. W. (1975). "Digital Signal Processing". Prentice Hall, Englewood Cliffs, New Jersey.
45. Orchard, H. J. and Temes, G. C. (1968). Filter design using transformed variable. *I.E.E.E. Trans. Circuit Theory* **CT-15**, **4**, 385–407.
46. Otnes, R. K. (1968). Recursive band-pass digital filter. *I.E.E.E. Proceedings* **56**, **2**, 207–8.
47. Ozaki, H. and Kasami, T. (1960). Positive real functions of several variables and their application to variable network. *I.E.E.E. Trans. Circuit Theory* **CT-7**, **3**, 251–60.
48. Peled, A. and Liu, B. (1976). "Digital Signal Processing". Wiley, New York.
49. Pistor, P. (1974). Stability criterion for recursive filters. *IBM J. Res. Div.* **18**, **1**, 59–71.
50. Rabiner, L. R., Graham, N. Y. and Helms, H. D. (1974). Linear programming design of IIR digital filters with arbitrary magnitude function. *I.E.E.E. Trans. Acoustics, Speech and Signal Processing* **ASSP-22**, **2**, 117–23.

51. Rader, C. M. and Gold, B. (1967). Digital filter design techniques in the frequency domain. *I.E.E.E. Proceedings* **55**, **2**, 149–71.
52. Rader, C. M. and Gold, B. (1969). "Digital Processing of Signals". McGraw-Hill, New York.
53. Sedlmeyer, A. and Fettweis, A. (1973). Digital Filters with true ladder configurations. *Int. J. Circuit Theory and Applications* **1**, 1–5.
54. Shanks, J. L., Treitel, S. and Justice, J. H. (1972). Stability and synthesis of two-dimensional recursive filters. *I.E.E.E. Trans. Audio Electroacoustics* **AU-20**, **2**, 115–28.
55. Steiglitz, K. (1970). Computer aided design of recursive digital filters. *I.E.E.E. Trans Audio Electroacoustics* **AU-18**, **2**, 123–9.
56. System/360 Scientific Subroutine Package (360A.-CM-03X) (1968). Version III, programmers manual, IBM Data Processing Division, White Plains, New York, Document 420-0205-3.
57. Tosino Koga (1966). Synthesis of finite passive n-port with prescribed two-variable reactance matrices. *I.E.E.E. Trans. Circuit Theory* **CT-13**, **1**, 31–52.
58. Wupper, H. and Meerkötter, K. (1975). New active filter synthesis based on scattering parameters. *I.E.E.E. Trans. Circuit and Systems* **CAS-22**, **7**, 594–602.

Chapter 4

Quantization Effects and Noise in Digital Filters

IVA Introduction

In Chapters 2 and 3 methods were presented to determine the necessary coefficients of required digital filters. In such methods the coefficients were implicitly assumed to have infinite precision, with the implication that the systems considered were sampled but not quantized in amplitude. In the implementation of digital filters, however, it is necessary to introduce some finite precision not only on the coefficients of the filters but also on the input samples and in addition on the results of arithmetic operations within the filters.

Firstly any input signal has to be quantized, that is to say the samples of the input signal have to be represented with a finite number of bits. Secondly the coefficients of the filter have to be represented with a number of bits compatible with the word length of the computer or indeed with special purpose hardware to be used in the operation. Finally some approximation is involved during the filtering operation too. In fact the digital filter implementation consists of a sequence of multiplications and additions, and when we multiply two numbers the result is generally longer than the operands. Since it is obviously impracticable to construct hardware with registers growing in length, it is necessary to introduce some approximation on the result of these operations.

The problem of error behaviour in digital filtering operations is very involved and is essentially a nonlinear problem, because such quantizers as needed in the above description are obviously nonlinear devices. However if the quantization is not very coarse it is possible to use a linear model for the system behaviour and a statistical model for the errors with most signals normally encountered in filtering applications. This is the case we consider in this chapter. It is primarily intended to give some criteria to the user for the choice of cells of arithmetic to be used in the applications, and it is by no means intended as a development of the theoretical background in any detail.

The only nonlinear effects we consider are concerned with (a) the limit cycle phenomenon which can be present when the signal in the filter structure occupies only few quantization steps and (b) overflow oscillations.

Most of the development of the error behaviour of filters is made having in mind hardware implementation or a software implementation with minicomputers, where memory and speed considerations can be very stringent.

This is also the reason why only a reference to the floating point error behaviour is made. In fact we do not consider the problem of hardware floating point implementations where the minimization of the word length can be very important and in most minicomputers floating point implementations the quantization noise problems are not very stringent. However some studies of this argument can be found in the literature.[13,14,16,17]

IVB Number representation

Before studying the error behaviour of digital filters it is necessary to describe how the numbers used in the implementation are represented. Two representations are mainly used: the fixed point representation and the floating point representation.

In the fixed point case several variations in the representation are possible, differing in the binary point position (right or left justified numbers) and in the negative number representation.

If the right justified fixed point form is used, each number is represented by a sequence of positive powers of two. To clarify what we mean let us consider an arithmetic operation with 16 bits (15 for the magnitude and 1 for the sign), which is common, for example, in all the minicomputers. Any of the possible 32768 ($=2^{15}$) sequences of 15 bits, having positive sign (sign bit 0) represents a number whose magnitude can be expressed by means of the relation

$$N=\sum_{i=0}^{M-1} d_i 2^i \tag{4.1}$$

where d_i is a binary variable assuming only the values 0,1 and $M=15$.

On the other hand in the left justified fixed point case when the binary point is assumed at the left of the most significant bit, any of the above considered sequences represents a number whose value is given by

$$N = \sum_{i=1}^{M} d_i 2^{-i} \tag{4.2}$$

that is a number in the range $(1-2^{-M}, 0)$.

These definitions can be expressed by saying that the right justified fixed point numbers represent integers while the left justified fixed point numbers represent fractions.

The main difference between the two representations is in the multiplication operation. The multiplication of two fractions is a fraction and therefore it cannot exceed the length of the registers if the least significant bits are eliminated during the operation.

Obviously overflow in the addition operation can occur with either of the two forms. Hence particular care must be given to the overflow problems in the implementation of digital filters.

In the floating point representation the sequence of binary digits which represents the number, is divided into two sub-sequences. The first, named *mantissa*, is a normally left justified signed fixed point number which represents the normalized magnitude of the number. The second, known as the exponent, represents the signed exponent of the power of two by which the mantissa has to be multiplied to obtain the value of the number. Accordingly the number is given by the relationship

$$N = \pm m 2^{\pm e} \tag{4.3}$$

It is evident that the floating point representation is more convenient when a great dynamic range of numbers has to be represented, because few extra bits in the exponent can be used to accommodate a great range of representable numbers.

Another item to consider is the representation of negative numbers. The simpler representation is that of sign magnitude representation. In this case one of the bits is reserved for the sign (normally 0 positive, 1 negative) while the rest of the digits constitute the positive magnitude of the number. The second representation examined here is the one based on complements. Two types of complements are possible in the binary base, the 1-complement and the 2-complement. Of these two we consider the 2-complement which is the most commonly used representation. In this case positive numbers are represented as in the sign magnitude case (sign 0) whilst negative numbers are represented in the integer case as

$$N_{\text{neg}} = 2^M - \sum_{i=0}^{M-1} d_i 2^i \tag{4.4}$$

or in the fractional case as

$$N_{\text{neg}} = 1 - \sum_{i=1}^{M} d_i 2^{-i} \tag{4.5}$$

where M is the number of binary digits of the systems.

Thus, for example, in the 16 bit system considered before, the number 1 is represented as 0 000 0000 0000 0001 whereas the number -1 is represented by $2^M - 1$, that is 1 111 1111 1111 1111. It can be noted that in this case too, negative numbers are identified by the sign bit 1, but the representation of the rest of the number is completely different from that in the sign magnitude representation. Some comments on the 2-complement representation are in order.

Firstly the representation is circular, that is, if we start from zero and keep adding 1 to the preceding number, we obtain all the positive numbers from 0 to $2^M - 1$. Then adding again 1 we obtain a negative number because the sign bit becomes equal to 1 and its magnitude corresponds to 2^M. Continuing on with the unity increment we obtain all the negative numbers to -1, then 0, that is the number from which the operation was started.

A consequence of this is a very important feature of this number system, namely that if the result of an addition of more than two positive and negative numbers is such as to be representable with the available number of binary digits, then the result is exact irrespective of the order of the addition operations, even if overflow takes place in a sequence of operations.

A second very important feature of the 2-complement representation is that subtraction can be obtained as addition of the minuend and the 2-complement version of the subtrahend. In fact, with reference for example to integer numbers, the addition of N_1 to the 2-complement version of N_2 as defined by (4.4) corresponds to $N_1 + 2^M - N_2$ and since the representation is modulo 2^M, this is equivalent to $N_1 - N_2$.

Thus it is possible using a 2-complement arithmetic representation to perform both addition and subtraction using only an adder. The problem is now the computation of the 2-complement version of a number. From the definitions (4.4) and (4.5) it seems that it is necessary to perform a subtraction to obtain the 2-complement. However more efficient methods exist. For example it is easy to see that the 2-complement of a number can be computed by complementing all the bits in its binary representation (sign included) and then adding 1 to this result.

Between the two number representations presented in the previous considerations, the fixed point one is simpler from the implementation point of view. The sum of two numbers is simple to implement and the multiplication can be obtained without overflow problems if the fixed point numbers represent fractions. The main problems for the application of this type of arithmetic to the digital filter implementation are in its limited dynamic range. Hence particular care has to be taken in the normalization of the variables in the filter structure to avoid overflow.

Using a floating point representation, some of the problems related to the dynamic range constraints can be avoided, with at most some extra bits in the representation. However the implementation of this type of arithmetic is more difficult and generally it is slower than the fixed point one. In fact to add two numbers in the floating point system it is necessary first to normalize them so that they have the same exponent and then to perform the addition. The result of addition has to be normalized again if necessary. Moreover a floating point multiplication is really the combination of two arithmetic operations, that is the product of the mantissas and the addition of the exponents.

It is important also to observe that in the fixed point representation, once overflow in the addition operations has been avoided through suitable normalizations of the signals, an error is introduced only in the multiplications. In the floating point representation on the other hand some error is introduced also in the additions, because in the normalization process some bits of one of the numbers are lost.

The increase in complexity in floating point arithmetic results in the preference of fixed point arithmetic in hardware implementations of digital filtering operations. In this case the economy obtainable in the computing hardware and memory and in addition the increase in speed possible can be more important than the increase in design work involved in (a) the choice of the best structure for the filter, (b) the computation of the overflow constraints and (c) the study of the error behaviour of the structure.

In software implementation on general purpose computers and minicomputers and in applications where the processing speed is not of the utmost importance a floating point implementation can give a greater flexibility in the synthesis procedures.

We observe finally here that in most minicomputers real numbers are represented as floating point numbers of 32 bits, 8 for the exponent (sign included) and 24 for the 2-complement mantissa. This allows the representation of 6-7 decimal digits and a good dynamic range, that is $2^{-128} \leq N \leq 2^{127}$.

Integer numbers are in general represented as right justified fixed point numbers. It can be observed that if fixed point arithmetic is used in the implementation using a high level language such as FORTRAN, the maximum data and coefficient length is limited by the fact that an overflow in the multiplication occurs if the sum of the bits of the two operands is greater than 15 bits (sign excluded).

Fortunately however most of the fixed point arithmetic units used in minicomputers compute really all the 32 bits that result in the multiplication of two 16 bit numbers. This allows the user to represent the variables as left justified fractions to a maximum of 15 bits plus sign and, after the multiplication, to choose the 16 most significant bits of the product using a lower level programming language.

IVC Truncation and rounding

The approximation with a limited number of digits of variables, for any representation of the numbers, which are in principle continuous, has as a consequence the introduction of an error. This error depends on the number (M) of binary digits used in the representation, on the type of representation (fixed point, floating point), on the type of negative number representation and finally on the method used to eliminate the digits exceeding the register length.

Two different approaches can be considered for the last operation. The first consists of simply chopping off the exceeding bits. The second consists of the approximation of the number to the nearest level, that is to the same value as in the truncation case if the ignored fraction is less than half the magnitude of the least significant bit in the representation and to the same value incremented by 1 in the other case.

The first type of approximation is called truncation, whilst the second one is referred to as rounding.

Rounding is obviously more precise than truncation since the maximum error is equal to one half the quantization step. Thus if the number is represented by a fraction of M bits (sign excluded) the error in the approximation will be in the range

$$-\tfrac{1}{2}2^{-M} < e(n) \leq \tfrac{1}{2}2^{-M} \qquad (4.6)$$

The maximum error is greater in the truncation case, being equal to the quantization step, and its value (in sign) depends on the negative number representation. If the sign magnitude form is used, the error is then defined as the difference between the obtained value and the true value and it is between 0 and -2^{-M} if the number is positive, whereas the error is between 2^{-M} and 0 if the number is negative; that is the error is always of opposite sign to the number. If the 2-complement representation is used the error is not correlated with the sign of the number to be quantized but it is always negative and within the interval $(0, -2^{-M})$.

A more involved situation exists when the floating point representation is chosen. In this case the approximation (truncation or rounding) is performed only on the mantissa of the number. This means that the absolute magnitude of the error depends on the magnitude of the number, that is in this case relative errors instead of absolute errors are important. Therefore in the floating point case the errors introduced in the approximation are multiplicative instead of additive. Thus if α represents the value of a number to be approximated and α' the resulting floating point number, with ε defined as the relative error then we have

$$\frac{\alpha'-\alpha}{\alpha}=\varepsilon \tag{4.7}$$

and consequently

$$\alpha'=\alpha+\varepsilon\alpha=\alpha(1+\varepsilon) \tag{4.8}$$

If we use, for example, a rounding arithmetic the error in the mantissa is expressed by eq. (4.6) and hence

$$-2^{l}\frac{2^{-M}}{2}<\alpha'-\alpha\leq 2^{l}\frac{2^{-M}}{2} \tag{4.9}$$

if l is the value of the exponent.

A bound for ε can be obtained using (4.7) and the relation $2^{l-1}<\alpha<2^{l}$ for which the result is

$$-2^{-M}<\varepsilon\leq 2^{-M} \tag{4.10}$$

IVD Input signal quantization

As mentioned earlier the first type of error we have to consider when digital linear filtering operations are to be used, is associated with the quantization of the signal serving as input to process. Therefore particular care has to be taken as to the choice of a quantization step which is suitable both with respect to the characteristics of the input signal itself (i.e. dynamic range, signal to noise ratio, etc.) and with respect to the type of operation which is to be performed.

The quantization operation can be considered as the substitution of the input signal samples with the most significant M digits of their binary representation, according to the rules considered in the previous section. Hence any input sample can be represented in the form

$$x'(n)=x(n)+e(n) \tag{4.11}$$

where $x'(n)$ is the true value of the sample and $e(n)$ is the associated error.

To proceed with the analysis it is necessary to introduce some hypotheses concerning the behaviour of these errors. In general a statistical model is assumed, based on the assumptions that (i) the sequence $e(n)$ is a sample sequence taken from a stationary random process, (ii) the error sequence is uncorrelated with the sequence of exact values of the signal, (iii) the error samples are uncorrelated (white-noise process), and (iv) the probability distribution is uniform over the range of quantization errors.

Obviously these assumptions are to some extent arbitrary and it is very easy to construct immediately cases where they fail, as for example a constant signal or a sinusoidal signal sampled at a frequency which is a rational multiple of the sinusoidal frequency. In the first case all the errors $e(n)$ are equal whilst in the second case they constitute a periodic sequence, for which cases the above hypotheses are not valid. However Bennet[1] verified that these assumptions realistically represent most of the signals which one is likely to encounter. Thus this model for the quantization error behaviour has enormous practical value. Roughly it can be said that the model can be used when the signal behaviour is such to cross several quantization steps going from a sample to the other and at same time the number of quantization levels or equivalently of bits in the number representation is not too small.

A consequence of the fourth assumption is that the error probability distribution has the form presented in Fig. 4.1(a) for rounding arithmetic and is of the form presented in Fig. 4.1(b) for 2-complement truncation arithmetic. From these probability distributions it is easy to compute the mean or variance of the error signals. They are in fact

$$m = 0 \tag{4.12}$$

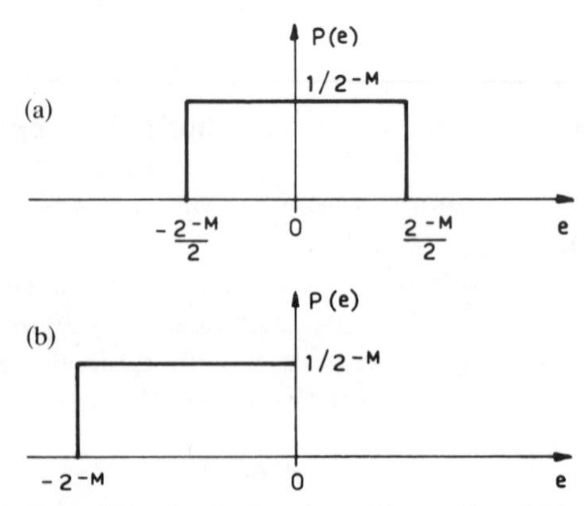

FIG. 4.1. Probability density functions: (a) rounding, (b) truncation.

$$\sigma^2 = \frac{2^{-2M}}{12} \tag{4.13}$$

for rounding, and

$$m = -\frac{2^{-M}}{2} \tag{4.14}$$

$$\sigma^2 = \frac{2^{-2M}}{12} \tag{4.15}$$

for 2 complement truncation.

From the above discussion it follows that the quantization process can be considered as the superposition on the infinite precision signal of a noise having mean values and variances expressed by (4.12), (4.13), (4.14), (4.15).

The amount of this error can be conveniently measured by the ratio, between the signal power and the noise power (signal-to-noise ratio, S/N). For example in the rounding case it is possible to write

$$\mathrm{S/N} = \sigma_x^2/\sigma_n^2 = \frac{\sigma_x^2}{2^{-2M}/12} = 12[2^{2M}\sigma_x^2] \tag{4.16}$$

and using a logarithmic measure we have

$$\mathrm{S/N\ dB} = 10 \log_{10} 12 + M 20 \log_{10} 2 + 10 \log_{10} \sigma_x^2 \tag{4.17}$$

It is interesting to observe that since $20 \log_{10} 2 \simeq 6$, the addition of one bit in the arithmetic increases the signal to noise ratio by approximately 6 dB.

The above exposition of the quantization process suggests some considerations about the choice of the quantization step to be used in the applications. This choice is determined by the necessary required precision, the noise present on the signal to be processed, and the operation that has to be performed on the signal.

The noise present on the signal introduces an upper limit on the number of quantization steps. Obviously it does not make any sense to use a small quantization step when much noise is present on the signal, because in this case we shall be quantizing the noise and not the signal. It is sufficient to quantize to a level that introduces an amount of quantization noise small in comparison with the noise present on the signal.

In the opposite case, when the noise is very small, the quantization step has to be chosen to obtain the desired quality at the output of the processing. In fact a degradation of the input signal quality can be caused by errors introduced in the processing. So the input signal quantization and the processing structure in terms of the algorithms employed, precision of the arithmetic used, etc. have to be chosen in a way which is compatible with the necessary required accuracy at the output.

IVE Effect of signal quantization on output noise

In the preceding section the methods describing the quantization process were discussed and a model of the noise behaviour was developed. Let us now consider what happens when the quantization noise superposed on the signal is processed by a digital filter.[9]

We assume in this case that the arithmetic operations in the filter are performed with infinite precision, that is to say that the only error in the process is the input quantization error. This is possible because the model we considered in the preceding section is a linear one, i.e. it is valid when the quantization step is small enough to make negligible all nonlinearities. This means that the different sources of errors in linear systems can be studied independently. It is very easy to compute the effect of the input quantization noise on the output of the filter by computing the output of the filter as a response to the input noise signal $e(n)$. Thus if $h(m)$ is the impulse response of the linear system its output when the input is the error sequence $e(n)$ is given by

$$\varepsilon(n) = \sum_{m=0}^{n} h(m)e(n-m) \tag{4.18}$$

The variance of $\varepsilon(n)$ can be written in the following form

$$\begin{aligned} \sigma_o^2(n) &= E\left[\sum_{m=0}^{n} h(m)e(n-m) \cdot \sum_{l=0}^{n} h(l)e(n-l)\right] \\ &= \sum_{m=0}^{n} \sum_{l=0}^{n} h(m)h(l)E[e(n-m)e(n-l)] \end{aligned} \tag{4.19}$$

and since $e(n)$ is assumed to be white we can write

$$\sigma_o^2(n) = \sum_{m=0}^{n} \sum_{l=0}^{n} h(m) \cdot h(l)\delta(l-m)\sigma_e^2 = \sigma_e^2 \sum_{m=0}^{n} h^2(m) \tag{4.20}$$

In the IIR case if the poles of the filter are in the stability region then $h(m) \to 0$ when $m \to \infty$ and a steady state value of the variance can be obtained in the form

$$\sigma_o^2 = \sigma_e^2 \sum_{m=0}^{\infty} h^2(m) \tag{4.21}$$

whilst in the FIR case eq. (4.21) can be written with the sum extended to the N non-zero values of the impulse response.

Using Parceval's theorem this result can be written also in the form

$$\sigma_o^2 = \frac{\sigma_e^2}{\pi} \int_0^{\pi} |H(e^{j\omega})|^2 \, d\omega \tag{4.22}$$

where $H(e^{j\omega})$ is the frequency transfer function of the system under consideration.

If we consider for example the first order system

$$y(n) = x(n) + ay(n-1) \tag{4.23}$$

its impulse response is given by

$$h(n) = a^n \tag{4.24}$$

and expression (4.21) can be directly evaluated in closed form yielding the steady state value

$$\sigma_o^2 = \sigma_e^2 \frac{1}{1-a^2} \tag{4.25}$$

It can be observed that if $a \simeq 1$, that is if the pole of the filter is near the unit circle, the input noise can be amplified by the system.

In the case of a second order section with two poles and no zeros we have

$$y(n) = x(n) - b_1 y(n-1) - b_2 y(n-2) \tag{4.26}$$

which has poles at the position $z = r\,e^{\pm j\theta}$ with $r = \sqrt{b_2}$ and $\cos\theta = b_1/2\sqrt{b_2}$. It is possible to compute again the result in closed form, obtaining

$$\sigma_o^2 = \sigma_e^2 \frac{1+r^2}{1-r^2} \frac{1}{r^4 + 1 - 2r^2 \cos\theta} \tag{4.27}$$

and if $r \simeq 1$, with the position $\rho = 1 - r$ it is possible to write

$$\sigma_o^2 \simeq \sigma_e^2 \frac{1}{4\rho \sin^2\theta} \tag{4.28}$$

This expression shows that the output noise depends on the distance of the poles from the unit circle and on the angular position of the poles. That is low frequency poles introduce much more noise in the output than high frequency ones.

IVF Quantization of the arithmetic operations: IIR case

As observed in the introduction it is always necessary not only to approximate the input data and the coefficients but also to introduce some quantization within the structure of the filter.

The filtering operation in general consists in a sequence of sums and products. Considering now a fixed point implementation and assuming that the coefficients are so normalized that no overflow can occur in the filter structure, an approximation has to be performed on the results of the products. The product of two fixed point numbers of N and M bits respectively is at most a number of $N+M$ bits. Consequently some quantization has to be performed to accommodate the results of the products in cells shorter than $N+M$ bits. Let us now study briefly how it is possible to solve the problem of prediction of the amount of error at the output of the filter as a consequence of the quantization operation. The problem in its general form is very involved and it is again basically a nonlinear problem. However if the length of the numbers which are used in the multiplication is not too short, then the same statistical model considered before can be used. Suppose that several sources of noise are present in the structure, one for each multiplication node.

With this representation the error behaviour can be studied by computing the amount of noise which is present at the output of the system as a result of the quantization of the results of products.

A multiplication node within the structure is represented as in Fig. 4.2, that is with the addition of a white-noise source, in cascade to the perfect product; this source represents the quantization error, having a variance of $\sigma_e^2 = 2^{-2M}/12$, where M is the number of bits in the representation. We observe that M can be greater than the number of bits in the input signal representation. This means that the variance of the error introduced in the approximation of products can be achieved, if necessary, with smaller value than that corresponding to the error introduced in the quantization of the input signal, simply by lengthening the memory registers.

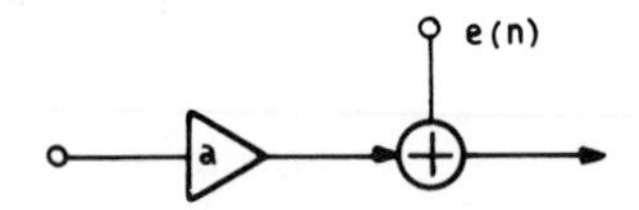

FIG. 4.2. Quantization noise model for a multiplier (fixed point).

When the above representation can be considered valid, the noise variance in the output due to every noise source in the structure can be easily computed using the same method of the preceding section, except that the impulse responses to the output from each point of the structure where the noise sources are considered to be injected have to be determined. This means that if the noise sources are L and we indicate with $h_k(m)$ the impulse response from the kth noise source to the output, the variance of the output due to any noise source in the steady state will be given by,

$$\sigma_{ok}^2 = \sigma_e^2 \sum_{m=0}^{\infty} h_k^2(m) \tag{4.29}$$

where σ_e^2 is the variance of the injected noise, and consequently the overall noise varance at the output is given by

$$\sigma_o^2 = \sum_{k=1}^{L} \sigma_{ok}^2 \tag{4.30}$$

Some general remarks can be made at this point. The first is that the noise behaviour of a recursive structure is very dependent on the particular structure which is used in the filtering implementation.[12] In fact the impulse response that the error sources see from the point where they have been injected to the output depends on the structure employed. This means that particular care has to be taken in the choice of filter structure which is important not only in the noise behaviour but also in the coefficient approximation problem.

A second observation which is related to the previous one is that not all the noise sources which are injected in the structure have the same importance. In fact if we consider for example a second order section in its canonical form, as shown in Fig. 4.3 the three error sources $e_1(n)$, $e_2(n)$, $e_3(n)$ are directly added to the output of the filter, whilst the sources $e_4(n)$ and $e_5(n)$ are injected in the feedback loop of the filter with the consequent possibility of amplification by the poles of the filter. A pole is of course a resonance in the frequency response and this can result in amplification of the noise as shown in the examples of the previous section.

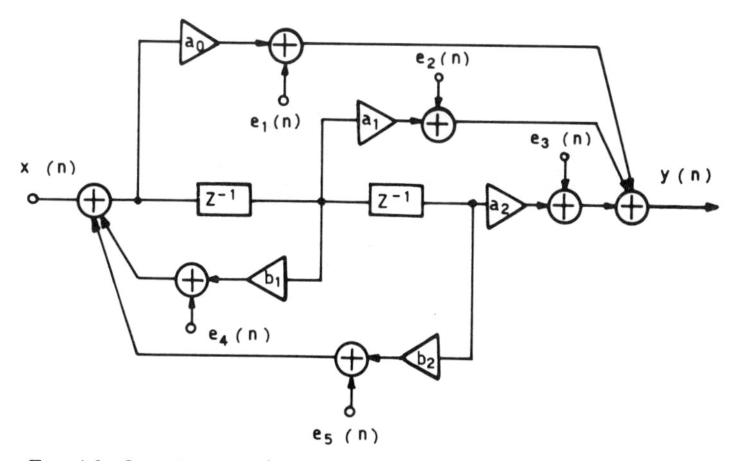

FIG. 4.3. Quantization noise model for a second order section (fixed point).

Let us now consider again the first order and second order sections of the previous section in order to study their behaviour with respect to this new type of error.

In the first order case the equivalent structure with the suitable noise source is shown in Fig. 4.4. It is evident from the figure that the error source due to the quantization of the product goes through the same path in the structure as the input signal. Hence the corresponding impulse

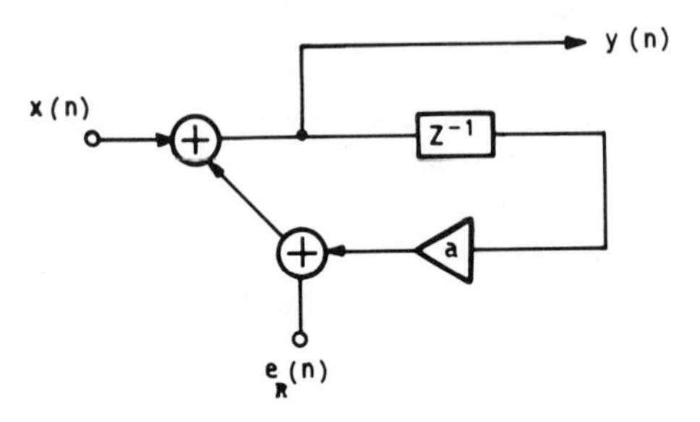

FIG. 4.4. Quantization noise model for a first order section (fixed point).

response is the same as far as the input noise source is concerned and thus the system behaves as if the input noise $e(n)$ signal were modified to be of the form

$$e_1(n) = e(n) + e_R(n) \tag{4.31}$$

In the second order section case the equivalent noise structure is shown in Fig. 4.5. In this case too the error sources $e_1(n)$, $e_2(n)$ encounter the same path to the output as the input signal and the results of the previous section can be used with the correct value of variance.

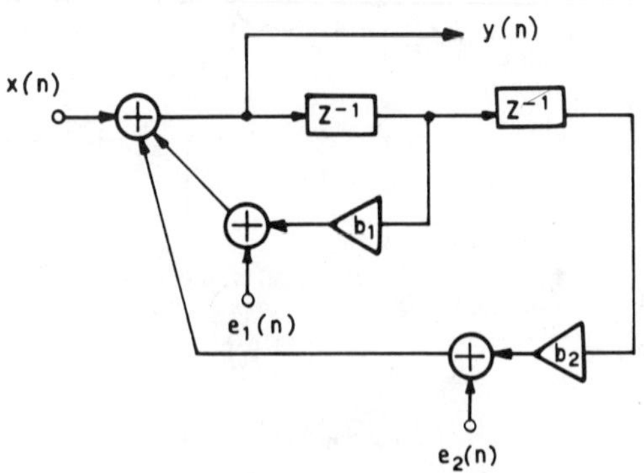

FIG. 4.5. Quantization noise model for a second order section with poles only (fixed point).

IVG Overflow constraints and noise-to-signal ratio

Using the model developed in Sections IVE, IVF it is possible to compute the variance of the output noise as a function of the input noise and of the noise introduced in the structure of the filter due to the approximation of the results of the arithmetic operations. But often one is more interested in the noise-to-signal (N/S) ratio of the output of the system as a function of the input noise-to-signal ratio. To compute the output (N/S) it is necessary to take into account not only the noise generated by the system but also the constraints introduced by the system on the input signal. In fact given an implementation with a register length of M bits, in general it is not possible to use for the signal all the dynamic range possible with this word length because it is necessary to avoid overflows in the structure.

Given a system with impulse response $h(n)$ the maximum possible output signal level is equal to

$$y_{\max} = x_{\max} \sum_{n=0}^{+\infty} |h(n)| \tag{4.32}$$

where $x_{\max}$ is the maximum value of the input. Hence by using a fixed point left justified implementation where overflow is possible only in the addition nodes, it is necessary for each addition node to compute the impulse response from the input to that node and to normalize the inputs to the node so as to avoid overflow. This obviously reduces the input signal power and increases the noise-to-signal ratio for a given word length, for a fixed output noise variance.

Unfortunately the overflow constraints and therefore the maximum noise-to-signal ratio which can be achieved at the output of a system depend on the form of the realization structure used for implementation. Hence it is not possible to give general results for the N/S ratio. Many experimental results and a discussion of various types of overflow constraints can be found in the literature.[11,12] It can be observed that the overflow constraints used earlier are too stringent in many applications. In many situations it is possible to make some assumptions about the behaviour of the input signal and to use less stringent constraints. For example if the input signal is a sinusoid or a very narrowband signal, the maximum output of the system can be at most equal to the amplification of the system at the frequency under consideration.

To clarify through an example the procedure outlined in this section let us consider the first order section (4.23) with white-noise input having a maximum amplitude of unity. To avoid overflow the input signal must be normalized to have a maximum amplitude of $1-|a|$ ($|a|<1$ for stability) and consequently the input signal variance is given by

$$\sigma_i^2 = \tfrac{1}{3}(1-|a|)^2 \tag{4.33}$$

Using eq. (4.25) the output signal variance is given by

$$\sigma_{os}^2 = \frac{1}{3}\frac{(1-|a|)^2}{1-a^2} \tag{4.34}$$

and using again eq. (4.25) for the quantization noise the output noise-to-signal ratio becomes

$$\frac{\sigma_{on}^2}{\sigma_{os}^2} = \frac{1}{4} 2^{-2M} \frac{1}{(1-|a|)^2} \tag{4.35}$$

To conclude the discussion we observe that in the cascade realization case the overflow constraints and the noise behaviour are functions also of the cell ordering and of the zero-pole pairing. In most filter design techniques the results can be expressed in terms of biquadratic cells, that is in terms of cells which can be obtained by choosing two complex conjugate but otherwise arbitrary couples of zeros and poles. Obviously the noise behaviour is different for different choices of zero-pole pairing. Studies have been carried out both with respect to the zero-pole pairing and also with respect to the cell ordering and some algorithms have been developed to produce an optimum ordering for different classes of signals.[12]

IVH Quantization of the arithmetic operations: FIR case

The problems discussed in the previous sections are less involved in the FIR filter case.

For the synthesis in the direct form structures,[4] for example, the noise introduced in the quantization of products does not pass through the structure of the filter, but it is simply added to its output. Hence if at a multiplication node a noise source $e(n)$ with variance σ_e^2 is introduced, the noise variance on the output is $N\sigma_e^2$ which is independent from the value of the coefficients and consequently independent from the type of filter (i.e. low-pass, high-pass, etc.) to be implemented. However the values of the coefficients have their effect on the output noise-to-signal ratio through the overflow constraints, which can be evaluated by means of the relation

$$y_{\max} = x_{\max} \sum_{n=0}^{N-1} |h(n)| \tag{4.36}$$

The situation is slightly more complicated if a cascade realization of the FIR filter is considered. In fact the noise at the output of any section then will pass through the successive sections in the cascade. If $h_i(n)$ is the unit sample response from the noise source $e_i(n)$ to the output it is possible to write the expression

$$\sigma_{oi}^2 = 3\sigma_e^2 \sum_{n=0}^{N-2i} |h_i(n)|^2 \tag{4.37}$$

for the output noise variance due to $e_i(n)$ and the total noise variance is

$$\sigma_o^2 = \sum_{i=1}^{L} \sigma_{oi}^2 \tag{4.38}$$

where L is the number of second order sections. From eq. (4.37) it is evident that the output noise depends on the ordering of the sections.[2,3]

These considerations are obviously valid for a nonrecursive implementation of FIR digital filters. If a recursive implementation, for example the frequency sampling structure, is used, it is necessary to study the noise behaviour in the same way as in the preceding sections.

IV I Error behaviour of 2-D filters

In the 2-D case the problem of the error behaviour can be described in the same way as in the 1-D case, except that the theory is much more involved for an analytical study due to the increase in dimension. It is not realistic to present here all the details associated with the error behaviour of such systems. However from a conceptual point of view the steps of the analysis and general study are not dissimilar to those of the 1-D case.[15]

The first step is the study of the effect of the noise superposed on the input signal on the noise variance at the output, which can be obtained by means of the relationship

$$\sigma_o^2 = \sigma_e^2 \sum_{m=0}^{\infty} \sum_{n=0}^{\infty} |h_{mn}|^2 \tag{4.39}$$

where h_{mn} is the impulse response of the two-dimensional filter. Then the study of noise due to the quantization of the results of the operations can be performed. This involves the computation of the impulse response $\{h_{m,n}^{ij}\}$ from every multiplication node to the output and every noise source of this type contributes to the output noise variance with a term of the form

$$\sigma_{ij}^2 = \sigma_e^2 \sum_{m=0}^{\infty} \sum_{n=0}^{\infty} |h_{m,n}^{ij}|^2 \tag{4.40}$$

whilst the average output variance is given by the sum of all the noise sources injected in the structure, that is

$$\sigma_o^2 = \sum_i \sum_j \sigma_{ij}^2 \tag{4.41}$$

Finally the computation of the output noise-to-signal ratio involves the computation of the overflow constraints to normalize the input signals to every summation node in order to avoid overflow.

IVJ Limit cycles and overflow oscillations

The study of the preceding sections was based on the assumptions of uncorrelated errors between themselves and also with respect to the input sequence. These are valid when the dynamic range of the signals is such that changes between one sample to the other are of several quantization steps. However there are some cases when this is not true. One such case occurs when the input of the filter becomes zero. In this case the output tends to decrease until the values of the output samples become small. At this point the errors introduced in the filter structure, associated with quantization of the results of operations within the filter, can result in a sustained non-zero output.

As an example let us consider the first order difference equation of the form

$$y(n)=x(n)-0.9y(n-1) \tag{4.42}$$

with zero input and an initial condition $y(-1)=10$, $x(0)=0$. Assuming an arithmetic which rounds to the nearest integer level, the output assumes the values $-9, 8, -7, 6, -5, 5, -5, 5, \ldots$, and the output cannot reach a value less than 5.

The general study of this kind of phenomenon is quite complex. However some useful results can be obtained with a simple and intuitive derivation due to Jackson.[10] Let us consider a first order operation of the form

$$y(n)=x(n)-ay(n-1) \tag{4.43}$$

where the result of the product $ay(n-1)$ is rounded to the nearest integer. If the coefficient a is equal to 1, that is if the pole of the first order system lies on the unit circle, the system holds the output to the value assumed by $y(n-1)$ when the input becomes zero. What we are looking for then is the range of initial conditions $y(n-1)$ for which the coefficient a is practically equal to 1 due to the rounding effect. This interval is commonly called the deadband of the filter.

The deadband is easily obtainable in this case by observing that to have a non-decreasing output k it is necessary to satisfy the disequality

$$k \leq |a|k+0.5 \tag{4.44}$$

for in this case the output, after the rounding, is equal to the input of the right band side of eq. (4.44) and it cannot be equal to $k+1$ because $|a|$

must be less than 1 in order to have a stable filter. From eq. (4.44) it is evident that the deadband is the region between $-k$ and k where k is the largest integer satisfying the relation

$$k \leq \frac{0.5}{1-|a|} \tag{4.45}$$

If a is negative the limit cycle is of constant amplitude and of the same sign, whilst if a is positive the limit cycle is of constant amplitude and alternating sign. If $|a|<0.5$ no limit cycle can occur.

A similar analysis can be performed for a second order section of the form

$$y(n)=x(n)-b_1y(n-1)-b_2y(n-2) \tag{4.46}$$

with a rounding arithmetic. In this case a deadband exists and it is the region from $-k$ to $+k$ where k is the largest integer satisfying the relation

$$k \leq \frac{0.5}{1-b_2} \qquad 0 \leq b_2 \leq 1 \tag{4.47}$$

which is the condition required to have the poles practically on the unit circle.

The value of the b_1 coefficient and the rounding of the corresponding product has its effect on the shape of the limit cycle, that is on its frequency.

A second nonlinear oscillatory effect can occur when an overflow is generated inside the structure of the filter.[7] Due to the circular nature of the 2-complement arithmetic system when an overflow occurs, the output of the corresponding node switches from the maximum of the positive scale to the minimum of the negative scale or vice versa. This of course causes very large oscillations in the output of the filter. Two methods can be used to prevent this phenomenon. The first is to calculate the overflow constraints in their most stringent form (4.32) and then ensure that an overflow cannot occur. The second consists of using saturating arithmetic, that is, arithmetic operations which in the case of an overflow substitute in the result the maximum value of the same sign of the addends.

IVK Quantization of the coefficients

As already mentioned deviation of the filter coefficients from their nominal values produces a deviation in the characteristics of the transfer function. In fact it can produce rather adverse effects when such coefficients are associated with the poles of the transfer function. The following examples show this effect.[6]

Consider the transfer function given by,

$$H(z)=(a_0+a_1z^{-1}+a_2z^{-2})/(1+b_1z^{-1}+b_2z^{-2}) \tag{4.48}$$

If we let $\rho\ e^{\pm j\theta}$ be the z-plane positions of the poles of eq. (4.48) then

$$\rho=\sqrt{b_2} \tag{4.49}$$

$$\cos\theta=b_1/2\sqrt{a_2} \tag{4.50}$$

Then for small changes in the b_1 and b_2 the pole positions shift such that

$$\Delta\rho=\frac{\partial\rho}{\partial b_1}\cdot\Delta b_1+\frac{\partial\rho}{\partial b_2}\cdot\Delta b_2 \tag{4.51}$$

or

$$\Delta\rho=\frac{1}{2\rho}\cdot\Delta b_2$$

and

$$\Delta\theta=-\frac{\Delta b_1}{2\rho\sin\theta}+\frac{\Delta b_2}{2\rho^2\tan\theta} \tag{4.52}$$

It may be observed that $\Delta\rho$ changes drastically for ρ near the unit circle whereas $\Delta\theta$ changes drastically for θ near zero.

An important point to note is that the changes in the positions of the poles are highly dependent on the structure used for the realization of given transfer function. For example, let eq. (4.48) describe a second order all-pass filter.

Then if the pole positions on the z-plane are $\rho\ e^{\pm j\theta}$ we have

$$\begin{aligned} a_1&=b_1=-2\rho\cos\theta\\ a_0&=b_2=\rho^2\\ a_2&=1 \end{aligned} \tag{4.53}$$

One structure which is normally used to realize such a transfer function is the biquadratic form given in Fig. 4.6.

Hence for small changes in b_1 and b_2 we have correspondingly the changes in ρ and θ given by eq. (4.51) and (4.52).

Let us now rewrite the transfer function of eq. (4.48) under the all pass constraints of eq. (4.53) in the form

$$\begin{aligned} H_A(z)&=[H_B(z)+\alpha]/[1+\alpha H_B(z)]\\ H_B(z)&=z^{-1}\cdot(z^{-1}+\beta)/(1+\beta z^{-1}) \end{aligned} \tag{4.54}$$

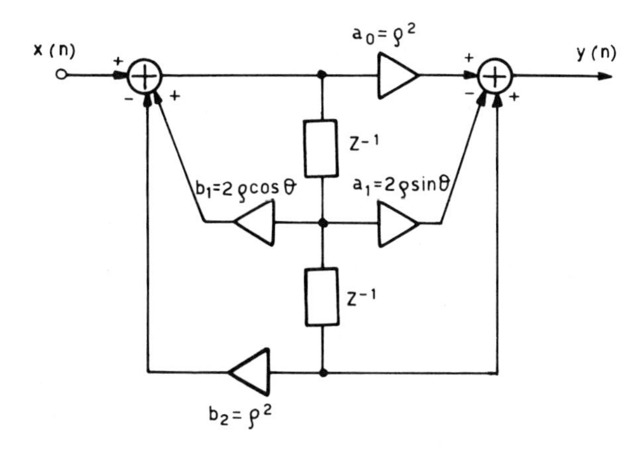

FIG. 4.6. Block diagram of a biquadratic section.

so that

$$\rho = \sqrt{\alpha}$$
$$\cos\theta = \beta(1+\alpha)/2\sqrt{\alpha} \tag{4.55}$$

The structure we shall use is the one given in Fig. 4.7. In this case we have

$$\Delta\rho = \frac{1}{2\rho}\cdot\Delta\alpha$$
$$\Delta\theta = \frac{(\rho^2-1)\cos\theta}{2\rho^2(1+\rho^2)}\cdot\Delta\alpha + \frac{1+\rho^2}{2\rho}\cdot\Delta\beta \tag{4.56}$$

It follows therefore that for this particular structure large errors in $\Delta\theta$ are not associated with small angles as was the case in the previous structures. In general it can be said that the structure used for the realization of

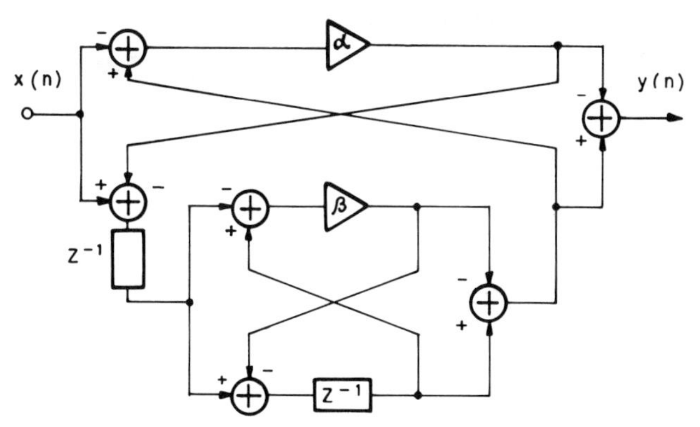

FIG. 4.7. Alternative implementation of a biquadratic section.

a given transfer function is of paramount importance as far as errors due to quantization are concerned.[8]

Let us now consider the deviation of a given transfer function $G(z)$ from its nominal value as a result of changing a particular multiplier of value β to $\beta + \Delta\beta$. We shall use the compensating theorem for signal flow diagrams since digital filters are signal flow graphs where some transmittances have complex values owing to their frequency dependent nature.

The digital filter is represented as a box as shown in Fig. 4.8 with β made accessible.

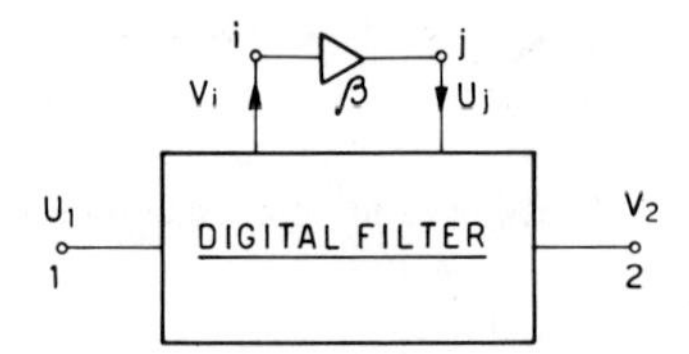

FIG. 4.8. Representation of a digital filter with the coefficient β made acessible.

If β is removed completely then it can be seen that the box has two inputs (U_1, U_j) and two outputs (V_2, V_i) and hence assuming linearity we can write

$$\left.\begin{aligned} V_2 &= T_{21}U_1 + T_{2j}U_j \\ V_i &= T_{i1}U_1 + T_{ij}U_j \end{aligned}\right\} \tag{4.57}$$

where

$$T_{21} = \left.\frac{V_2}{U_1}\right|_{U_j=0} \tag{4.58}$$

In terms of a signal flow diagram we have the representation of Fig. 4.9.

With reference to the signal flow graph, if $U_j = V_i \cdot \beta$ then we have

$$\frac{V_i}{U_1} = T_{i1}/(1 - T_{ij}) \tag{4.59}$$

$$\frac{V_2}{U_1} = T_{21} + \beta T_{i1}T_{2j}/(1 - \beta T_{ij}) \tag{4.60}$$

If for the same signal flow graph we changed β to β' then

$$V_i \to V_i'$$

$$U_j \to U_j'$$

$$V_2 \to V_2'$$

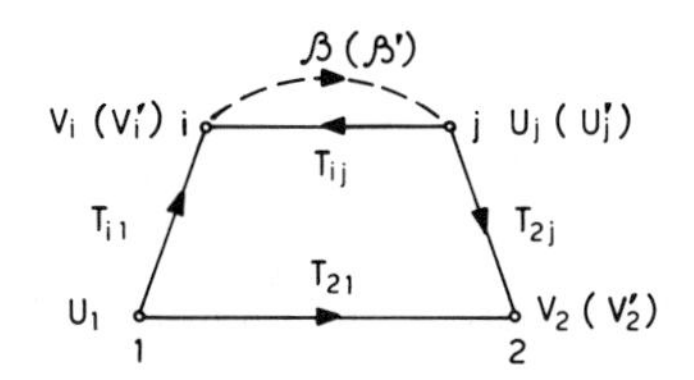

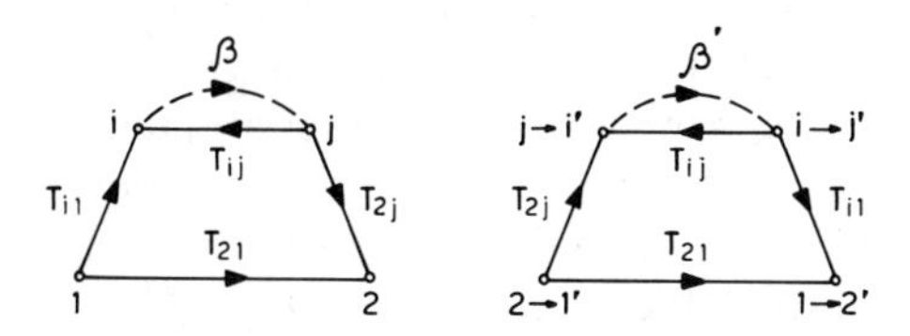

FIG. 4.9. Flow diagram of the filter of Fig. 4.8.

This can be appreciated from eq. (4.57) where the dual-input, dual-output linear system has U_j changed to U_j'; then all variables except U_1 will change to their primed values.

Therefore we shall have,

$$\frac{V_i'}{U_1} = T_{i1}/(1-\beta' T_{ij}) \tag{4.61}$$

and

$$\frac{V_2'}{U_1} = T_{21} + \beta' T_{i1} T_{2j}/(1-\beta' T_{ij}) \tag{4.62}$$

From (4.60) and (4.62) we obtain[6]

$$\frac{V_2' - V_2}{U_1} = T_{i1} T_{2j}(\beta' - \beta)/(1-\beta T_{ij})(1-\beta' T_{ij}) \tag{4.63}$$

Equation (4.63) is a general equation relating the change in the output $(V_2' - V_2)$ when the multiplier β changes to β'. If we consider an infinitesimal change in β i.e. $\beta \to \beta + \mathrm{d}\beta$ then we have

$$\frac{\mathrm{d}V_2}{U_1} = [T_{i1} T_{2j}/(1-\beta T_{ij})^2]\, \mathrm{d}p \tag{4.64}$$

and since $V_2/U_1 = H_{21}(z)$ the transfer function of the signal flow (or digital filter) then

$$\frac{\partial H_{21}(z)}{\partial \beta} = [T_{i1}/(1-\beta T_{ij})] \cdot [T_{2j}/(1-\beta T_{ij})] \tag{4.65}$$

or

$$\frac{\partial H_{21}(z)}{\partial \beta} = H_{i1}(z) \cdot H'_{2j}(z) \tag{4.66}$$

where $H_{i1}(z)$ is the transfer function of signal flow graph (or digital filter) from 1 to i and $H_{2j}(z)$ is the transfer function $T_{2j}/(1-T_{ij})$ which can be interpreted as the transfer function from j to 2 when $U_1 = 0$ or equivalently $H'_{2j}(z)$ is the transfer function from 1 to i in the adjoint signal flow graph.

The above result of eq. (4.66) is very well known in classical analogue networks and has also been derived for digital filters.[5,8] The result of eq. (4.63), however, is more general in that one can have any change in the multiplier value which need not be infinitesimally small. Once again in eq. (4.63) one can interpret the pair $T_{i1}/(1-\beta T_{ij}) = H_{2j}$ and $T_{2j}/(1-\beta' T_{ij}) = H'_{2j}(z)$ or the pair $T_{i1}/(1-\beta' T_{ij}) = H'_{i1}(z)$ and $T_{2j}/(1-\beta T_{ij}) = H_{2j}(z)$ in the same way as above.

References

1. Bennet, W. R. (1948). Spectra of quantized signals. *Bell System Techn. J.* **27**, **3**, 446–72.
2. Chan, D. S. K. and Rabiner, L. R. (1973). Theory of roundoff noise in cascade realizations of finite impulse response digital filters. *Bell System Techn. J.* **52**, **3**, 329–45.
3. Chan, D. S. K. and Rabiner, L. R. (1973). An algorithm for minimizing roundoff noise in cascade realizations of finite impulse response digital filters. *Bell System Techn. J.* **52**, **3**, 347–85.
4. Chan, D. S. K. and Rabiner, L. R. (1973). Analysis of quantization errors in the direct form for finite impulse response digital filters, *I.E.E.E. Trans, Audio Electroacoustics* **AU-21**, **4**, 354–66.
5. Colonna, S. *et al.* (1972). Determination of sensitivity functions of digital filters. *I.E.E.E. Trans. Circuit Theory* (correspondence) **CT-19**, **5**, 538–9.
6. Constantinides, A. G. (1973). Some sensitivity aspects concerning digital filters. I.E.E. Colloquium on Sensitivity Theory and Tolerance Design of Circuits, 1973/3.
7. Ebert, P. M., Mazo, J. E. and Taylor, M. G. (1969). Overflow oscillations in digital filters. *Bell System Techn. J.* **48**, **9**, 2999–3020.
8. Fettweis, A. (1972). On the connection between multiplier word length limitation and roundoff noise in digital filters. *I.E.E.E. Trans. Circuit Theory* **CT-19**, **5**, 486–91.
9. Gold, B. and Rader, C. M. (1969). "Digital Processing of Signals", Chapter 4. McGraw-Hill, New York.
10. Jackson, L. B. (1969). An analysis of limit cycles due to multiplication rounding in recursive digital filters. Proceedings: 7th Allerton Conference on Circuit System Theory, 69–78.
11. Jackson, L. B. (1970). On the interaction of roundoff noise and dynamic range in digital filters. *Bell System Techn. J.* **49**, **2**, 159–84.

12. Jackson, L. B. (1970). Roundoff noise analysis for fixed-point digital filters realized in cascade and parallel form. *I.E.E.E. Trans. Audio Electroacoustics* **AU-18**, **2**, 107–22.
13. Kan, E. P. F. and Aggarwal, J. K. (1971). Error analysis of digital filters employing floating-point arithmetic. *I.E.E.E. Trans. Circuit Theory* **CT-18**, **6**, 678–86.
14. Liu, B. and Kaneko, T. (1969). Error analysis of digital filters realized with floating point arithmetic. *I.E.E.E. Proceedings* **57**, **10**, 1735–47.
15. Ni, M. D. and Aggarwal, J. K. (1974). Two-dimensional digital filtering and its error analysis. *I.E.E.E. Trans. Computers* **C-23**, **9**, 942–54.
16. Sandberg, I. W. (1967). Floating-point-roundoff accumulation in digital filter realization. *Bell System Techn. J.* **46**, **8**, 1775–91.
17. Weinstein, C. J. and Oppenheim, A. V. (1969). A comparison of roundoff noise in floating point and fixed point digital filter realization. *I.E.E.E. Proceedings* (correspondence) **57**, **6**, 1181–3.

Chapter 5

Evaluation of Digital Filter Coefficients

VA Introduction

In the previous chapters some methods for designing 1-D and 2-D digital filters have been described. In this chapter some practical information is given as to the choice of the techniques most useful for particular applications and also some quantitative evaluation in the complexity of the filtering problem is presented.

There are several ways of formulating a filtering problem where in general some variables in the filtering operation are fixed due to the character of the problem whilst some others can vary in order to meet the filtering specifications.

For example in the low-pass filtering operation the design problem can be presented in the following alternative forms:

(1) Given the width of the transition band ($\Delta\omega = 2\pi\,\Delta F$) and the values of the maximum admissible deviation in the passband δ_1 and/or in the stopband δ_2, to find the minimum value of N (the order of the filter) necessary to meet these specifications.

(2) Given the width of the transition band ΔF and the maximum value N of the order of the filter, to find the minimum deviation in the passband δ_1 and/or in the stopband δ_2 obtainable.

(3) Given the maximum admissible deviation in the passband δ_1 and/or the stopband δ_2, and the value of N, to find the minimum transition bandwidth ΔF for which the specifications are met.

In general design methods are formulated according to one of the above schemes. To relate the solution of one formulation to the other formulations of the problem it is necessary to study the nature of the approximation procedure. Sometimes it is not possible to have analytical relations between the variables involved in the approximation and therefore it is necessary to rely on empirical relations, design graphs and tables and perhaps on iterative design procedures. To implement some of the design techniques discussed in this chapter, in a very simple and direct form, some FORTRAN programs which have been run and tested on a minicomputer Data General Eclipse S200 are presented in Appendix 3. The purpose of their inclusion is to give to the reader who wishes to try applying digital filtering techniques in his specific area of interest, the possibility of testing the suitability of digital filtering techniques in his applications without a great deal of preliminary work.

VB 1-D FIR design techniques

VB1 The window method

The most general method for designing FIR digital filters is the window method. In fact filters with arbitrary frequency responses can be designed if their impulse response sequence (that is the Fourier series of the periodic frequency response) is known. Sometimes this Fourier series cannot be evaluated in an analytical form, but it can be estimated in a fast way by means of the discrete Fourier transform.[6] In fact the frequency response to be synthesized can be sampled at a very small sampling interval so as to avoid aliasing in the time domain. Then this sequence of frequency samples can be inverse-transformed to produce the samples of the impulse response. At this point a small number of these samples near the origin will be used where the aliasing is negligible by truncating the sequence using windows described in Chapter 2.

The filters obtained this way are not optimum in that the window method does not satisfy any filter optimality constraint. However this design technique can be very useful in applications where the speed and the simplicity of the design is the fundamental constraint, even though an increase in the number of coefficients necessary to meet some set of specifications may occur, in comparison with the corresponding optimum filter.

The analytical derivation of the samples of the impulse response is very simple for the most commonly used forms of filter i.e. for the low-pass, high-pass, band-pass, band-stop filters, differentiators and the Hilbert transformers.

In these cases due to the simple form of the function $H(e^{j\omega})$, it is easy to evaluate the coefficients of the impulse response by evaluating directly the

Type	N odd	N even
low-pass	$h(0)=2f_t$ $h(n)=\frac{\sin 2\pi n f_t}{n\pi}$	$h(n)=\frac{\sin 2\pi(n-\frac{1}{2})f_t}{(n-\frac{1}{2})\pi}$
high-pass	$h(0)=1-2f_t$ $h(n)=-\frac{\sin 2\pi n f_t}{\pi n}$	$h(n)=\frac{1}{(n-\frac{1}{2})}[\sin (n-\frac{1}{2})\pi-\sin 2\pi(n-\frac{1}{2})f_t]$
band-pass	$h(0)=2(f_2-f_1)$ $h(n)=\frac{1}{\pi n}[\sin 2\pi n f_2-\sin 2\pi n f_1]$	$h(n)=\frac{1}{(n-\frac{1}{2})\pi}[\sin 2\pi(n-\frac{1}{2})f_2-\sin 2\pi(n-\frac{1}{2})f_1]$
stopband	$h(0)=1-2(f_2-f_1)$ $h(n)=\frac{1}{\pi n}[\sin 2\pi n f_1-\sin 2\pi n f_2]$	$h(n)=\frac{1}{(n-\frac{1}{2})\pi}[\sin \pi(n-\frac{1}{2})\pi-\sin 2\pi(n-\frac{1}{2})f_2+\sin 2\pi(n-\frac{1}{2})f_1]$
differentiator	$h(0)=0$ $h(n)=-\frac{1}{\pi n^2}[\sin 2\pi n f_c-2\pi n f_c\cdot\cos 2\pi n f_c]$	$h(n)=\frac{1}{\pi^2(n-\frac{1}{2})^2}[\sin 2\pi(n-\frac{1}{2})f_c-2\pi(n-\frac{1}{2})f_c\cos 2\pi(n-\frac{1}{2})f_c]$
Hilbert transformer	$h(0)=0$ $h(n)=\frac{1}{\pi n}[\cos 2\pi n f_1-\cos 2\pi n f_2]$	$h(n)=\frac{1}{\pi(n-\frac{1}{2})}[\cos 2\pi(n-\frac{1}{2})f_1-\cos 2\pi(n-\frac{1}{2})f_2]$

TABLE 5.1

coefficients of the Fourier series of its periodic repetition, according to the relation

$$h(n)=\frac{1}{2\pi}\int_{-\pi}^{\pi} H(e^{j\omega})\, e^{j\omega n}\, d\omega \tag{5.1}$$

In Table 5.1 the impulse responses of the above mentioned filters are summarized and written in the noncausal form. Based on the impulse responses of Table 5.1 a simple FORTRAN program is presented in Appendix 3, for the design of FIR digital filters using the window technique. Referring to the listing of the program for the detailed instructions, the simplicity of method can be seen directly and correspondingly the speed of the window design technique is evident. Two windows are considered in the program, that is the Lanczos and the Weber-type (Cappellini) windows. However it is very easy to introduce other windows. For example, if the Kaiser window is to be used, there is a very efficient routine for the evaluation of I_0 Bessel function due to Kaiser in Rabiner.[14] Let us now consider how the window method can be used to satisfy the sets of specifications considered in the introduction and correspondingly what are the relations between the design parameters; that is, the number of the impulse response samples (N), the maximum passband and stopband deviation (δ_1), the transition bandwidth (defined as the normalized interval between the point where the frequency response becomes equal to $1-\delta_1$ and the point where the frequency response is equal to δ_1) and finally the type of used window function.

The ripple in the filter response depends obviously only on the type of correction function which is used.

In fact, as observed in Chapter 2, the ripple of the windowed filter depends on the amplitude of the sidelobes of the chosen window and they are obviously fixed when the window has been chosen.

The maximum passband and stopband fluctuations for the Lanczos-type, Weber-type (Cappellini) and Kaiser windows as a function of the window parameters are shown in Fig. 5.1 ($\delta=\delta_1=\delta_2$). For Weber–Cappellini (WEBER–CA) window the data of Table 2.2 were used.

Once the type of window to be used has been chosen and hence the value of its parameter to produce the required deviation, when the maximum acceptable ripple is one of the input parameters of the design problem, then two further parameters are yet free to vary, that is N and ΔF. These two variables are related by an uncertainty relation of the form

$$N\cdot\Delta F=k \tag{5.2}$$

where k does not depend on the cutoff frequency of the low-pass filter, but it depends on the type of window used and on the value of the window parameter.

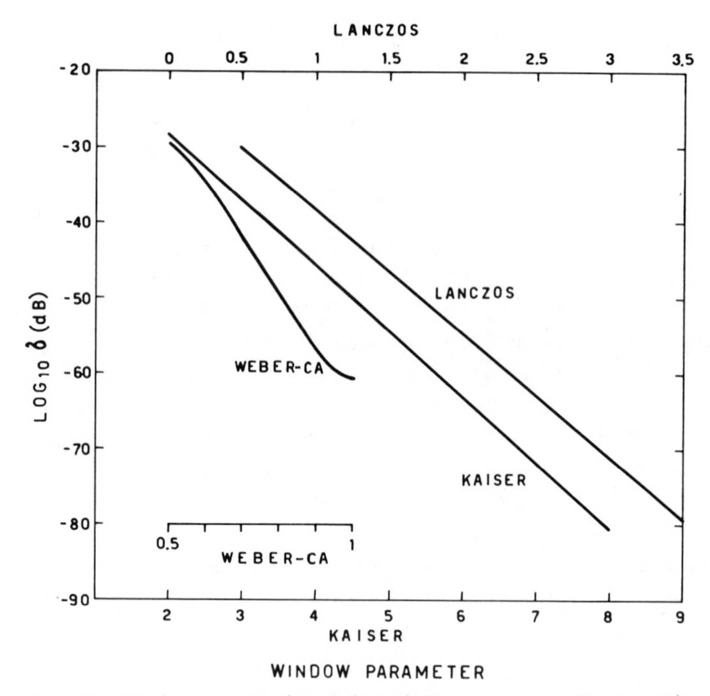

FIG. 5.1. Maximum fluctuation against window parameters ($\delta = \delta_1 = \delta_2$).

Design curves can be constructed where, for every window, the product $(N-1)\Delta F$ is plotted against the value of the maximum ripple, which depends only on the value of the parameter of the window. Using these plots which are shown in Fig. 5.2 it is possible to determine for every window, the product $(N-1)\Delta F$ corresponding to different values of δ_1.

Thus with the aid of the two sets of curves presented in Fig. 5.1 and Fig. 5.2 it is possible to satisfy the three types of formulation of the digital filter design problem considered in the introduction.

In the first case (ΔF and δ_1 fixed), the curves of Fig. 5.1 can be used to find the parameter of the correction function and then the curves of Fig. 5.2 can be used to find the minimum N to meet the specifications. In the second case (N and ΔF are fixed) the curves of Fig. 5.2 can be used to find for every window the minimum δ_1 obtainable, corresponding to the fixed value of $(N-1)\Delta F$. At this point the curves of Fig. 5.1 can be used to obtain the parameter of the window to produce the ripple δ_1. In the third case (N and δ_1 fixed), the curves of Fig. 5.2 can be used to obtain the value of $(N-1)\Delta F$ corresponding to the desired ripple, and the value of ΔF is directly obtained since N is fixed. The value of the window

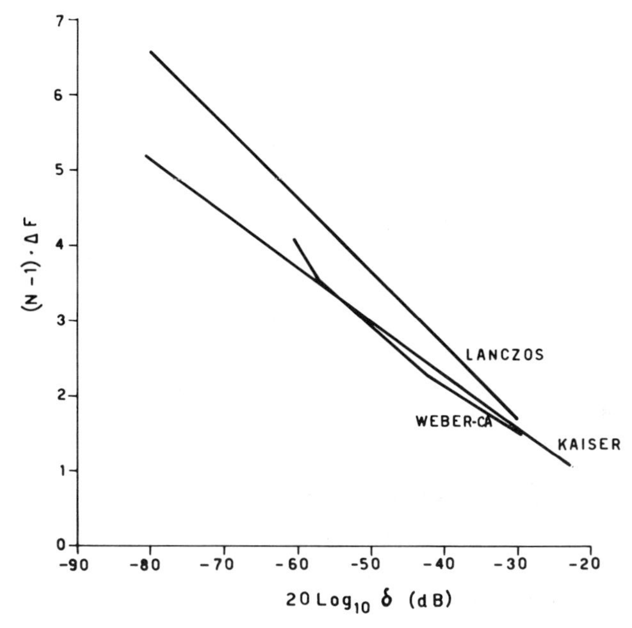

FIG. 5.2. Product $(N-1)\,\Delta F$ against the maximum fluctuation ($\delta=\delta_1=\delta_2$).

parameter can then be found using the curves in Fig. 5.1. The above considerations have been developed with reference to the low-pass filter design, where the number of variables to be considered is not too high. However they can be extended to other types of filters. For example the design of a band-pass filter can be considered roughly as the design of two low-pass filters, one for every transition band, with equal bandwidth and equal passband and stopband ripples. Thus the filter design can be dimensioned on the equivalent low-pass filter design. Fig. 2.8 is an example of the use of the program in Appendix 3.

VB2 The frequency sampling method

The second design method, the implementation of which is discussed here in some detail, is the frequency sampling technique. This technique can be very useful, if the resulting FIR filters are implemented recursively as described by eq. (2.86). In addition this method is particularly suitable when very narrow band filters are to be designed, because in this case most of the frequency samples are zero or indeed when several different filtering operations are to be performed on the same signal. In this last case in fact the comb filter and the second order sections are the same for all the filters and only the weights at the output of the second-order sections change. A program for the design of frequency sampling filters using the linear programming algorithm is presented in Appendix 3. It can be used to design low-pass, high-pass, band-pass, band-stop filters, differentiators and Hilbert transformers, using the sampling points of type 1 filters according to Fig. 2.13.

The specification of the bands has to be formulated in terms of the sampling points and not in terms of the normalized frequency. That is the number N of sampling points in the interval 0–1 of the normalized frequency has to be an input data to the program, whilst two other integers N_1 and N_2 represent the extrema of the passband. Obviously if low-pass, high-pass filters and differentiators have to be designed only one extremum of the band-pass has to be specified and only the value of N_1 is used. It has to be observed that N_1 and N_2 really are the points where the transfer function is equal to the passband value and hence the coefficients to vary in the optimization are outside the range of points specified by N_1 and N_2. Finally when two transition bands are present in the transfer function, the coefficients in these transition bands are assumed to be symmetric. The optimization is performed using the APMM routine in the IBM SSP (Ch. 3) and the stopband error is minimized. As an example the filters in Fig. 2.15(a) and (b) have been designed with this program. The output of the program corresponds to the values of the variable coefficients in the transition bands, which are used to minimize the stopband error. These values, in addition to the passband ones, along with their phase angles to have linear phase filters with real impulse responses of suitable symmetries, can be directly used to implement the filter using the recursive structure of (2.86). If the coefficients of the corresponding nonrecursive structure are needed, they can be computed using the inverse discrete Fourier transform.

A routine is needed for the APMM program, which is used to compute the values of the interpolation functions. In this case the interpolation functions are those appearing in the relationships of (2.97) and (2.99), and the routine FINTA in Appendix 3 is used to compute their values. If filters defined by type 2 frequency samples have to be designed, this routine must be changed and some further simple changes must be made in the program to define the new grid of interpolated ascissas.

Finally we observe that in the differentiator design, the filter is defined in the frequency region which starts from 0 and ends at some frequency $F1$ specified by N_1, while in the Hilbert transform case, the designed filter is basically a passband filter. In fact looking at the form of the interpolated frequency response as in eqs (2.97) and (2.99), it is evident that the only difference between a band-pass Hilbert transformer and a band-pass filter is in the phase factor which multiplies the real part of the transfer function.

Thus in the design, where only the real part of the frequency response is considered, the Hilbert transformer is indistinguishable from the band-pass filter.

From the computational point of view we observe that the linear programming routine APMM is rather sensitive to the accuracy of the computation. The program presented here uses single precision arithmetic, because it has been implemented on a minicomputer with the consequent memory limitation. Thus problems of precision can be present when the number of variables increases with the corresponding decrease in the value of stopband ripple. In these cases the program has to be run using double precision arithmetic. Finally the relations between the variables of the design must be studied so as to have at least a rough estimate of their values to solve filter design problems. The low-pass filter design is considered for simplicity, but the results can be easily extended to other forms of filters.

Using Table 5.2, which describes the behaviour of the stopband deviation in terms of the number of variable coefficients in the approximation, it is very easy to study the behaviour of the filters, due to the fact that two parameters, that is the number of coefficients (frequency samples) and the transition bandwidth, are directly related when the number of frequency samples in the transition band is fixed.

Hence the three cases considered in the introduction can be solved in the following manner.

If δ_2 and ΔF are fixed, the number of variable coefficients N_v to produce a ripple δ_2 is chosen using Table 5.2. At this point the frequency samples are so spaced that N_v samples fit in the ΔF interval and the number N of coefficients is directly obtained. When N and ΔF are fixed the number of variable samples is directly obtained by considering the number of frequency samples which are in the transition band. At this point δ_2 can be estimated from Table 5.2. In the third case, when N and δ_2 are fixed, the minimum transition bandwidth results directly from the number of variable coefficients necessary to have a ripple δ_2.

Number of variable coefficients	Range of variation of maximum stopband error[a]
1	−44 to −54 dB
2	−65 to −75 dB
3	−85 to −95 dB

[a] A stopband attenuation exceeding these limits is generally obtained when the filter to be designed is very narrow-band or very wide-band.

TABLE 5.2. Low-pass design frequency-sampling technique.

VB3 Optimum filters

Let us now discuss a technique based on the application of the Remez exchange algorithm for the design of optimum FIR filters according to the Chebyshev criterion (minimax) described in Chapter 2.

In Appendix 3 a computer program due to McClellan[8] is given, for the design of optimum FIR filters using the Remez algorithm. The program is very general and it can be used to design multiple passband stopband filters, differentiators and Hilbert transformers, using, as error function to minimize, the weighted error norm defined by eq. (2.134).

The program is modular and flexible and it can be easily modified to design filters with arbitrary frequency responses. In fact the forms of the function to be approximated and of the weighting function are specified by two routines (EFF and WATE) which can be replaced to define arbitrary functions as and when required.

Referring to Appendix 3 for an explanation of the use of the program, it is observed that we are interested in the parameters which must be fixed at the start of the program. The main advantage of the formulation of the approximation problem considered, in comparison with other optimum filter design procedures, is that the transition bands of the filter to be designed can be precisely fixed. Once the transition bands have been fixed and the number of functions to be used in the approximation, which is equal to the number of coefficients of the resulting filter, has been chosen, the program minimizes the error functional and computes the coefficients of the filter for which the maximum error in the approximation bands is minimum.

Then the main advantage of the program lies in the fact that it is possible to fix the edges of the transition bands, whilst the main drawback is in the fact that it is necessary to fix *a priori* the number of approximation functions. Thus it is not possible to satisfy directly sets of specifications like the ones corresponding to the first and third cases considered in the introduction of this chapter. Obviously these requirements can be met using the program in its original form by means of an iterative procedure. For example in case (1) the program can be used iteratively starting from a low value of N and increasing it to a value for which the requirements are satisfied. This procedure, though in principle possible, is in practice viable only in very simple cases, due to the fact that the procedure is relatively slow, in spite of the efficiency of the algorithm and its implementation.

Unfortunately analytical relations between the relevant parameters in the design have not been found even in the simple case of the design of low-pass filters. However some empirical relations have been found in the low-pass filter case between the following parameters

(1) the performance measure D_p defined as

$$D_p = (N-1)\,\Delta F$$

(2) the maximum passband deviation δ_1
(3) the maximum stopband deviation δ_2.
For example, if one defines the parameter D as[10]

$$D(\delta_1, \delta_2) = [a_1(\log_{10}\delta_1)^2 + a_2\log_{10}\delta_1 + a_3]\log_{10}\delta_2 + [a_4(\log_{10}\delta_1)^2 + a_5\log_{10}\delta_1 + a_6] \tag{5.3}$$

where

$$a_1 = 5.309\ 10^{-3}$$
$$a_2 = 7.114\ 10^{-2}$$
$$a_3 = -4.761\ 10^{-1}$$
$$a_4 = -2{\cdot}66\ 10^{-3}$$
$$a_5 = -5.941\ 10^{-1}$$
$$a_6 = -4.278\ 10^{-1}$$

and the parameter $f(\delta_1, \delta_2)$ as

$$f(\delta_1, \delta_2) = b_1 + b_2(\log_{10}\delta_1 - \log_{10}\delta_2) \tag{5.4}$$

where

$$b_1 = 11.01217$$
$$b_2 = 0.51244$$

then the following empirical relation

$$D(\delta_1, \delta_2) = (N-1)\Delta F + f(\delta_1, \delta_2)(\Delta F)^2 \tag{5.5}$$

can be written.

Using this relation it is possible to find at least a starting point for a procedure to compute the value of one of the parameters, when the other parameters are fixed, according to our formulations of the design problem given in the introduction. If δ_1, δ_2 and ΔF are specified and N is to be found, obviously D and f can be obtained using (5.3) and (5.4) relations, whilst N can be found from the (5.5). Thus this value of N together with ΔF and $k = \delta_1/\delta_2$ can be chosen as input parameters of the program producing as a result the values of δ_1 and δ_2 corresponding to the imposed constraints on N and ΔF and also the corresponding coefficients. If δ_1 and δ_2 are greater than the specifications the program can be iterated with an increased value of N, whilst if δ_1 and δ_2 are less than the specifications the design can be stopped or it can be iterated decreasing the value of N.

If ΔF and N are imposed by the problem, two cases are possible, the first occurring when $k = \delta_1/\delta_2$ is fixed, the second where one of the parameters δ_1 and δ_2 is specified. In the first case the program can be used directly producing the values of δ_1 and δ_2. In the other case the non-fixed parametef between δ_1 and δ_2 can be estimated using eq. (5.5) and the obtained value can be used to evaluate k. Finally if δ_1, δ_2 and N are specified, the starting value of ΔF can be estimated from relationship (5.5) and it can be used as an input parameter in the design.

The program presented here is very general in the sense that it is possible to use it to design multiple stopband and passband filters. However when filters with more than two bands are to be designed, particular care must be taken in the choice of a suitable set of input parameters.

The solution of the approximation problem is basically in terms of the alternation problem (Chapter 2, Section II I). Thus the filter must have at least $(N+3)/2$ extrema if N is odd and at least $N/2+1$ extrema if N is even. This minimum value corresponds to the number of extrema which are constrained to be equal to the maximum of the deviation in the approximation problem. However if one considers the maximum number of extrema which are possible in the approximation, they are equal to[13]

$$N_e \leq (N+1)/2 + (2NB-2) \tag{5.6}$$

for N odd

$$N_e \leq N/2 + (2NB-2) \tag{5.7}$$

and N even respectively, where NB is the number of bands in the 0–0.5 interval of the normalized frequency, and the first band is assumed to start in the point $F=0$ whilst the last band is assumed to end at $F=0.5$. Now the program constrains only the minimum number of extrema, and the other extrema, which are possible according to the relationships of (5.6) and (5.7), are unconstrained. They can occur either in the optimization bands and in this case they are less than the maximum deviation or they can occur in the unspecified regions (transition bands). In this case they are completely unconstrained and can give oscillations which cannot be accepted. Thus the case can occur in which a filter is obtained which meets the specifications as far as the deviation in the approximation bands are concerned, but too great errors occur in the transition bands. In Fig. 5.3 an example of this behaviour is shown,[13] where a stopband filter with the following band edge frequencies 0.0, 0.14375973, 0.16533943, 0.37032451, 0.41679744, 0.5 is designed. It is evident from the figure that a very high ripple is present in the second transition band.

If one now considers the low-pass design case relation (5.6), for example, for the N odd case, the number of extrema can assume only two values,

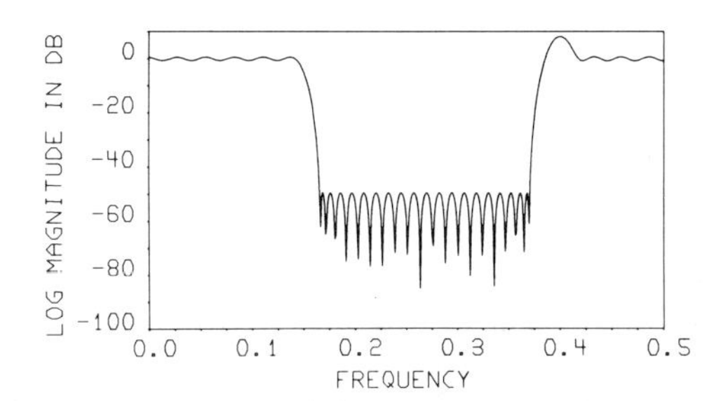

FIG. 5.3. Example of a multiband filter having a ripple in the transition band.

namely $(N+3)/2$ and $(N+5)/2$. Hence at most one of the maxima is not constrained in the approximation procedure and it is not very probable that it will occur in the transition band. When the number of bands increases the number of possible unconstrained extrema increases and correspondingly the probability of having a maximum in the transition band increases.

Unfortunately the number of parameters which are involved in the design of multiband filters is too great to be able to find relations describing the behaviour of the optimum filters, after a careful choice of the parameters is made to avoid the undesired maxima in the transition bands. However many simulations have been made and some procedures have been proposed to choose the input parameters of the algorithm.

The procedure to find the suitable values of the input parameters starts by considering the multiple-band filter design as a composite design of several low-pass filters, one for every transition band. For every equivalent low-pass filter the transition band and the required ripple in the two approximation bands are assumed to be known. Thus by means of relation (5.5) values of N can be obtained, one for every equivalent low-pass design. At this point the normal procedure for designing the multiband filter is to choose the higher value between these N_i values, to meet the most stringent of the specifications. But it has been observed in the simulations that the undesired transition band ripples are more likely to occur when the values of the N_i are very different, whilst an acceptable frequency response can in general be obtained if the values N_i are close, i.e. when

$$N_1 \simeq N_2 \simeq \ldots \simeq N_i.$$

The problem is therefore to find a method to modify the specifications so as to obtain equivalent low-pass filter designs with the same values of N. Several methods have been investigated and the one which is probably the most convenient is based on the modification of the transition bandwidths to obtain equivalent low-pass filters having the same values of N. The procedure can be summarized as follows.

(1) The multiband filter design is divided into several low-pass designs, one for every transition band.
(2) Using the relationship of (5.5) the values N_i for these filters are obtained.
(3) Using the greater value between the N_i, the relationship of (5.5) is now used to obtain a new set of values of the band edges, corresponding to narrower transition bands, which are compatible with the values of ripples.
(4) These new values of the band edges are used as input parameters of the design program.
It can be observed that the modification of the starting parameters is in the direction of imposing more severe specifications to the filter. Thus the procedure cannot interfere with the specifications of the problem.

In the case of Fig. 5.3 the N_i have values of 75 and 35. To equalize the N_i values the upper stopband edge has to be raised to 0.3958. The corresponding frequency response is shown in Fig. 5.4.

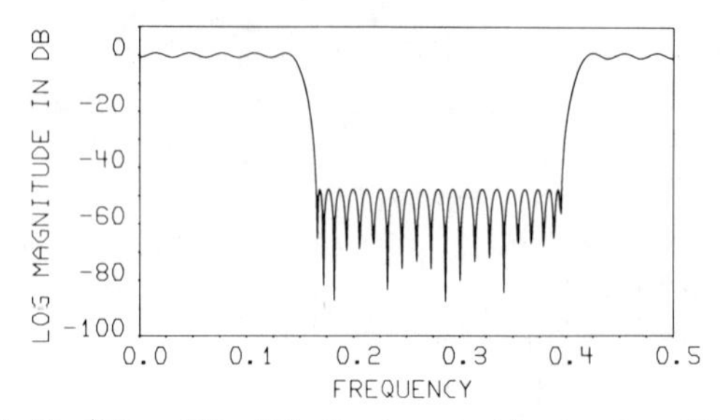

FIG. 5.4. The filter of Fig. 5.3 after the transition band modification.

Extensive simulations have been done also for the design of optimum differentiators and Hilbert transformers and design curves and tables, representing the approximation error as a function of ΔF with N as parameter or as a function of N with ΔF as a parameter, have been presented by Rabiner and Shafer.[11,12]

VC 2-D FIR filter design

The design of 2-D FIR filters, as observed in Chapter 2, is a problem much more difficult computationally than the 1-D design, with the only exception being the window method.

Here we wish to discuss briefly the implementation of the McClellan transformation[7] from 1-D to 2-D filters, which can be very useful when, for example, circularly symmetric filters are to be designed. Using this technique the design is performed in 1-D and then the transformation is applied. If the 1-D design is made using the method to design optimum filters discussed in the previous section, filters very near to 2-D optimum obtainable from linear programming techniques can be determined, if they are sufficiently narrow band.

A program is presented in Appendix 3 for designing 2-D filters using this procedure, that is to perform the mapping according to the relations of (2.160) to (2.166). The program accepts as input the coefficients of the 1-D filter and produces the matrix of coefficients of the 2-D filter.

The main problem of this procedure from the computational point of view is precision in the computation of the coefficients of the expansion in terms of Chebyshev polynomials. These coefficients have in fact a great dynamic range.

The written program is capable of accepting a transformation relation as that of (2.164) represented by four coefficients. Thus, for example, it can be used also to design fan filters. Recently a procedure[9] has been presented for the approximation of arbitrary contours in the transformation by means of functions of the type (2.164) with a greater number of coefficients. Obviously the program can be used only to implement transformations which can be written in the form (2.164) while modifications have to be made if more complex mappings are used.

A filter designed with the above program is shown in Fig. 2.20.

VD 1-D IIR filter design

In this section some practical considerations about the methods studied in Chapter 3 for the design of IIR digital filters are presented. In particular the bilinear transform is considered, and its application to the digitalization of rational transfer functions of classical analogue filters, defined on the s plane or in the framework of the wave digital filter approach.

VD 1 Review of classical analogue filters

Before going into the details of the digitalization methods let us briefly review the main properties of the common families of analogue low-pass filters.

The first family of filters considered is the Butterworth one, which consists of filters which are maximally flat, that is of filters having $2N-1$ derivatives of the squared magnitude, for an Nth order filter, zero at the origin. Their squared magnitude response has the form

$$|H(j\Omega)|^2 = \frac{1}{1+(j\Omega/\Omega_c)^{2N}} \tag{5.8}$$

where Ω_c is the cutoff frequency at which the magnitude response has an amplitude of $1/\sqrt{2}$ (or in a logarithmic scale: -3 dB). These filters with a passband and stopband monotonic behaviour have poles on the s-plane equally spaced on a circle of radius Ω_c. For N odd there are poles at the points corresponding to angle 0 and π, whilst in the N even case the first pole occurs at an angle $\pi/2N$.

Chebyshev filters have an equiripple behaviour which minimizes the maximum error in one of the two bands, that is either the passband or the stopband. In the other band in which the minimax condition is not applied the behaviour is monotonic. The first class of filters (equiripple passband) is referred to as type 1 Chebyshev filters, whilst the second as type 2 Chebyshev filters. The squared magnitude response of type 1 is written in the form

$$|H(j\Omega)|^2 = \frac{1}{1+\varepsilon^2 V_N^2\left(\frac{\Omega}{\Omega_c}\right)} \tag{5.9}$$

where V_N is a Chebyshev polynomial of order N. Such polynomials can be obtained by means of the recursive relation

$$V_{N+1}(x) - 2xV_N(x) + V_{N-1}(x) = 0 \tag{5.10}$$

with

$$V_1(x) = x \qquad V_2(x) = 2x^2 - 1 \tag{5.11}$$

or equivalently by means of the relationships

$$V_N(x) = \begin{cases} \cos(N\cos^{-1} x) & |x| \leq 1 \\ \cos h(N\cosh^{-1} x) & |x| > 1 \end{cases} \tag{5.12}$$

The squared magnitude function oscillates between 1 and $1/(1+\varepsilon^2)$ in the passband, being equal to 1 at $\Omega = 0$ point if N is odd or equal to $1/(1+\varepsilon^2)$ at the same point if N is even. Ω_c is the cutoff frequency, the point of the transition band where the filter reaches the value

$$1/(1+\varepsilon^2)$$

The poles of the filter lie on an ellipse which is completely determined when N, ε, and Ω_c are known. The ellipse is in fact tangent to the circles (Butterworth circles) of radii $a \cdot \Omega_c$ and $b \cdot \Omega_c$ where

$$b, a = \tfrac{1}{2}[(\sqrt{\varepsilon^{-2}+1}+\varepsilon^{-1})^{1/N} \pm (\sqrt{\varepsilon^{-2}+1}+\varepsilon^{-1})^{-1/N}] \tag{5.13}$$

and the axis b of the ellipse lies on the imaginary axis of the s plane. The poles can be easily determined by drawing the radii corresponding to the two Butterworth filters. The ordinate of the ellipse pole is equal to the ordinate of the pole on the larger circle, and the ascissa is equal to the ascissa of the pole on the smaller circle. Functional iteration techniques have also been used for the design of Chebyshev digital filters.[1]

A third and very important family of filters is that of elliptic filters. An elliptic filter has equiripple behaviour in the stopband and in the passband and it is an optimum filter, in the sense that for a given order and a given ripple specification, it has the narrower transition bandwidth.

The squared magnitude response is given by

$$|H(j\Omega)|^2 = \frac{1}{1+\varepsilon^2 sn^2(\Omega, k_1)} \tag{5.14}$$

where ε is a parameter related to the passband filter ripple equal to $1 \pm 1/(1+\varepsilon^2)$, $k_1 = \varepsilon/\sqrt{A^2-1}$ a parameter related to the stopband ripple $1/A^2$ and sn is the Jacobi elliptic function. The mathematical theory of elliptic functions is very involved and will not be presented here. However it can be found in the literature[5] and also a FORTRAN program to design digital elliptic filters has been published.[4] Also an algebraic approach to elliptic digital filter design through functional iteration is available[2] producing filters of order 2^n where n is a positive integer.

VD2 Design of digital IIR filters by means of the bilinear transform

A very simple method for designing a digital filter starting from an analogue filter is based on the bilinear transform as shown in Chapter 3. Its application is as follows. Starting (i) from the critical frequencies of the required digital filter, (ii) from the maximum deviation in the passband and from the minimum attenuation in the stopband, the critical frequencies of the analogue filter are obtained by means of the relationship (3.3)

$$\Omega = k \tan \frac{\omega T}{2} \tag{5.15}$$

This expression represents the mapping of the frequency axis caused by the bilinear transform, as shown in Chapter 3. At this point the problem consists essentially of designing an analogue filter, which can be obtained from general tables[16] developed for designing analogue filters. Thus we obtain the order of the filter necessary to meet the specifications and correspondingly the positions of the poles (and zeros in the elliptic filter case). From knowledge of the zero-pole positions, it is possible to construct the transfer function of the filter on the s plane and the digital filter can be obtained by means of the simple algebraic substitution

$$s \rightarrow k \frac{1-z^{-1}}{1+z^{-1}} \tag{5.16}$$

However digital filters of the Butterworth and Chebyshev type defined by the relations (5.8) and (5.9) can be designed directly in the discrete domain.

Consider for example a Butterworth filter defined by the s plane relation

$$H(s)H(-s)=\frac{1}{1+(-s^2)^N} \tag{5.17}$$

By applying the bilinear z-transform it is possible to obtain the z transfer function

$$H(z)H(z^{-1})=\frac{1}{1+\left[-\left(k\frac{1-z^{-1}}{1+z^{-1}}\right)^2\right]^N} \tag{5.18}$$

from which by letting $z^{-1}=\mathrm{e}^{-j\omega}$ we have

$$|H(e^{j\omega})|^2=\frac{1}{1+\left[k^2\tan^2\dfrac{\omega}{2}\right]^N} \tag{5.19}$$

where the constant k can be chosen to yield an amplitude equal to $1/\sqrt{2}$ for $\omega=\omega_c$ that is from the relation

$$k^2\tan^2\frac{\omega_c}{2}=1 \tag{5.20}$$

thus yielding

$$k^2=1/\tan^2\left(\frac{\omega_c}{2}\right) \tag{5.21}$$

On substituting this value for k in the relation (5.19) the following result is obtained

$$|H(\mathrm{e}^{j\omega})|^2=\frac{1}{1+\left[\dfrac{\tan\omega/2}{\tan\omega_c/2}\right]^{2N}} \tag{5.22}$$

This is therefore the general form of the squared magnitude function of the filter obtained applying the bilinear transform to the analogue Butterworth filter.

The same procedure can be repeated for the Chebyshev filter, obtaining the digital filter squared magnitude function

$$|H(\mathrm{e}^{j\omega})|^2=\frac{1}{1+\varepsilon^2V_N^2\left(\dfrac{\tan\omega/2}{\tan\omega_c/2}\right)} \tag{5.23}$$

The poles of these filters can be easily obtained,[3] and they can be expressed in the form

$$u_m = \frac{1 - x_m^2 - y_m^2}{(1 - x_m^2) + y_m^2} \tag{5.24}$$

$$v_m = \frac{2y_m}{(1 - x_m^2) + y_m^2}$$

where

$$\begin{aligned} x_m &= a \tan\frac{\omega_c}{2} \cos\frac{m\pi}{N} \\ y_m &= b \tan\frac{\omega_c}{2} \sin\frac{m\pi}{N} \end{aligned} \qquad m = 0, 1, \ldots, 2N-1 \tag{5.25}$$

if N is odd, and

$$\begin{aligned} x_m &= a \tan\frac{\omega_c}{2} \cos\frac{2m+1}{2N}\pi \\ y_m &= b \tan\frac{\omega_c}{2} \sin\frac{2m+1}{2N}\pi \end{aligned} \qquad m = 0, 1, \ldots, 2N-1 \tag{5.26}$$

when N is even; a and b are equal to 1 in the Butterworth case. Thus when the parameters of the design are known, which are N, ω_c and ε in the Chebyshev case, then the coefficients of the filter can be obtained by computing the pole positions by means of either (5.25) or (5.26). Construction of the rational z transfer function from the zero and pole positions is achieved by chosing the N poles in the stability region. It should be noted that when the analogue filter has poles only, the corresponding digital filter has poles and zeros. From the relationship of (5.18) and the corresponding relationship for the Chebyshev filters it is in fact easy to verify that a digital filter obtained with this technique has a zero of order N at $z = -1$.

The problem now is to investigate the relationship between the order of the filter N, the passband deviation δ and the transition bandwidth ΔF defined by the cutoff frequency F_c and the frequency F_A at which the squared magnitude frequency response is less or equal to $1/A^2$.

Let us now consider the three types of design specifications mentioned in the introduction. In the first case (ΔF and δ fixed) the design procedure has to start with the evaluation of the order of the filter necessary to meet the specifications in terms of the desired attenuation, transition bandwidth and passband deviation. The passband deviation can be controlled in the case of Chebyshev type 1 filters by ε. In any case, having defined F_c and F_A (that is the transition bandwidth), the desired value $1/A^2$ of $|H(e^{j\omega})|^2$ at F_A and ε in the Chebyshev case, it is possible to determine N iteratively, starting from a first order filter and increasing the order of the filter to the point where the attenuation at F_A is greater than the desired value. At this point the design is completely determined. The functional iterative techniques of Constantinides[1,2] are suitable for this purpose but they tend to over-design the filter specifications. In the second case (N and ΔF fixed) the design is completely determined for the Butterworth filter case by obtaining directly the value of the attenuation at F_A. In the Chebyshev case a trade-off can be considered between the minimum deviation in the passband and the maximum attenuation at F_A.

In the third case (N and δ fixed) the filter is completely specified and the transition bandwidth is directly obtainable during the design procedure. A computer program is presented in Appendix 3 which designs Butterworth and type 1 Chebyshev filters by means of the above relations and computes the coefficients of their cascade structures. The inputs to the program are the critical frequencies F_c and F_A (on the normalized frequency scale), the value of the desired attenuation of the filters at F_A, and the value of the maximum passband ripple if a Chebyshev filter is to be designed. The order of the filter is computed iteratively. Two examples of filter design with this program are shown in Fig. 5.5 and Fig. 5.6.

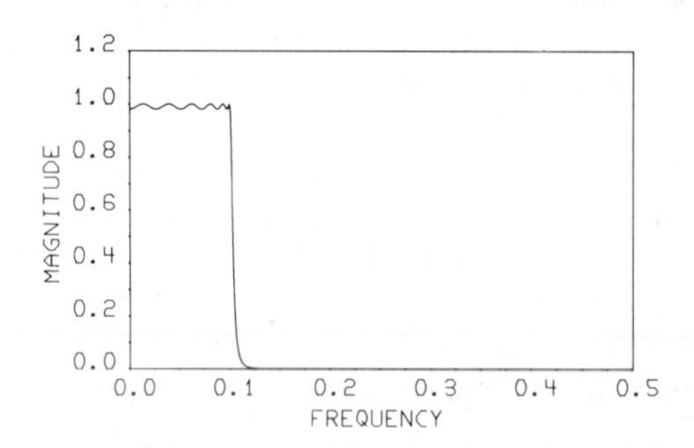

FIG. 5.5. Example of IIR Chebyshev filter: $N = 12$, $F_c/F_s = 0.1$.

Assuming a realization structure by means of second order sections, only even order filters are designed by the program. However it is quite simple to modify the program to design odd order filters by replacing eq. (5.26) with eq. (5.25) and introducing a first order section in the structure.

The program presented here can be used to design only low-pass filters. However to design other types of filters such as band-pass, high-pass and band-stop it is possible to start with the design of a normalized low-pass

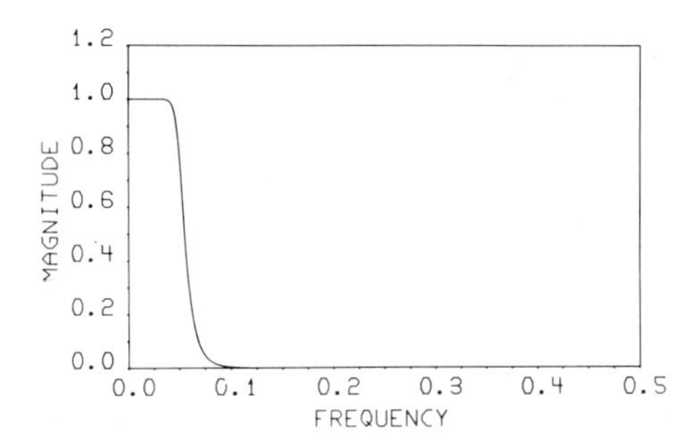

FIG. 5.6. Example of IIR Butterworth filter: $N=8$, $F_c/F_s=0.05$.

filter and then to apply the appropriate frequency transformation of Chapter 3 as shown in Table 3.3. A simple routine (TRASF) to perform these transformations is presented in Appendix 3. This routine can be used to transform second order sections obtained in the design procedure to produce second or fourth order sections if band-pass and band-stop filters are required to be designed. It is essential to note that the transformations operate by re-aligning the frequency axis and hence prewarping of the transition band of the normalized low-pass filter is in general necessary.

The design technique outlined in the above considerations allows the design of recursive digital filters starting from their zero-pole pattern, with their consequent representation in terms of second order sections. However as shown in Chapter 3, this is not the only way for designing digital filters by starting from analogue structures. In fact it is possible to start from analogue ladder structures and to transform these to the digital domain using the bilinear transform, by means of one of the different approaches considered in Chapter 3. The design problem in this case reduces to the following steps for the low-pass filters:

(1) evaluation of the order of the filter which can be made iteratively as above;

(2) determination of the values of the components of the corresponding normalized filter. This can be done using general design tables[16] or analytical relations for the Butterworth on Chebyshev cases;

(3) evaluation of the multipliers of the digital filters recurring to one of the methods presented in Chapter 3.

VE 2-D IIR filter design

In the case of 2-D IIR filters, the techniques available to us are too cumbersome and unwieldy to be presented here in the form of simple programs for their design. In fact in general one has to have several programs available for the 2-D IIR filter design as follows:

(a) the design program itself incorporating the design approach;

(b) a stability test program to be used if the design method, as is almost invariably the case, does not guarantee stability;

(c) a stabilization program to be used if the stability test happened to be unsuccessful.

A program to stabilize filters using the Hilbert transform technique is in Read.[15] This can be very useful even if some counter examples have been found, where the technique was shown to be unsuccessful. In general a stabilization method produces stabilized filters having an infinite number of coefficients and thus a truncation has to be introduced. This means that it is necessary to run a stability test after a stabilization procedure has been applied.

As a final general comment we observe that the problem of the design of 2-D IIR filters is currently one of the most extensively studied problems and important improvements in the design techniques are expected in the near future.

References

1. Constantinides, A. G. (1967). Synthesis of Chebychev digital filters. *Electronics Letters* **3**, **3**, 124–6.
2. Constantinides, A. G. (1967). Elliptic digital filters. *Electronics Letters* **3**, **6**, 255–6.
3. Gold, B. and Rader, C. M. (1969). "Digital Processing of Signals". McGraw-Hill, New York.
4. Gray, A. H., Jr. and Markel, J. D. (1976). A computer program for designing digital elliptic filters. *I.E.E.E. Trans. Acoustic, Speech and Signal Processing* **ASSP 24**, **6**, 529–38.
5. Guillemin, E. A. (1957). "Synthesis of Passive Networks". Wiley, New York.
6. Helms, H. D. (1968). Nonrecursive digital filters: design methods for achieving specifications on frequency response. *I.E.E.E. Trans. Audio Electroacoustics* **AU 16**, **3**, 333–42.
7. McClellan, J. H. (1973). The design of two-dimensional digital filters by transformations. Proceedings: 7th Annual Princeton Conference Information Sciences and Systems, 247–51.
8. McClellan, J. H., Parks, T. W. and Rabiner, L. R. (1973). A computer program for designing optimum FIR linear phase digital filters. *I.E.E.E. Trans. Audio Electroacoustics* **AU 21**, **6**, 506–26.
9. Mersereau, R. M., Mecklenbrauker, W. F. G. and Quatieri, T. F. (1976). McClellan transformations for two-dimensional digital filtering: I design. *I.E.E.E. Trans. Circuits and Systems*, **CAS 23**, **7**, 405–14.
10. Rabiner, L. R. (1973). Approximate design relationship for low-pass FIR digital filters. *I.E.E.E. Trans. Audio Electroacoustics* **AU 21**, **5**, 456–60.

11. Rabiner, L. R. and Shafer, R. W. (1974). On the behaviour of minimax relative error FIR digital differentiators. *Bell System Tech. J.* **53**, **2**, 333–61.
12. Rabiner, L. R. and Shafer, R. W. (1974). On the behaviour of minimax FIR digital Hilbert transformers. *Bell System Tech. J.* **53**, **2**, 363–90.
13. Rabiner, L. R., Kaiser, J. F. and Shafer, R. W. (1974). Some considerations in the design of multiband finite impulse response digital filters. *I.E.E.E. Trans. Acoustic, Speech and Signal Processing* **ASSP 22**, **6**, 462–72.
14. Rabiner, L. R. and Gold, B. (1975). "Theory and Application of Digital Signal Processing". Prentice-Hall, Englewood Cliffs, New Jersey.
15. Read, R. R. and Treitel, S. (1973). The stabilization of two-dimensional recursive filters via the discrete Hilbert transform. *I.E.E.E. Trans. Geoscience Elec.* **GE 11**, **3**, 153–60.
16. Saal, R. (1961). The design of filters using the catalogue of normalized low-pass filters. Telefunken AG.

Chapter 6

Software Implementation of Digital Filters

VIA Introduction

The techniques described in Chapters 2, 3 and 5 were aimed at the design aspects of digital filters, that is at the determination of the coefficient vectors (1-D) or matrices (2-D) to meet some set of specifications. Here we discuss the algorithms which can be used to implement a filtering operation on a general purpose digital computer.

In Appendix 4 simple FORTRAN routines are listed, which implement some of the algorithms described in this chapter. They have been written as a direct implementation of the algorithms presented in this chapter, using floating point number representation. The programs included here are not meant to be very efficient either from the speed or from the length of code point of view, but they are useful working routines from which further development is possible, if needed. These routines have been written and tested using a Data General Eclipse S 200 minicomputer and the FORTRAN V compiler.

Obviously to implement digital filtering operations using a minicomputer it is necessary to consider some constraints regarding the memory, the operation speed, the types of available peripherals, which are less stringent when a larger machine is used. Thus some of the considerations in this chapter are suggested by our experience as minicomputer users and are consequently less stringent for a general purpose large computer user.

However minicomputers have been so widely diffused within the data processing application field, that in our opinion this approach can be of general interest.

VIB Some general comments on implementation comparison

One of the first items in the discussion of digital filter implementation methods is the choice of a figure of merit to facilitate a comparison of the relative efficiency of possible implementations.

In this efficiency comparison the most commonly used figure of merit is the number of multiplications, which have to be performed, while the time necessary for the additions and the control of the operations (loop control, indices computation, etc.) are considered negligible.

This is a reasonable figure of merit when a hardware implementation is considered. In this case, in fact, the main constraint is the multiplication time, whilst the other operations of control and additions can be performed with dedicated hardware which runs concurrently with the multiplier.

If a software implementation is to be used, particular care has to be taken in the problem of algorithm comparison. In fact some general statements about the efficiency of different methods are always valid (for example, FFT FIR digital filtering is definitely more efficient than a direct convolution when the number of coefficients is high), but the trade-off between the methods is a function of several conditions such as:

(1) fixed point or floating point implementation
(2) availability of an hardware fixed point and/or floating point arithmetic unit
(3) high or low level programming language.

If a fixed point implementation is considered and a fixed point arithmetic hardware is not available, then most of the time is really used in the evaluation of products. But if a hardware fixed point arithmetic unit is available, although multiplication time is again greater than addition time, it is nevertheless of the order of only few memory cycles longer. Therefore, few control instructions can have an execution time of the same order as multiplication time. This means, for example, that the use of symmetries in the impulse response of FIR filters, for the purpose of decreasing the number of multiplications, cannot be as convenient as it would seem, if this use implies a more complex structure for the program.

In a floating point implementation, if a dedicated hardware is not available for the execution of floating point operations, the operation times are of the order of some hundreds of microseconds. This implies that the complexity of the control part of the algorithm is not significant in comparison with the operation time, thus making convenient the use of symmetries for the reduction of the number of required multiplications.

The situation is different if a floating point hardware is present. In this case the operation times are of the order of few memory cycles with the consequence that the algorithm complexity becomes very important.

As an example let us consider the case of the Eclipse S 200 minicomputer, used in the development of the programs. The floating point addition time is at the worst case 2.4 μs, whereas the multiplication time is 3.9 μs. Thus only five more instructions in the control part of the algorithm due to the more complex index computation necessary to exploit symmetries in the impulse response, are sufficient to reduce to zero the advantage of reducing the number of multiplications.

In this last case the programming language used is also of importance. When a low level language is used, such as the ASSEMBLER, it is possible to optimize the structure of the program and thus reduce to a minimum the number of instructions needed to be executed in the control part of the algorithm.

VIC Techniques for the implementation of 1-D FIR filters

VIC1 The direct convolution

A simple algorithm to implement 1-D FIR filters is the direct convolution in the time domain, that is the direct evaluation of the convolution sum which defines the filter,

$$y(n)=\sum_{k=0}^{N-1} h(k)x(n-k) \tag{6.1}$$

Very simple routines can be written to implement eq. (6.1), even if they are not very efficient from the point of view of processing speed. Two different forms can be considered. The first assumes the availability of only one sample of the signal at every sampling interval and consequently it features the shift register operation. In the second form it is assumed a batch of data is to be operated upon and hence the routine is organized as a convolution operation sliding on this given batch of data.

As an example in Appendix 4 the FORTRAN code of a very simple routine (FUNCTION FLTR) of the first type is presented. It requires N cells for the impulse response and $N+1$ cells for the data register, which have to be cleared before starting the operation.

Also a routine of the second type (SUBROUTINE FLTRM) is presented in Appendix 4. It needs N cells for the impulse response in the reverse order and it operates on the entire vector of data. The output samples are memorized in the same vector of the input data, ignoring the initial and final transients of $N-1$ points.

A reduction of the number of multiplications can be obtained if some information is available about the symmetries of the impulse response. In this case the number of multiplications N can be reduced to $N/2$ (N even) or to $N/2+1$ (N odd) adding or subtracting the two samples which have to be multiplied by the same coefficient. For example, if the impulse response is even and N is even, then eq. (6.1) can be rewritten in a simpler and more convenient form for the above purposes

$$y(n)=\sum_{k=0}^{N/2-1} h(k)\cdot[x(n-k)+x(n-N+1+k)] \tag{6.2}$$

Some measurements on processing time have been performed with the two routines presented above, obtaining estimates of the time $t_p=CN$ necessary to produce an output sample and also of the value of C which can be interpreted as a program dependent multiplication time. The results for some values of N are shown in Table 6.1. Remembering the value of multiplication time for the computer used here, it is evident that the *true* multiplication time is heavily dependent on the structure of the algorithm considered for the filter.

N	t_p(FLTR)	t_p(FLTRM)	C(FLTR)	C(FLTRM)
21	1 ms	0.7 ms	≃46 μs	≃33 μs
31	1.44	1.04		
51	2.33	1.66		
71	3.16	2.25		
101	4.66	3.33		

TABLE 6.1

VIC2 Filtering in the frequency domain

When the memory is not the main constraint to the implementation problem, a method, which in most cases is much faster than the direct convolution, can be used for the implementation of digital filtering operations. It is based on the transformation of the convolution to a product in the frequency domain, using the circular convolution theorem of the discrete Fourier transform. The efficiency of this method relies essentially on the efficiency of the FFT algorithms.[6,11]

Assume that we have a signal $\{x(n)\}$, $n=0,\ldots,M-1$, which has to be filtered by means of a FIR filter whose impulse response is $\{h(n)\}$, $n=0,\ldots,N-1$. If L is the minimum power of 2 greater than the sum $M+N-1$, it is possible to perform the filtering in the frequency domain using the following procedure.

(1) Construct a new signal $\{\tilde{x}(n)\}$, $n=0,\ldots,L-1$, where

$$\begin{cases}\tilde{x}(n)=x(n) & n=0,\ldots,M-1\\ \tilde{x}(n)=0 & n=M,\ldots,L-1\end{cases} \tag{6.3}$$

(2) Construct a new impulse response $\{\tilde{h}(n)\}$, $n=0,\ldots,L-1$ defined as

$$\begin{cases} \tilde{h}(n)=h(n) & n=0,\ldots,N-1 \\ \tilde{h}(n)=0 & n=N,\ldots,L-1 \end{cases} \tag{6.4}$$

(3) Compute the FFT of $\{\tilde{x}(n)\}$ obtaining the sequence $\{\tilde{X}(k)\}$, $k=0,\ldots,L-1$
(4) Compute the FFT of $\{\tilde{h}(n)\}$ obtaining the sequence $\{\tilde{H}(k)\}$, $k=0,\ldots,L-1$
(5) Compute the complex product of the two sequences $\{\tilde{H}(k)\}$ and $\{\tilde{X}(k)\}$, obtaining a new sequence $\{Y(k)\}$, $k=0,\ldots,L-1$
(6) Compute the IFFT of the sequence $\{Y(k)\}$ obtaining a sequence $\{y(n)\}$, $n=0,\ldots,L-1$, which, according to the circular convolution theorem, is the circular convolution of $\{x(n)\}$ and $\{h(n)\}$.

The circularity of the discrete convolution is the reason why in step (1) of the above procedure L was defined as the minimum power of two greater than the sum of M and N and not as the maximum between M and N. In fact to obtain the noncircular convolution of two sequences a guard space has to be created of a length which is equal to the number of points of the shorter between the two sequences minus 1.

Let us now explain how this procedure can be more convenient from the processing time point of view than the direct convolution described before, using as a figure of merit the number of multiplications.

The convolution of two real sequences of length M and N respectively requires a number of real multiplications equal to

$$N_c = M \cdot N \tag{6.5}$$

The filtering performed in the frequency domain, on the other hand, assuming we know the Fourier transform of the impulse response of the filter, requires two FFT computations of L points plus L complex multiplications. Thus the number of required complex multiplications is given by

$$N_{mc} = 2\left(\frac{L}{2}\log_2 L\right) + L \tag{6.6}$$

whilst the number of real multiplications is given by

$$N_{mr} = 4L\,(\log_2 L + 1) \tag{6.7}$$

It can be observed that the worst condition in the frequency domain filtering occurs when the number M differs from a power of 2 by an amount which is less than the value of N, so that $L \simeq 2M$, whilst the best condition is when $M+N-1$ is less than a power of 2 but very near to it. The computation of the relationships (6.5) and (6.7) for the best conditions in the frequency domain approach shows a crossing over point between the two methods (direct convolution and FFT) when the number of coefficients of the filter is of the order of 40.

A further increase in processing speed by a factor of approximately 2 can be obtained by using the fact that the signal to be filtered is real. In this case the L point FFT can be obtained by a suitable recombination of the $L/2$ point transform of an artificial complex sequence defined as

$$x_c(n) = x_r(n) + jx_i(n) \qquad n = 0, \ldots, \frac{L}{2} - 1 \tag{6.8}$$

where

$$\begin{aligned} x_r(n) &= \tilde{x}(2n) \\ x_i(n) &= \tilde{x}(2n+1) \end{aligned} \qquad n = 0, \ldots, \frac{L}{2} - 1 \tag{6.9}$$

After the computation of the sequence $\{X_c(k)\}$, $k = 1, \ldots, L/2 - 1$, the Fourier transforms of $\{x_r(n)\}$ and $\{x_i(n)\}$, $\{X_r(k)\}$ and $\{X_i(k)\}$ respectively can be obtained by means of the relations[4] ($X_c(L/2) = X_c(0)$)

$$\begin{aligned} X_r(k) &= \frac{X_c(k) + X_c^*(L/2 - k)}{2} \\ X_i(k) &= \frac{X_c(k) - X_c^*(L/2 - k)}{2j} \end{aligned} \qquad k = 0, \ldots, \frac{L}{2} - 1 \tag{6.10}$$

and from $\{X_r(k)\}$ and $\{X_i(k)\}$, the sequence $\{X(k)\}$, $k = 0, \ldots, L - 1$ can be derived using the expressions

$$\begin{aligned} X(k) &= X_r(k) + e^{-j2\pi k/L} X_i(k) & k &= 0, \ldots, \frac{L}{2} - 1 \\ X(k) &= X_r\left(k - \frac{L}{2}\right) + \mathrm{e}^{-j2\pi k/L} X_i\left(k - \frac{L}{2}\right) & k &= \frac{L}{2}, \ldots, L - 1 \end{aligned} \tag{6.11}$$

which result directly from the decimation in time FFT algorithm.

As observed in Section VIB, the comparison of the number of multiplications is only a very rough estimate of the relative algorithm efficiency. The right comparison must really be done by considering also the kind of computer used and the programming complexity of the algorithm. The time to process M data in the direct convolution case is given by

$$t_M = c_1 N \cdot M \tag{6.12}$$

where c_1 takes into account the actual implementation, while the time to compute a FFT transform can be written in the form

$$t_F = c_2(L \log_2 L) \tag{6.13}$$

where again c_2 depends on the particular routine used. By considering for example the routines presented in Appendix 4, c_1 is approximately 46 μs using the function FLTR and approximately 33 μs using the subroutine FLTRM, whereas the values of c_2 vary from about 71 μs to approximately 85 μs for L varying from 4096 to 64, as shown in Table 6.2.

L	4096	2048	1024	512	256	128	64
$\sim c_2$ (μs)	71	71	73	73	76	80	85

TABLE 6.2

VIC3 The select-save and the overlap-add methods

The main problem using the algorithm of Section VIC2 is that to perform the filtering operation it is necessary to compute the FFT of two sequences of length L, equal to the minimum power of 2 greater than $(M+N-1)$. This has two effects. The first is that the memory is not used in an efficient way, because two complex vectors of length L are needed. The second is that if $N \ll L$ a waste of time results in the computation of an L point transform on the N points of the impulse response. Moreover, often it is necessary to process a long sequence, which cannot all be contained in the available memory. In this case it is necessary to read in the memory a batch of data, process them, then write out the output samples, iterating the procedure until the whole input signal has been processed. This can be achieved by filtering subsets of the input samples by means of the frequency domain approach and recombining the obtained sub-sequences. This sectioning can be performed by two different procedures, referred to as the select-save and overlap-add methods, which differ in the way the difficulties due to the circularity of the convolution are circumvented.

The select-save method is organized in accordance with the following steps.

(1) Given the value of N, that is the length of the impulse response, a value for L, corresponding to the number of points on which the Fourier transforms are computed, is chosen. This choice is generally done according to a study,[6] results of which are included at the end of this section and which gives the optimum value of the FFT length as a function of the value of N. If the optimum value is too large for the available memory a suboptimum solution can be chosen with decreasing efficiency in the algorithm.

(2) The FFT of the filter impulse response is computed, by constructing a new impulse response $\{\tilde{h}(n)\}$, $n=0,\ldots,L-1$, defined as

$$\begin{cases} \tilde{h}(n)=h(n) & n=0,\ldots,N-1 \\ \tilde{h}(n)=0 & n=N,\ldots,L-1 \end{cases} \tag{6.14}$$

(3) L points of the input signal $\{x_i(n)\}$ are selected and the Fourier transform $\{X_i(k)\}$, $k=0,\ldots,L-1$ is computed.
(4) The complex product $\{Y_i(k)\}=\{\tilde{H}(k)\cdot X_i(k)\}$ is computed for $k=0,\ldots,L-1$.
(5) The IFFT of $\{Y_i(k)\}$ is computed producing the sequence $\{y_i(n)\}$, $n=0,\ldots,L-1$.

If now we consider the fact that the discrete convolution is circular and that the impulse response is defined as being different from zero only in the interval $0-(N-1)$ with $N<L$, it is evident that the first $N-1$ samples of $\{y_i(n)\}$ with $n=0,\ldots,N-2$ are wrong, whilst the points from $N-1$ to $L-1$ are samples of the noncircular convolution. Thus this procedure produces at every step of the iteration $L-N+1$ points of the desired convolution from $(N-1)$ to $L-1$.

It is evident that the right convolution for the M points of the input signal can be obtained if the successive sections of the input signal are overlapped for $N-1$ points. In fact even if the first $N-1$ samples obtained at every step of the iteration are wrong, the contribution to the convolution of the corresponding $N-1$ samples has been obtained in the preceding iteration. This allows us to drop the first $N-1$ points obtained in the ith iteration and to recombine the other $L-N+1$ with the $L-N+1$ points obtained in the $(i-1)$th iteration. This procedure is sketched in Fig. 6.1. In Appendix 4 a routine which implements this algorithm is presented. This routine has as input a sample vector of M points, the values of M, N and L, the coefficient vector and some other vectors used for the transform computation. The routine has been written to exploit the fact that the signal is real in order to halve the number of steps in the algorithm. Several properties of the DFT transform can be used to obtain the same result: for example it is remembered that an algorithm exists to transform a real sequence of L points using a complex transform of $L/2$ points or that the DFTs of two real signals can be computed using one of them as the real part and the other as the imaginary part in a complex transform, and then using the relations (6.10). It can be shown also that the convolutions of two real sequences $\{x_1(n)\}$, $\{x_2(n)\}$, $n=0,\ldots,L-1$ with the same real sequence $\{h(n)\}$, $n=0,\ldots,N-1$ can be obtained using only one convolution step.

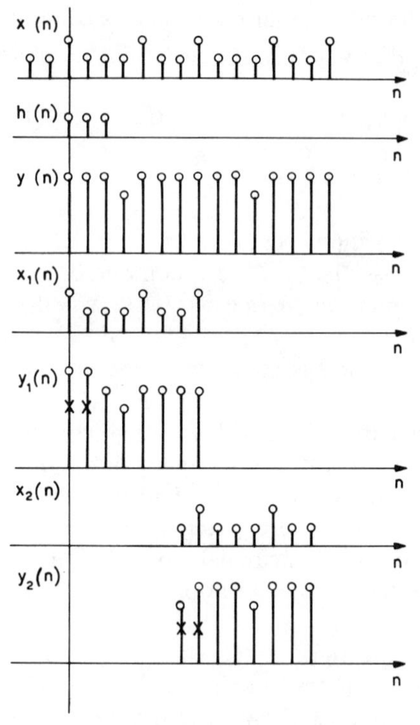

FIG. 6.1. The select-save algorithm in 1-D.

Let $\{\tilde{H}(k)\}$, $k=0,\ldots,L-1$ be the FFT of the sequence $\{h(n)\}$ defined as in the step 2 of the procedure described above, and let us form the complex sequence

$$x(n)=x_1(n)+jx_2(n) \qquad n=0,\ldots,L-1 \tag{6.15}$$

whose discrete Fourier transform is given by

$$\begin{aligned}X(k)&=\sum_{n=0}^{L-1} x(n)\exp\left(-j\frac{2\pi kn}{L}\right)=\sum_{n=0}^{L-1} x_1(n)\exp\left(-j\frac{2\pi kn}{L}\right)\\&\quad+j\sum_{n=0}^{L-1} x_2(n)\exp\left(-j\frac{2\pi kn}{L}\right)\\&=X_1(k)+jX_2(k)\end{aligned} \tag{6.16}$$

On using the discrete convolution theorem a sequence $\{y(n)\}$, $n=0,\ldots,L-1$ can be obtained by inverse transforming the product of $X(k)$ and $\tilde{H}(k)$, which corresponds to the convolution of $\{x(n)\}$ with $\{\tilde{h}(n)\}$. It can then be written as

$$y(n)=y_1(n)+jy_2(n)=\sum_{k=0}^{L-1} X(k)\cdot \tilde{H}(k)\exp\left(j\frac{2\pi kn}{L}\right)$$
$$=\sum_{k=0}^{L-1}[X_1(k)+jX_2(k)]\tilde{H}(k)\exp\left(j\frac{2\pi kn}{L}\right)$$
$$=\sum_{k=0}^{L-1} X_1(k)\tilde{H}(k)\exp\left(j\frac{2\pi kn}{L}\right)+j\sum_{k=0}^{L-1} X_2(k)\tilde{H}(k)\exp\left(j\frac{2\pi kn}{L}\right) \tag{6.17}$$

Now in eq. (6.17) above the terms $\sum_{k=0}^{L-1} X_1(k)\tilde{H}(k)\exp[j(2\pi kn/L)]$ and $\sum_{k=0}^{L-1} X_2(k)\tilde{H}(k)\exp[j(2\pi kn/N)]$ are real, since they are the inverse transforms of the products of transforms of two real sequences. Therefore it is possible from eq. (6.17) to write the following relationships for the real and imaginary parts of the sequence $y(n)$

$$\left.\begin{aligned} y_1(n)&=\sum_{k=0}^{L-1} X_1(k)\tilde{H}(k)\exp\left(j\frac{2\pi kn}{L}\right)\\ y_2(n)&=\sum_{k=0}^{L-1} X_2(k)\tilde{H}(k)\exp\left(j\frac{2\pi kn}{L}\right)\end{aligned}\right\}\quad n=0,\ldots,L-1 \tag{6.18}$$

and due to the convolution theorem $y_1(n)$ is equal to the convolution of $x_1(n)$ and $\tilde{h}(n)$ while $y_2(n)$ is equal to the convolution of $x_2(n)$ and $\tilde{h}(n)$.

Obviously this property allows us to process two of the previously considered segments in only one iteration, thus halving the number of iterations of the procedure, which appeared necessary to process M points of the input signal. The second method used to segment the filtering operation is the overlap-add method. With reference to the description of the select-save method, the steps 1–5 are the same for this case but for step 3 where the signal selection is performed. In this case step 3 consists of the selection of sub-sequences $\{\tilde{x}(n)\}$, $n=0,\ldots,L-1$ of the input signal obtained in the following way

$$\begin{cases}\tilde{x}(n)=x(n) & n=0,\ldots,L-N\\ \tilde{x}(n)=0 & n=L-N+1,\ldots,L\end{cases} \tag{6.19}$$

and therefore having only $L-N+1$ elements different from zero. But $\{h(n)\}$ has only N elements different from zero and hence by using an L point FFT a noncircular convolution can be obtained. As a result at every step of the procedure L points of the convolution are obtained showing a triangular effect at the extrema, due to the fact that the impulse response, in its shifted version has not overlapped with the signal in the first $N-1$ points and that the last $N-1$ points of the convolution contain input samples which are equal to zero, according to the definition of the L point input signal segment. When the next segment of L points is processed, the same effects are observed. The first $N-1$ points in the new iteration do not contain the contribution of the last samples of the previous segment. However this contribution was computed in the previous step and was contained in the last $N-1$ points. So by adding the last $N-1$ points of the ith iteration to the first $(N-1)$ points of the $(i+1)$th iteration, it is possible to reconstruct the right convolution in this part. This procedure can be visualized with reference to Fig. 6.2.

It can be observed at this point that given a filter with an N point impulse response and using transforms of length L only $L-N+1$ points are filtered

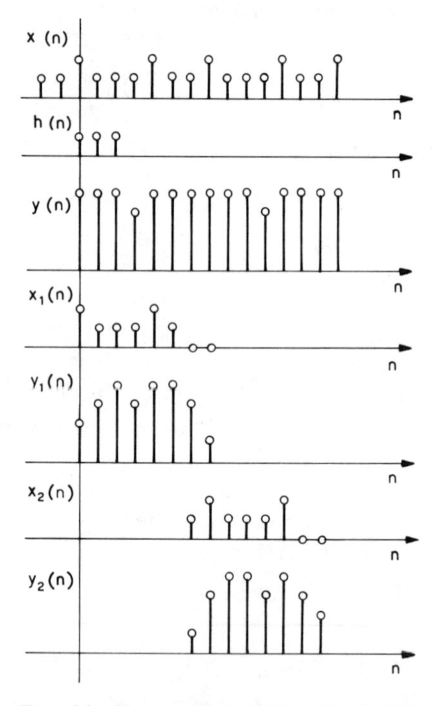

FIG. 6.2. The overlap-add algorithm in 1-D.

at any iteration of the procedure. Therefore to filter M points of the input signal, $M/(L-N+1)$ iterations are necessary. If the number of operations at every step is given by (6.7) the total number N_{tot} is then given by

$$\frac{M}{L-N+1}[4L(\log_2 L+1)] \tag{6.20}$$

that is, given M and N, it is a function of L.

The problem of computation of the value of L for which the number of operations is minimum has been considered by Helms[6] and his results are shown in Table 6.3, for $N \leq 1000$.

Length of filter	Optimum value of L
<11	32
11–17	64
18–29	128
30–52	256
53–94	512
95–171	1024
172–310	2048
311–575	4096
576–1000	8192

TABLE 6.3

It can be seen from the table that the optimum value of L tends to become high when the length of the filter increases. Fortunately the number of operations is not a very rapidly increasing function of L. So that when the available memory is not sufficient for the use of the optimum value, a suboptimum value can be used without a drastic decrease in efficiency. For example, if a value $L \simeq 2N$ is chosen, the number of operations increases only by a factor 4/3.

To compare this filtering procedure to the direct convolution some tests have been performed with respect to measurement of time, t_p, necessary to produce an output point and the corresponding equivalent multiplication time, that is t_p/N. Some results are shown in Table 6.4; it can be observed that by using the routines given in this section the cross point between the two methods is around 20 coefficients.

N	t_p(FLTRF)	t_p/N
21	$\simeq$0.76 ms	$\simeq$36.3 μs
31	0.81	26.1
51	0.88	17.3
71	0.905	12.7
101	1.12	11.1

TABLE 6.4

VID Techniques for the implementation of 1-D IIR filters

In the previous section the problem of FIR filter implementation was considered and the following result was found. Only when the number of coefficients in the filtering operation is very small is it convenient to perform the filtering by means of the direct convolution. When IIR filters have to be used, in general they are defined by a moderate number of coefficients. Hence it is possible to implement them efficiently using the difference equation which defines the filter; that is of the form

$$y(n) = \sum_{k=0}^{N-1} a(k)x(n-k) - \sum_{k=1}^{M-1} b(k)y(n-k) \tag{6.21}$$

or using a cascade or parallel combination of second order sections.

In Appendix 4 a simple routine FILRI to implement the IIR filtering operation in its cascade form is presented. It uses 6 memory cells for the coefficients and 3 for the states and it needs 5 multiplications and 5 additions for every second order section in the structure. The memory of the filter has to be cleared at the beginning of the filtering operation and the routine is organized in such a way as to accept one point at the input and to produce one point at the output.

The implementation method considered is a general one in the sense that any IIR digital filter can be implemented using a routine performing the operations in the (6.21) or in a cascade or parallel combination of second order sections. In fact for any design procedure, it is possible to find the poles and zeros of the filter and from their position a finite difference equation of the type (6.21) or a cascade of first and second order equations can be written. Moreover the z-transform corresponding to the (6.21) can be re-expressed to produce a parallel implementation of the same digital filter.

However some design methods, such as the wave digital filter approach, are intended to map to the digital domain a topology of components that has been found optimum from some point of view (sensitivity) in the continuous analogue case. As it may be seen from Chapter 3, these methods start from a structure in the continuous domain (ladder filters) and replace every analogue component by a digital structure conserving in this manner the position of the corresponding component in the continuous filter case.

From these considerations it is evident that it is not convenient to implement the filters obtained with this method using eq. (6.21), but rather it is necessary to implement them in terms of cascades of elementary sections resulting from the transformation of the continuous domain components. Obviously it is not possible in this case to give a general implementation routine, because the topology of the digital equivalent of any component depends on the form of the continuous filter structure, the transformation method, the type of component.

As an example, however, let us consider the case of a low-pass filter of the LC ladder doubly terminated type, obtained by means of the Constantinides two-port approach using the scattering parameter description. The very simple routines presented in Appendix 4, even though not for general application, are representative of the implementation complexity of the single sections of the wave digital filter structures.

These routines, which are appropriate for the case of the design from the generator, can be used to implement the forward path (COND) and backward path (CON1) of the structure equivalent to the shunt capacitor and the forward path (ZL11) and the backward path (ZLLL) of the series inductor. To implement the entire filter the COND and ZLL1 routines have to be called alternatively starting from the generator to compute the forward wave, and then CON1 and ZLLL have to be called alternatively to compute the reflected wave from the load to the generator. Such routines will simulate polynomial type LC ladder filters. In the case of filters with tuned circuits some transformations are possible which will modify the routines and further details on the nature of these transformations are to be found in the references.[2,3]

VIE Techniques for the implementation of 2-D FIR filters

VIE1 The direct convolution

The problem of implementation of an efficient filtering operation is of fundamental importance in the 2-D case, where the use of direct convolution can result in over-long processing times, as is apparent from the following example. Let us assume that we have an image of 512×512 points and a filter with an impulse response of 32×32 points. Direct convolution, without the use of any symmetry properties of the impulse response, results in a number of products which is equal to $512 \times 512 \times 32 \times 32$. Even if we assume that we have a floating point arithmetic unit which performs a multiplication in 10 μs, the products correspond to a computing time of 44.74 minutes. Adding to this the time necessary to implement the control of the algorithms and to read the data in the memory if, as in the minicomputer case, the entire image does not fit into the available memory, the processing time becomes prohibitively excessive in most applications.

If the computer employed has a sufficient memory to contain the matrix to be processed, the filtering operation using the direct convolution method is straightforward as in the 1-D case, and it is equivalent to a double sum sliding over the matrix to be processed. Unfortunately in most cases the matrix is too large to fit into the computer memory and hence it has to be memorized on an external medium and portions of it have to be read in and fed to the filtering routine individually. Assuming that we have to process a matrix $M \times M$ with an $N \times N$ filter, then the operation using the direct convolution method can be organized in the following way:

(1) A matrix of N rows and $M+2N-2$ columns is constructed, with the first $N-1$ rows equal to zero and the last row equal to the first row of the matrix to be filtered, with guard spaces of $N-1$ points at the extrema.

(2) The row Nth is filtered.

(3) The filtered row is written on an external medium.

(4) All the rows of the submatrix in memory are shifted upward by 1 position, that is the first row is eliminated, the second row is transferred to the position of the first one and so on.

(5) A new row is read from the external medium and is written in the Nth row of the matrix in memory.

The above procedure is iterated going to step 2 until the matrix is completely processed.

A routine to implement the above algorithm is presented in Appendix 4. In this form it assumes that we have the matrix available on a random access medium (disk), even though it can be used with any sequential access medium, because the records (rows of the matrix) are read sequentially (from the first row to the last or from the last one to the first one). The routine needs the coefficients of the filter and some information about the conditions at the edges: that is the number of zero rows and columns that are necessary if one wishes to obtain the transient regions at the edges.

This routine (AFILT) is examined for use also in cascade with the routine BFILT, which will be explained in a later section, in order to implement recursive filtering operations. This is the reason why some parameters are present to control the angle of the data matrix from which the convolution is initiated. This is useful in recursive filtering where, in order to obtain a zero-phase transfer function, it is necessary to filter the matrix four times starting from the four different angles.

This direct method is not very efficient from the speed point of view as already shown in 1-D case. However some improvements in the filtering step of the above procedure in terms of number of multiplications can be obtained using the symmetries of the impulse response of the filter. The more convenient method from this point of view is when linear phase circularly symmetric filters are used. In this case a symmetry between the quadrants and also a symmetry with respect to the diagonals of the quadrants exist. This reduces by a factor of approximately 8 the number of products necessary to produce an output point.

VIE2 Filtering in the frequency domain

A second and more efficient method for performing two-dimensional FIR filtering operations is based on the frequency domain approach. The procedure is the same in principle to the one used in the 1-D case, as examined in Section VIC2; it can be summarized in the following way, where it is assumed we have an input matrix $\{x(n_1, n_2)\}$, $n_1 = 0, \ldots, M_1 - 1$; $n_2 = 0, \ldots, M_2 - 1$ of $M_1 \times M_2$ points and a filter $\{h(n_1, n_2)\}$, $n_1 = 0, \ldots, N_1 - 1$; $n_2 = 0, \ldots, N_2 - 1$ of $N_1 \times N_2$ points:

(1) The 2-D FFT $\{\tilde{X}(k_1, k_2)\}$, $k_1 = 0, \ldots, L_1 - 1$; $k_2 = 0, \ldots, L_2 - 1$ of the matrix $\{\tilde{x}(n_1, n_2)\}$ is computed, where $\{\tilde{x}(n_1, n_2)\}$ is defined as

$$\begin{aligned} \tilde{x}(n_1, n_2) &= x(n_1, n_2) \qquad && n_1 = 0, \ldots, M_1 - 1 \\ & && n_2 = 0, \ldots, M_2 - 1 \\ \tilde{x}(n_1, n_2) &= 0 && n_1 = M_1, \ldots, L_1 - 1 \\ & && n_2 = M_2, \ldots, L_2 - 1 \end{aligned} \tag{6.22}$$

where L_1 and L_2 are the minimum powers of 2 greater than $M_1 + N_1 - 1$, $M_2 + N_2 - 1$ respectively.

(2) The 2-D FFT $\{\tilde{H}(k_1, k_2)\}$, $k_1 = 0, \ldots, L_1 - 1$, $k_2 = 0, \ldots, L_2 - 1$ of the filter impulse response (or point spread function) $\{\tilde{h}(n_1, n_2)\}$, $n_1 = 0, \ldots, L_1 - 1$; $n_2 = 0, \ldots, L_2 - 1$ is computed where

$$\begin{aligned} \tilde{h}(n_1, n_2) &= h(n_1, n_2) \qquad && n_1 = 0, \ldots, N_1 - 1 \\ & && n_2 = 0, \ldots, N_2 - 1 \\ \tilde{h}(n_1, n_2) &= 0 && n_1 = N_1, \ldots, L_1 - 1 \\ & && n_2 = N_2, \ldots, L_2 - 1 \end{aligned} \tag{6.23}$$

(3) The complex product of $\{\tilde{X}(k_1, k_2)\}$ and $\{\tilde{H}(k_1, k_2)\}$ is evaluated to produce the sequence $\{Y(k_1, k_2)\}$, $k_1 = 0, \ldots, L_1 - 1$; $k_2 = 0, \ldots, L_2 - 1$.

(4) The IFFT of the sequence $\{Y(k_1, k_2)\}$ is computed to produce a sequence $\{y(n_1, n_2)\}$, $n_1 = 0, \ldots, L_1 - 1$; $n_2 = 0, \ldots, L_2 - 1$, which, due to the 2-D convolution theorem, is the convolution of the $\{x(n_1, n_2)\}$ and $\{h(n_1, n_2)\}$ sequences. The problem connected with the circularity of the convolution has been circumvented by augmenting the input matrix with the suitable number of zeros.

As a comparison with the direct method it can be observed that the time necessary to filter any image of $M_1 \times M_2$ points with a filter of $N_1 \times N_2$ coefficients can be written as:

$$t_c = K_1 N_1 N_2 M_1 M_2 \tag{6.24}$$

while the time necessary to perform the same operation using the frequency domain approach is given by:

$$t_f = K_2 L_1 L_2 (\log_2 L_1 + \log_2 L_2) \tag{6.25}$$

where K_1 and K_2 are constants dependent on the computer used for these operations and on the program employed for the implementation.

Some observations about the above procedure can be made to minimize the processing time and the memory allocation, due the fact that in most cases the matrix to be filtered is real. The discrete Fourier transform of an L point real sequence has a real part which is symmetric with respect to the origin and therefore it is also symmetric with respect to the $L/2$ point; the imaginary part is antisymmetric with respect the same points. This implies that the real part has $L/2+1$ independent values, whereas the imaginary part has $L/2-1$ independent values (for the antisymmetry the points 0 and $L/2$ are equal to zero). So the transform of any L point real sequence is specified by only L real values and can be written in the same memory positions as the starting sequence. When all the rows are transformed a matrix of the same dimension as the starting matrix is obtained and each row contains the $L/2+1$ real parts and $L/2-1$ imaginary parts of the transform of the corresponding row. If now we consider the columns, we have $L/2+1$ independent columns of which two (the first and the $(N/2+1)$th) are real. Obviously the transforms of the two real columns can be computed and written in the same columns, in accordance with the criterion described before, while the $L/2-1$ complex columns can be transformed without any problem and the transforms rewritten in the same places as the untransformed columns. This implies that, by using the symmetries of the transforms of real signals, the transform of a real matrix can be entirely contained within the same amount of memory as the original starting matrix.

A second observation to be made is that in the first step of the procedure, that is in the row transformation, it is necessary to compute FFTs of real sequences. Thus a reduction in processing time can be obtained through the general methods of real signal transformation discussed earlier. If, for example, a complex transform is used to process together two rows, the number of 1-D FFTs to be performed to transform a matrix with L_1 rows and L_2 columns is $L_1/2+L_2/2$ and not, of course, L_1+L_2.

However it must be mentioned that the preceding observations about processing time are based on the assumption that the matrix to be processed is all contained in memory and is therefore readily available.

When the matrix is not in memory but on an external medium, such as a disk, particular care has to be taken to minimize the number of accesses on the disk. The main problem in this case is that, assuming we have the matrix written on the disk by rows, when it may be simple to access the rows of the matrix in the first step of the procedure (row transformation), it is not easy to access the columns on the disk, because for every column it is necessary to go through the entire matrix. To minimize the time needed for reading in the columns in the second part of the transformation, it is more convenient first to transpose the matrix on the disk and then to read it again by rows. Several methods are employed to minimize the time necessary to transpose a matrix on the disk, see for example Delcaro and Sicuranza.[5] Their efficiency is obviously a function of the portion of the matrix which is accommodated within the central memory of the computer.

A second drawback of the frequency domain approach is that, as in 1-D, the 2-D discrete convolution is in fact a circular convolution. Therefore to avoid errors at the edges of the filtered matrix, it is necessary to augment the matrix to be filtered with a number of rows and columns which must be equal to the filter length minus one. The consequence of this is that if the number of rows and columns M of the matrix, assumed here to be square, is such that $(M+N-1)$ is just greater than some power of 2, it is necessary to transform an image with $2M$ rows and $2M$ columns, with a corresponding loss in efficiency.

VIE3 The select-save and the overlap-add methods

A third method which can be used to perform FIR filtering operations is a segmentation method which is a direct extension of select-save or overlap-add methods considered for 1-D case. Let us now briefly consider the two-dimensional equivalent of the select-save method, to filter an input matrix $\{x(n_1, n_2)\}$, $n_1=0, \ldots, M_1-1$; $n_2=0, \ldots, M_2-1$, with a filter $\{h(n_1, n_2)\}$, $n_1=0, \ldots, N_1-1$; $n_2=0, \ldots, N_2-1$.

Having chosen the dimension $L_1 \times L_2$ (with L_1, L_2 as powers of 2) of the submatrix to be processed at every step, every matrix is processed in the frequency domain as in the previous section. Due to the circularity of the two-dimensional convolution and due to the fact that a guard space of zero values has not be created, the results are exact only for $n_1 \geq N_1-1$, $n_2 \geq N_2-1$, that is in the cross-hatched region of Fig. 6.3 indicated by ①. Therefore the next segment to be considered in the direction, for example, of increasing n_1, that is a new segment with elements of the same rows of the preceding one and of different columns, must start from the column $L_1-(N_1-1)$. Due to the circularity of the convolution, in the second step there will be some erroneous values but these have already obtained in the previous step as correct and a new portion of $(L_1-N_1+1) \cdot (L_2-N_2+1)$ points contiguous with the preceding set of points is obtained.

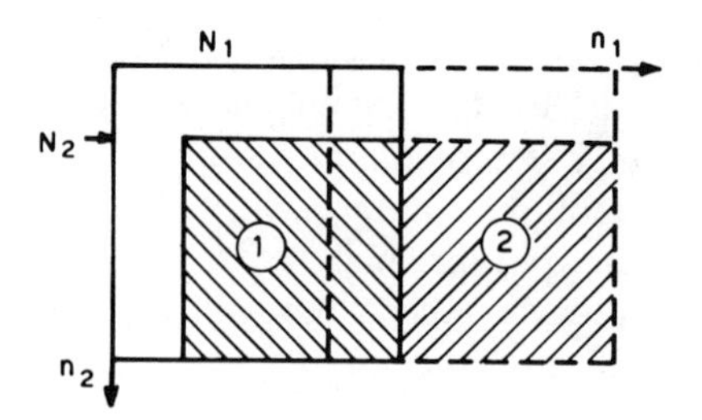

FIG. 6.3. The select-save algorithm in 2-D.

Using this procedure it is possible to obtain the entire image filtered in a number of steps which depend on the length of the filter and on the size of the two-dimensional transform used. A study similar to the one performed in 1-D case, and aimed at the choice of the best dimension for the segments to be used in the processing has been done by Hunt.[7] The time to perform a step in the select-save procedure is given by

$$t=\gamma L_1 L_2(\log_2 L_1+\log_2 L_2) \tag{6.26}$$

and the number of points processed in every section is in fact $(L_1-N_1+1)\cdot(L_2-N_2+1)$. Thus the time necessary to process the entire input matrix is given by

$$t=\gamma L_1 L_2(\log_2 L_1+\log_2 L_2)\frac{M_1 M_2}{(L_1-N_1+1)(L_2-N_2+1)} \tag{6.27}$$

and therefore the problem is to minimize the ratio

$$\alpha=\frac{L_1 L_2(\log_2 L_1+\log_2 L_2)}{(L_1-N_1+1)(L_2-N_2+1)} \tag{6.28}$$

The results are shown in Table 6.5 for $N_1, N_2 \leq 50$. It is evident from this table that for moderate length of the impulse response, the size of segments tends to become somewhat large. If it is not possible to have enough memory to allocate under these optimum values and a lower value has to

	N_2	10	20	30	40	50
N_1	10	128,128	128,256	128,256	128,512	128,512
	20		256,256	256,512	256,512	256,512
	30			512,512	512,512	512,512
	40				512,512	512,512
	50					512,512

TABLE 6.5

be used, the decrease in efficiency of the method can be computed from the value of α for optimum L_1, L_2 and the value of α for the chosen values of L_1 and L_2. Unfortunately, as it can be easily verified, the decrease in efficiency using suboptimum values of L_1 and L_2 is greater in the 2-D case in comparison with the 1-D case.

The general advantage of this approach in comparison with the direct frequency domain approach lies in the following points:

(1) the time necessary to compute the spectrum of the filter is reduced;

(2) the memory allocations necessary for the filter is reduced;

(3) only a segment of the input matrix has to be present within the memory at every step. The implication here is that if the size of the segments is chosen to be accommodated within the memory of the used computer, even if the values of L_1 and L_2 are not the optimum ones, the access to the external medium is reduced to read in the input signals and to write out the results.

The above considerations have been developed for the 2-D equivalent of the select-save method. A method can be also developed which is equivalent to the overlap-add and which differs from the above case in the manner in which the reading and writing of the data on the disk is performed.

VIF Techniques for the implementation of 2-D IIR filters

For the 2-D IIR filter implementation, which is generally performed by means of direct evaluation of the difference equation defining the filter, most of the points made with regard to the FIR direct convolution can be repeated, but for the fact that in this case the coefficients are less in number.

A simple routine for the implementation of the recursive part of the filtering operation is presented in Appendix 4 (subroutine BFILT), whereas any nonrecursive part which may be present can be always implemented by means of the subroutine AFILT.

This routine, which performs the quadrantal filtering operation, assumes that we have the matrix to be filtered on an external random access medium. In terms of memory requirements it needs one matrix with the number of rows equal to the dimension of the recursive part of the filter and the number of columns equal to the dimension of the rows of the matrix to be processed. Two guard spaces of zero samples are required, of which one constitutes the set of initial conditions and the second the tails in the filtering operations as shown in Fig. 6.4 ($H_2 = M_2$). These tails are necessary if the final purpose of the filtering operation is a zero-phase processing of the input matrix. In this case, using quadrantal filters, it is

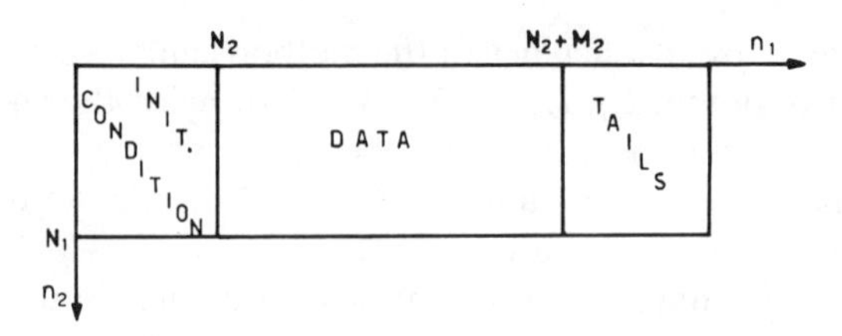

FIG. 6.4. Working memory for the subroutine BFILT ($H_2 = M_2$).

necessary to process the matrix four times, with the same filter but starting from the four different angles. Obviously to obtain a true zero-phase processing operation it would be necessary to perform four filtering operations with indices ranging from zero to infinite, since the impulse response of the filter is infinite. In practice filtering has to be performed to the point where the output is negligible. The tail region is used to this purpose.

The mechanism of the filtering operation is the same as in AFILT routine, but for the fact that in this case the convolution is performed on the output matrix, while only a row of the input matrix is present at any time in memory and only one point of it is used to obtain an output point. When an output row has been computed, it is written on an external medium and then it is used within the routine to construct the matrix of output rows to be used in the successive operations, in the same way as in the AFILT case where the new input row was used. The routine has in addition some controls to simulate the four different quadrant filters. The filtering operation is really performed in the same way whilst it is the matrix to be filtered which is fed to the routine in a suitable way. With the input matrix memorized by rows, the processing scheme is as follows:

First quadrant: the rows are read and filtered with the BFILT routine. An output matrix with longer rows (tails) than the starting one is obtained. The same is true also for the following steps.

Second quadrant: the rows are accessed sequentially but they are transposed before the filtering operation commences.

Third quadrant: the rows are read in reverse order, that is from the last one to the first one.

Fourth quadrant: the rows are read in reverse order and transposed before filtering.

If the filter has in addition a nonrecursive part, the filtering operation can be ended by using in cascade the AFILT routine.

Finally it can be observed that the access to the external medium is sequential from the first row to the last, or vice versa in all steps and hence it can be implemented with a sequential access memory system.

VIG Use of non-sinusoidal functions

Certain computational advantages are apparent if non-sinusoidal functions such as Walsh or Haar are substituted for Fourier functions in the FIR case as described by Beauchamp.[1] Speed of computation is increased and memory requirements are reduced. It is also easier to implement such procedures in small computers since complex arithmetic procedures are not needed.

The use of such functions implies certain limitations noted elsewhere[9] due to the inapplicability of arithmetic convolution such that a simple relationship between transform products and convolution of two series does not exist. Consequently it is necessary to make use of the classical methods of direct Wiener filtering.[10]

Considering the one-dimensional case, the input vector $x(t)$ is assumed to consist of additive zero-mean signal, $s(t)$ and noise $n(t)$, uncorrelated with each other. Utilizing a transformation operation, A consisting of a matrix of N by N values then:

$$\begin{aligned} X(f) &= A \cdot s(t) + A \cdot n(t) \\ &= S(f) + N(f) \end{aligned} \tag{6.29}$$

The resulting vector, $X(f)$ may be multiplied by an N by N matrix of filter weights, G and inversely transformed to produce a filtered output:

$$y(t) = A^{-1} \cdot G \cdot Ax(t) \tag{6.30}$$

G is chosen to provide the best mean-square estimate of the signal component $s(t)$ of the input vector, $x(t)$ (Fig. 6.5).

The transform procedure can employ the FFT, FWT (Fast Walsh transform) or FHT (Fast Haar transform).

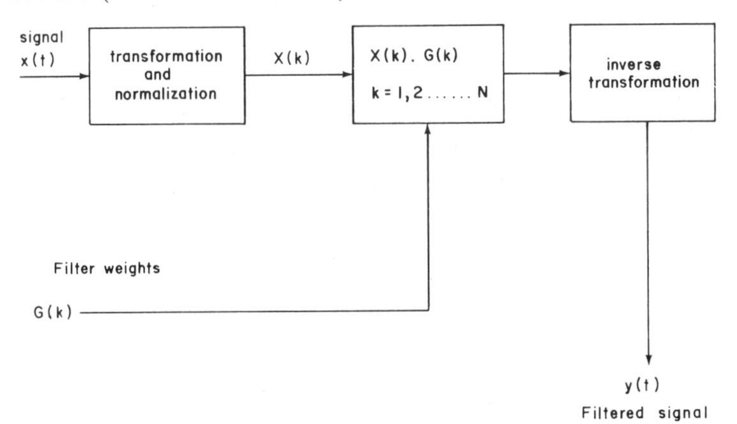

FIG. 6.5. Block diagram of transformation application.

The FWT and FHT involve only addition and subtraction of real numbers and are consequently easier to implement and faster in performance than the FFT which requires multiplication of complex numbers.[1]

In its simplest form for one-dimensional filtering G can consist of a string of ones followed by zeros corresponding to a total of N items in the transformed signal. An example is shown in Fig. 6.6 for the case of

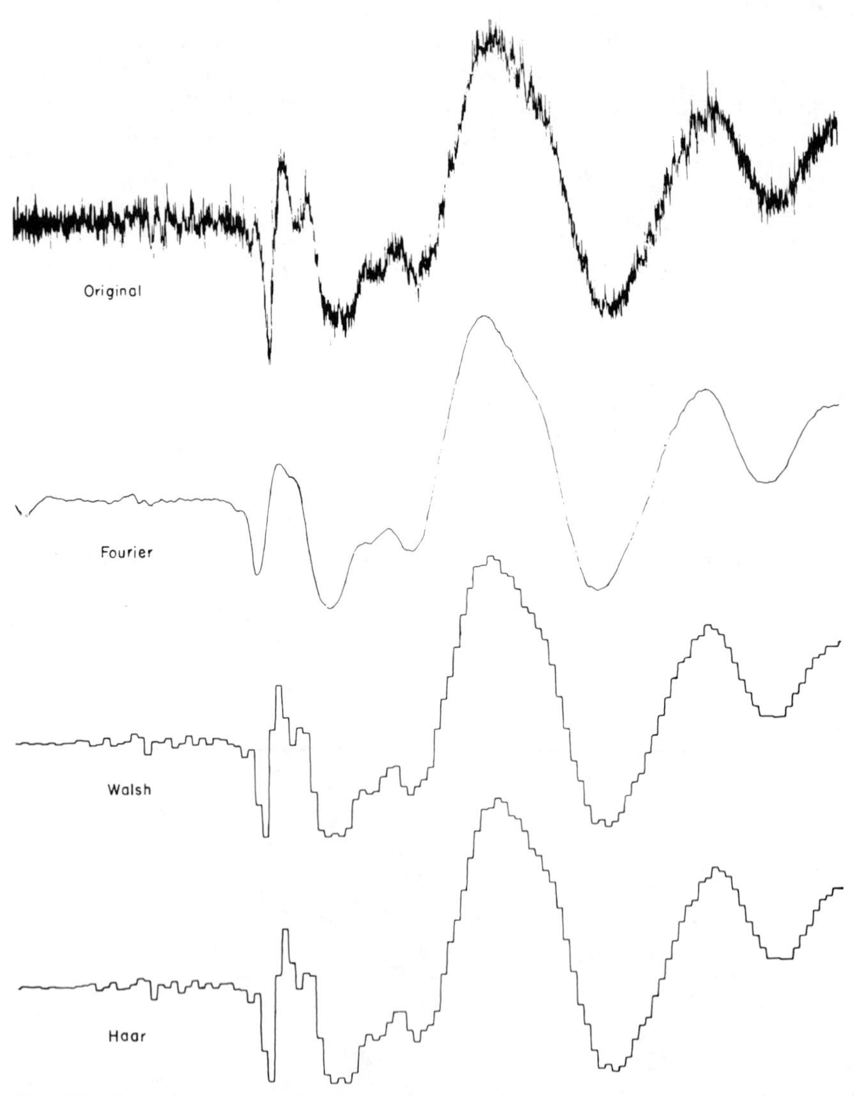

FIG. 6.6. Example of low-pass filtering by using Fourier, Walsh and Haar transformations.

low-pass filtering of a noisy transient. In this case Walsh and Haar series having only 256 terms were used which results in a stepped representation of the filtered signal. This is a characteristic of Walsh or Haar filtering; which also occurs in two-dimensional filtering, producing a chequer-board effect, unless a high order series is used.

This example employs a vector filter matrix, containing non-zero components along the diagonal only. Scalar filtering, where non-zero components can occur throughout the matrix, will produce improved results with fewer matrix terms. However there are design problems in producing the necessary matrix for a given filter frequency performance and multiplication by the filter matrix is slower, demanding up to N^2 multiplications. A reduction in computation time is obtained if many of the off-diagonal coefficients are made zero and efficient design procedures for this sub-optimal case are available.[8]

References

1. Beauchamp, K. G. (1975). "Walsh Functions and their Applications". Academic Press, London and New York.
2. Constantinides, A. G. (1976). Design of digital filters from LC ladder networks. *I.E.E. Proceedings* **123**, **12**, 1307–12.
3. Constantinides, A. G. (1978). Wave digital filters. *Alta Frequenza* **47**.
4. Cooley, J. W., Lewis, P. A. W. and Welch, P. D. (1969). The Finite Fourier Transform. *I.E.E.E. Trans. Audio Electroacoustics* **AU-17**, **2**, 77–85.
5. Delcaro, L. G. and Sicuranza, G. L. (1974). A method for transposing externally stored matrices. *I.E.E.E. Trans. Comp.* **C-23**, **9**, 967–70.
6. Helms, H. D. (1967). Fast Fourier Transform method of computing difference equations and simulating filters. *I.E.E.E. Trans. Audio Electroacoustics* **AU-15**, **2**, 85–90.
7. Hunt, B. R. (1972). Minimizing the computation time for using the technique of sectioning for digital filtering of pictures. *I.E.E.E. Trans. Comp.* **C-21**, **11**, 1219–22.
8. Kahveci, A. E. and Hall, E. L. (1972). Frequency domain design of sequency filters. 1972 Proceedings: Applications of Walsh Functions, Washington D.C., AD 744650.
9. Pichler, F. R. (1970). Some aspects of a theory of correlation with respect to Walsh harmonic analysis. Report R-70-11, Department of Electrical Engineering, Maryland University, AD 714596.
10. Pratt, W. K. (1972). Generalized Wiener filtering computation techniques. *I.E.E.E. Trans. Comp.* **C-21**, **7**, 636–41.
11. Stockham, T. G. (1966). High speed convolution and correlation. Proceedings: AFIPS Conference, **28**, 229–44.

Chapter 7

Hardware Implementation of Digital Filters

VIIA Introduction

As shown in the previous chapter any digital filtering implementation consists of a series of multiplications of samples of the input and/or output signals by constants and of additions of these products. Thus the main building blocks which have to be present in a digital filtering hardware are the following:

(a) memory cells,
(b) adders,
(c) multipliers,
(d) a programming unit which controls the sequence of operations.

The purpose of this chapter is not to give an exhaustive discussion of the hardwares suitable for digital filtering in terms of processor structure, input–output, microprogramming, machine languages and related topics. Rather we shall attempt to give an idea of the complexity and speeds obtainable from the arithmetic hardware, the core of any signal processing system, and to determine regions of applicability of hardware digital filtering techniques and the problems associated with these.

In addition a description is given of some integrated circuits currently available and suitable for digital filtering applications.

What is perhaps required in the general case is a commercially available integrated second order section or some other structure implementing a basic section in some new approach as, for example, a wave or linearly transformed digital building block.

One implementation of a second order section is currently available, as shown in a following section, but it is of limited diffusion. It is hoped that in the near future such kinds of integrated circuits will be made generally available to the user, to produce a large scale use of digital filtering operations.

VIIB Considerations on the general structure of digital filtering hardware

A general block diagram of a digital filtering hardware is presented in Fig. 7.1 and, as shown, it is composed of the following building blocks:

(a) a memory for the input samples;

(b) a memory for the output samples if IIR operations have to be performed;

(c) a coefficient memory;

(d) an arithmetic unit, which is capable of performing multiplications and additions;

(e) a control unit for the control of the sequence of operations.

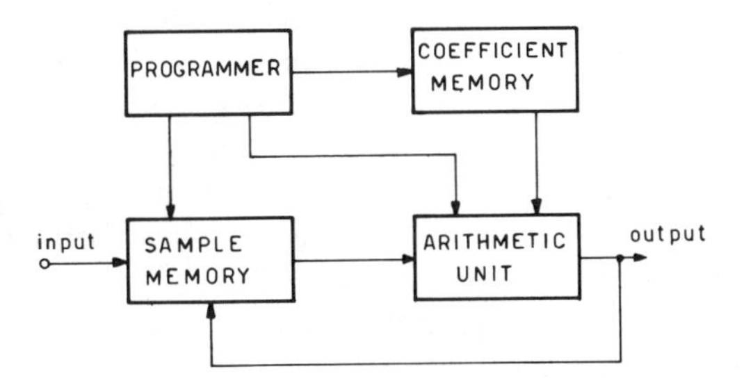

FIG. 7.1. Block diagram of a typical digital filtering processor.

Several different possibilities are now available for the implementation of the individual parts of this block diagram, which differ in speed of operation, level of integration, power consumption, etc. It is not possible to discuss here all these possibilities and all the structures which have been proposed for the implementation of digital filtering hardwares.[1,2] Therefore only some general topics are discussed which may be useful for users who are not accustomed to the problems of digital hardware implementation.

The first general and most powerful concept to be considered is that of multiplexing. There are a lot of applications where it is necessary to process several low frequency signals in parallel, as, for example, in the case of the spectral analysis performed by means of a bank of filters on a low frequency signal. In this case, if a fast arithmetic unit is available, all the filtering operations can be performed with the same arithmetic unit, which is used sequentially to perform the operations required for these different filters.

Thus processors for low frequency data can share the same arithmetic unit between all the necessary channels. In the opposite situation, when the signal to be processed is a very high frequency one, the operations to be performed can be shared between different arithmetic units, which work in parallel. This procedure can be in principle expanded to the point where a different arithmetic unit is used for every different operation to be performed. Between these two extrema, trade-offs can be considered differing in speed, complexity of the arithmetic hardware and complexity of the control part of the system. For example in several cases it may be more convenient to use several arithmetic units implemented using standard medium speed chips, instead of using only one arithmetic unit implemented with very sophisticated technologies, even if the parallelism in the operation increases the complexity of the control part.

As far as the control part is concerned, two possibilities can be considered. The first is its implementation as a logical network which generates all the signals necessary to control the processor. The second is to use a microprogrammed structure, where the processing unit is constructed to accept a set of instructions (microinstructions) and sequences of such instructions are memorized on a suitable, generally high speed, memory and executed during the processor operation. This method is much more flexible than the fixed network considered before because, obviously, the basic structure can be reprogrammed if different operations have to be performed by the processor, but in some cases it can be slower than the preceding one.

As a final remark it is necessary to point out the importance of the choice of the number representation, both in terms of type and in terms of length of the representation, and correspondingly the importance of a very accurate study of the problems related to the dynamic range of the implemented system. Thus facilities for suitable amplifications or attenuations of the signal must be provided before the addition nodes of a specific structure, and additional circuits may be necessary to control the overflow and saturate the channel if an overflow occurred so as to avoid the phenomenon of overflow oscillations as considered in Chapter 4.

It is obviously very simple to detect an overflow in the addition of two fixed point numbers. In fact an overflow can occur if the two numbers to be added have the same sign and hence it is only necessary to look at the sign bit after the operation. Obviously an overflow occurred only if the sign bit has changed. In this case a suitable circuit must generate a number which is equal to the full scale, having the sign of the two operands.

VIIC Memory implementation

In the previous section the general structure of a typical digital filtering processor was considered and it was observed that some memory is necessary to store the coefficients which define the filter and also to store the samples of the signals, which are used during the operation.

The memory circuits are amongst the most developed components in semiconductor technology and several alternatives are available, the choice of which depends on several factors, for example:

(a) the speed of operation. This is the main constraint and in general it determines the type of technology which is necessary to be used in the memory implementation;

(b) the type of access—random or sequential, serial or parallel;

(c) the programmability of the memory (RAM, ROM, EPROM).

The first requirement, which, as already mentioned, is fundamental for the choice of technology to be used and consequently the possible degree of integration, is the speed of operation, and naturally it is influenced by the bandwidth of the signal to be processed.

If the speed requirements turn out not to be too high, then MOS technology can be used, which has the greater degree of integration. Using this kind of technology it is in fact possible to find integrated circuits which contain up to 16 kbits of memory, with access times of the order of some hundreds of nanoseconds, with the only drawback that in such cases the chips tend to be dynamic, that is they need to be refreshed periodically to maintain the stored information. Thus particular care has to be taken in the refreshing operation, which must not interfere with the normal operation of the memory.

Static memories of 4 kbits are however becoming available and will be much more common in the near future. They will eliminate the problem caused by the refreshing operation above and the necessary additional hardware needed for it.

If a serial memory is to be used, shift registers using MOS technologies with capacities up to 2 kbits can be found with a maximum clock frequency of several MHz.

The main problem with these kinds of shift registers is that generally they require several clocks and they are dynamic. This means that they have a maximum acceptable interval between clock pulses which of course limits the low frequency use of the system. Hence a limitation is present on the minimum usable sampling frequency for the signal to be processed.

If such a limitation cannot be accepted, static MOS shift registers are also available, which have a smaller degree of integration (to 1 kbits) and a lower speed of up to some MHz but they have no limitations on the low frequency region.

When higher processing speed is necessary, bipolar memories are available both in the random access form and in the shift register form. In the first case, using TTL technologies, random access memories are available with circuit integration of up to 1024 bits per chip and with cycle times of the order of <50 ns. If sequential memory is needed, bipolar shift registers are available having maximum circuit integration of 256 bits and a maximum clock frequency of the order of 35–40 MHz.

Finally if in some part of the processor some very fast memory cells are needed, it is possible to use flip-flops realized by means of the ECL technology which have a maximum clock frequency of about 100 MHz.

An important new component in the digital processing implementation is the ROM (Read Only Memory) and the PROM (Programmable Read Only Memory). In this case memories with a high degree of integration (16–32 kbits in MOS, 8 kbits in bipolar) are available which can be written either at the production stage using suitable masks or in the field. These memories, which often have very small access times (<1 μs in MOS, <100 ns in bipolar), can thus be written with the coefficients of the filter to be implemented and read during the operation.

If some programmability of the values of the coefficients is needed, EPROM (Erasable Programmable Read Only Memories) can be used, which are essentially ROM memorics which however can be written and optically cleared (with ultraviolet radiation through a suitable window in the enclosure) more than once.

VIID Implementation of binary adders

The basic building block in any arithmetic unit is the adder. It can be used to perform additions and subtractions if a 2-complement representation is used and it is the fundamental arithmetic operator in the multiplier implementation.

A two bit adder is a combinational digital network with a truth table as follows:

A	B	S	C
0	0	0	0
0	1	1	0
1	0	1	0
1	1	0	1

where A and B are the two digits to be added, S is the sum and C is the carry.

If addition of two numbers of N digits has to be performed, summing any two digits of the numbers, it is necessary to consider the carry generated in the sum of the two less significant bits. So the adder becomes a three-input, two-output network, corresponding to the following truth table:

A	B	C_i	S	C_o
0	0	0	0	0
0	0	1	1	0
0	1	0	1	0
0	1	1	0	1
1	0	0	1	0
1	0	1	0	1
1	1	0	0	1
1	1	1	1	1

The network corresponding to the above truth table can be implemented, for example, in terms of elementary logical operators presented in Fig. 7.2. However chips performing this operation (full-adder-FA) on two bits or on more than two bits in a parallel structure, are commercially available with a wide range of speeds and power consumption.

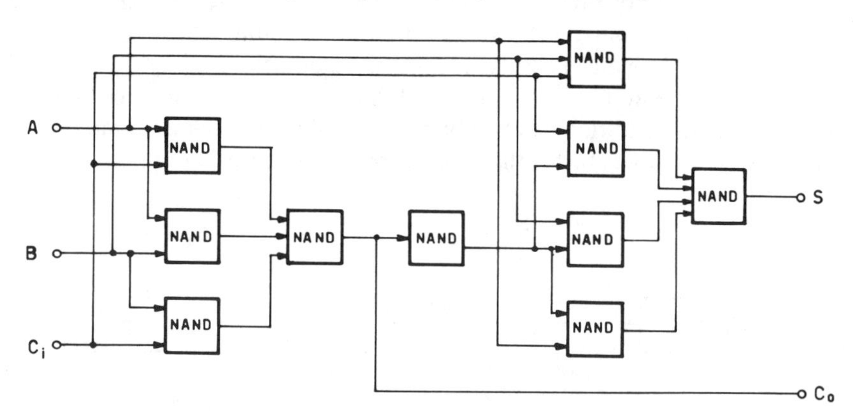

FIG. 7.2. Logical block diagram of a full adder.

If addition of two binary number of N bits has to be performed, several structures can be used, which differ in circuit complexity and speed of operation.

A simple adder for two N bit numbers is the serial one. It is arranged as shown in Fig. 7.3 having a full adder, a flip-flop to memorize the carry generated at the ith sum to be used in the $(i+1)$th step and a gate to reset the carry flip-flop at the end of the addition. To perform the addition of two N bit numbers with the serial full adder, digits having the same

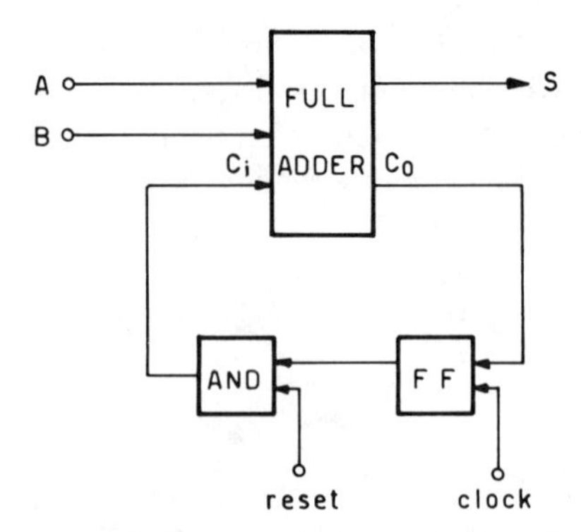

FIG. 7.3. Block diagram of a serial N bit adder.

position are presented sequentially to the circuit by which the sum and the carry to be used in the next step of the operation are produced.

This type of circuit is the simplest one to be used to perform the sum of two numbers but is obviously the slowest. In fact to perform the addition of two N bit numbers, the time necessary corresponds to N clock intervals.

Addition in only one clock interval can be obtained using a parallel structure. This can be achieved as follows. Assume that we have two numbers of N bits and N full-adders connected as in Fig. 7.4. In this case all N bits of the number are presented in parallel, the result is obtained in parallel and the addition can be obtained in only one clock interval, but this adder is not N times faster than the corresponding serial one.

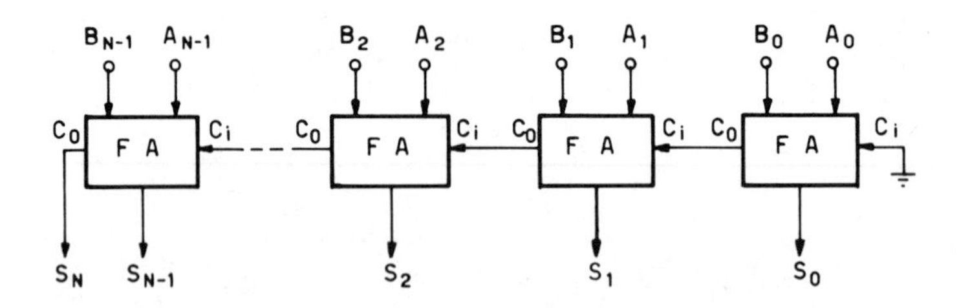

FIG 7.4. Block diagram of a parallel N bit adder.

This is explained by the fact that in many bit configurations the carry bit generated in the first cell has to propagate its effect through all the parallel adders to the last cell. This means that the carry has to propagate through $N-2$ cells before its effect can be sensed. For, as shown in Fig. 7.2 the propagation through every cell is a propagation through several gates, the total delay can be high. Hence, although it is true that in this case the result can be obtained in only one clock interval, the clock rate cannot be as high as in the serial case due to the delay involved in the carry propagation. Two methods however can be used to speed up the operation, that is to obtain a parallel structure which can be clocked as fast as the serial structure.

The first is to use a pipeline adder. In this structure, which is well suited for digital filtering operations, where the additions are in a regular sequence, a flip-flop is introduced between the adders for the carry to be stored and fed to the following cell in the next clock interval, as shown in Fig. 7.5. Thus there is no longer any possibility of performing an addition within a clock period, but the following advantages can be considered:

(1) at any clock pulse the signals have to propagate only through an adder, so that the clock frequency of the structure can be as high as in the serial case;

(2) if a sequence of additions, as in digital filtering case, has to be performed, N additions can be simultaneously in progress in the structure, if N is the number of the cells of the pipeline. At any step the first cell is adding the least significant bits of two numbers, the second cell is adding the second bits and so on. This means that even if every sum needs N clock intervals to be performed, when the pipeline is full a new sum is completed at every clock interval and the equivalent addition time is only one clock interval of a very fast clock.

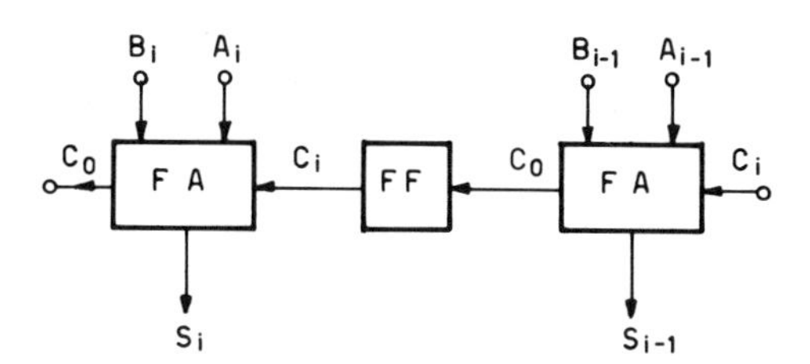

FIG. 7.5. Block diagram of a cell of a pipeline adder.

The second method to speed up the propagation of the carry, so as to increase the clock frequency and hence the number of additions in a fixed time interval in a parallel adder, is to use a look-ahead carry adder. In this case a second logical network is used to generate independently the carry to the sections of the parallel adder as a function of the bits of the numbers to be added. Using this method the carry for every cell of the adder structure can be computed in parallel with the sum evaluation and the carry at any cell can be fed without waiting for the propagation through the adder structure.

The carry generated at the *i*th stage of an adder is a function of the bits to be added and of the previous carry, and it can be expressed by the relation:

$$C_i = A_i \cdot B_i + (A_i \oplus B_i) \cdot C_{i-1} \tag{7.1}$$

where $\cdot$ represents a logical AND, + represents a logical OR and $\oplus$ represents a logical EX–OR.

Considering, for example, a four bit adder, the expressions for the carry can be written in the form:

$$\begin{aligned} C_1 &= A_1 \cdot B_1 + (A_1 \oplus B_1) \cdot C_0 \\ C_2 &= A_2 \cdot B_2 + (A_2 \oplus B_2) \cdot C_1 \\ C_3 &= A_3 \cdot B_3 + (A_3 \oplus B_3) \cdot C_2 \end{aligned} \tag{7.2}$$

and substituting for the carries we have:

$$\begin{aligned} &C_1 = A_1 \cdot B_1 \\ &C_2 = A_2 \cdot B_2 + (A_2 \oplus B_2) \cdot A_1 \cdot B_1 \\ &C_3 = A_3 \cdot B_3 + (A_3 \oplus B_3) \cdot A_2 \cdot B_2 + (A_3 \oplus B_3) \cdot (A_2 \oplus B_2) \cdot A_1 \cdot B_1 \end{aligned} \tag{7.3}$$

which correspond to only a three gate delay to generate the third carry.

The main problem, using this approach, is that the expression for C_i tends to become more and more complex when the number of stages increases. Hence in general some trade-off has to be accepted based on the division of the adder in several parts, using the look-ahead technique in each part, while the carry is allowed to propagate through different parts.

We mentioned before that a wide range of integrated circuits exists, which implement multiple full-adder functions. Let us now give an idea of the operation times obtainable by means of the more common integrated circuit families commercially available.

Using the most widely diffused family of digital integrated circuits, that is the TTL family and using, for example, 4 bit arithmetic units offered by most of the integrated circuits manufacturers, it is possible to construct a 16 bit parallel adder with ripple carry between the chips, having a typical addition time of 60 ns. If the same circuits are used with additional chips which implement the look-ahead carry technique for the carries between the chips, the addition time becomes as low as typically 36 ns. The addition time for the 16 bit case is lowered to about 20 ns if schottky TTL units are used, with the same chip count. Finally if ECL circuits are used, it is possible to have 1 bit adders with a typical delay of about 3 ns and look-ahead carry generators with a typical delay of about 2 ns.

VIIE Implementation of digital binary multipliers

The second fundamental arithmetic circuit needed in digital filter implementations is the multiplier. This circuit is the most critical both from the speed of operation and from the circuit complexity points of view for digital filter implementations. The availability of fast integrated multipliers is indeed the key point for the diffusion of digital filtering hardware in the applications side. It can be observed that the fundamental limit to the digital filtering operation speed is the multiplication time, which is really the minimum time to produce the output of a digital filter in the limiting case of a system having one multiplier per coefficient.

The simplest and more direct algorithm to implement a binary multiplication is the algorithm of successive additions, which is precisely the paper and pencil case.

As an example let us consider the product of two positive numbers, of which one (110) is chosen as the multiplicand and the other (101) as the multiplier. The operation can be written as:

$$\begin{array}{r} 110 \\ 000 \\ \underline{110} \\ 11110 \end{array} \tag{7.4}$$

corresponding to the algorithm of multiple shifts and additions of partial products. A circuit can be easily implemented to perform the above operation, except that in several cases additions are performed sequentially, that is by accumulating the successive partial products.

From the above considerations it is evident that the multiplication consists of the evaluation of partial products and of the accumulation of the shifted partial products.

The partial product evaluation is a very simple matter in the binary system. The multiplier digits can assume only two values, 0 and 1, and correspondingly the partial product can only be equal to zero if the digit is zero and equal to the multiplicand if the digit is 1. So the partial product evaluation corresponds to a gating of the multiplicand to let the multiplicand through if the multiplier digit is 1 or to produce a zero output if the digit is zero. This operation corresponds to a logical AND operation and hence the binary product corresponds to the logical AND operation.

Let us now consider the more commonly used multiplier structures assuming only positive numbers as inputs. In the following we shall consider the case of signed binary number multiplication.

VIIE1 The serial multiplier

The simplest multiplier which uses the successive addition algorithm can be implemented using the serial full-adder of Fig. 7.3, a logical AND circuit and a serial-in serial-out register, according to the block diagram presented in Fig. 7.6. The two numbers to be multiplied, A (multiplier) and B (multiplicand), are presented to the multiplier input (AND circuit) at two different clock frequencies, such that the clock of the multiplier is N times slower than the multiplicand clock, where N is the number of bits of the multiplicand. Thus for every bit of the multiplier the partial product is evaluated and a serial addition with the partial results stored in the register is performed. The gate in the path between the output of the register and the input of the adder is used to reset the partial sum at the beginning of the operation. The shift of 1 bit necessary for the algorithm is obtained automatically if the register has a length of $N-1$ cells.

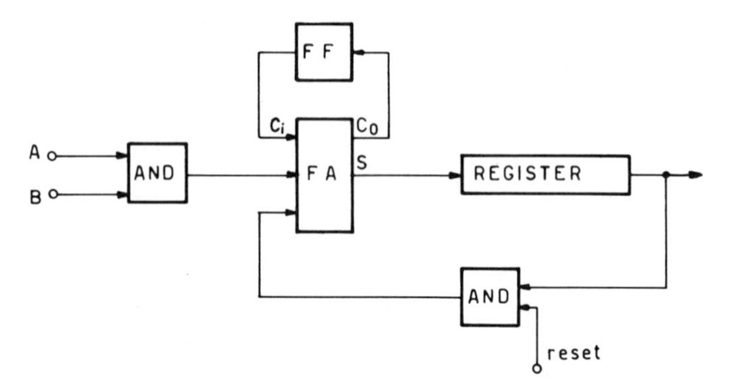

FIG. 7.6. Block diagram of a serial multiplier.

Finally as far as the operation time is concerned, it can be observed that the complete product of $M+N$ bits can be obtained in $M \cdot N$ clock intervals of the multiplicand and that this clock can be fast due to the small delay in the serial adder.

VIIE2 The serial-parallel multiplier

Using the same general algorithm as above a faster multiplier can be obtained by means of the parallel adder presented in the preceding section.

In this case the multiplier assumes the configuration of Fig. 7.7. N binary multipliers (AND gates) are present (one for every bit of the multiplicand), the gating signals for which are the bits of the multiplier as before. Moreover N full adders and a parallel-in parallel-out register of N bits to

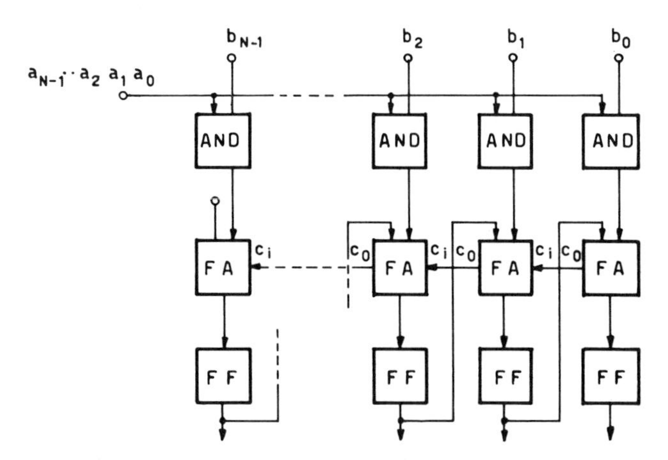

FIG. 7.7. Block diagram of a serial-parallel multiplier.

store the partial results are needed. The shift of partial results is obtained by means of a suitable connection from the outputs of the register to the inputs of the parallel adder.

The complete product of $M+N$ bits is obtained in M clock intervals, if M is the number of bits of the multiplier, and the time to perform the operation depends on the implementation of the parallel adder, whose propagation delay limits the maximum clock frequency possible.

This type of multiplier is called serial-parallel multiplier, because one of the numbers has to be available in parallel form, whilst the other is used in his serial form. A second form of the serial-parallel multiplier is possible, whose structure is very modular and can be easily modified to obtain a pipeline system.[4]

This structure can be easily described with reference to the example (7.4) and it implements an arithmetic circuit which executes the additions as a per column basis, that is in the same way as in the paper and pencil algorithm.

This serial-parallel multiplier has the form presented in Fig. 7.8. It is easy to verify by the example (7.4) that this structure is really capable of performing the multiplication by means of successive additions of columns of the shifted partial product matrix and that it is obviously a serial-parallel structure because one of the numbers must be presented to the multipliers in parallel form whilst the other has to be presented in serial form.

It can be observed that the preceding multiplier needed only M clock periods to obtain the final result whereas this kind of multiplier needs $N+M$ clock periods to produce the same result. Moreover it can be seen that in this case there is also the problem that the addition of N bits has to propagate through the array of adders, thus reducing the maximum

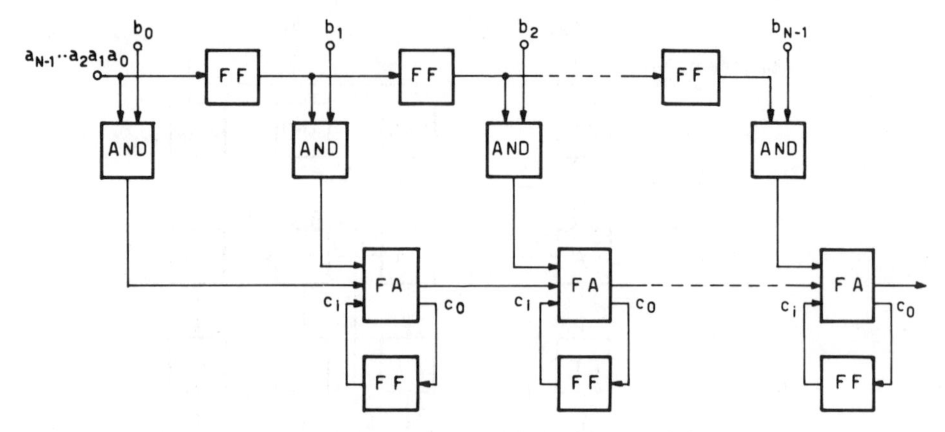

FIG. 7.8. Block diagram of a serial-parallel multiplier.

clock frequency possible. However the circuit presented here is ideally suited for a pipeline structure which can be obtained simply by introducing two additional flip-flops in every cell of the structure, one on the multiplier line and the other on the addition line, thus obtaining cells of the kind shown in Fig. 7.9. The maximum delay at any step of the operation is produced by the single full-adder, with the disadvantage that the number of clock intervals to produce the complete product is now doubled. Thus we have an increase in the maximum clock frequency which can be used, but we have also an increase in the number of clock pulses necessary to produce the result.

However it can be observed that using this type of multiplier both in pipeline and non-pipeline form, it is possible to have more than one

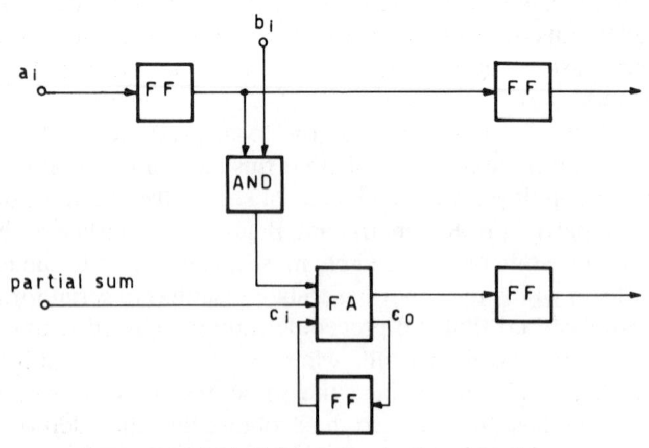

FIG. 7.9. Block diagram of a pipeline multiplier.

multiplication propagating through the structure, if, as normally is the case in digital filtering operations, it is not necessary to obtain all the $M+N$ bits of the product, but only a rounded or truncated result having a number of bits equal to those of one of the operands. In this structure this can be obtained very simply by propagating in a different line a rounding signal in synchronism with the least significant bit of the multiplier, thus not allowing the sum to propagate through the serial combination of adders for a number of clock pulses corresponding to the number of bits to be eliminated. The cell of the structure can be modified to obtain the form of Fig. 7.10, where the pipeline case is shown. The rounding signal propagating in synchronism with the least significant bit of the multiplier inhibits the propagation of the partial sum through the adder cascade until all the bits of the multiplier are contained in the multiplier. At this point the number is shifted out of the serial register and the bits of the product are generated in sequence. Due to the separation caused in the cascade of adders by the rounding signal, it is possible to have a second number propagating through the register before the preceding number has been completely shifted out. This reduces to $2M$ the number of clock intervals necessary to produce a result in the pipeline multiplier or to M, as in the previous serial-parallel multiplier, if the structure is not a pipeline.

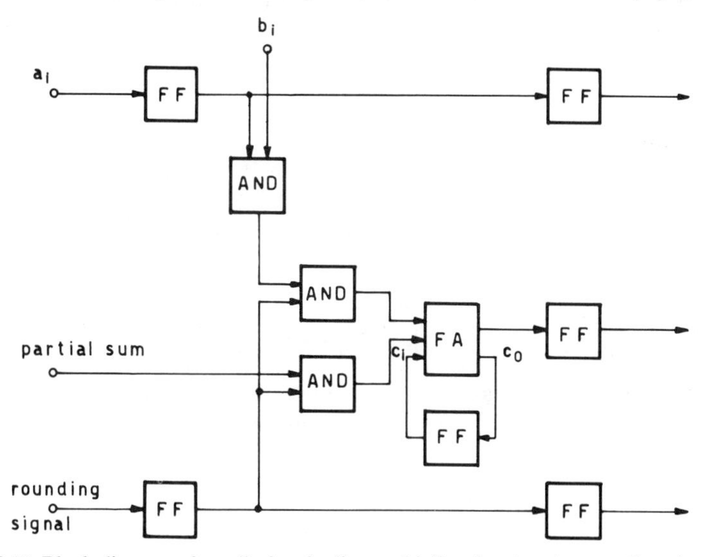

FIG. 7.10. Block diagram of a cell of a pipeline multiplier showing the rounding signal path.

Two further remarks can be made at this point. The first is that if one of the M bits is the sign bit, which must be treated by a different circuit because the multiplier accepts only positive numbers, the corresponding time interval can be used to obtain one bit more of the result to be used in the rounding operation.

The second remark is that in this serial-parallel multiplier, whilst the maximum length of the multiplicand is fixed by the number of cells in the structure, the number of bits of the multiplier is variable and consequently so is the number of clock intervals necessary to produce the result.

VIIE3 The 2-complement multiplier

All the points made so far have been based on the assumption that a sign-magnitude number representation is employed. Multiplication is then performed on the magnitudes only of the two numbers, whilst an additional circuit (logical EX–OR) computes the sign of the result. The sign can be reintroduced as the most significant bit in the product when the result is transferred out of the multiplier.

However if the two numbers to be multiplied are represented in 2-complement form, it is necessary either to convert negative numbers before their introduction in the multiplier or to modify the multiplier structure to accept numbers in the 2-complement representation.

The 2-complement representation of a left justified fixed point number A is as follows:

$$A = -a_0 + \sum_{k=1}^{N-1} a_k 2^{-k} \tag{7.5}$$

where N is the number of bits, and a_0 is equal to 0 for a positive number whereas for a negative number it is equal to 1.

The product of two 2-complement numbers A and B, both of N bits, can therefore be written in the form

$$AB = a_0 b_0 - a_0 \sum_{j=1}^{N-1} b_j 2^{-j} - b_0 \sum_{k=1}^{N-1} a_k 2^{-k} + \sum_{k=1}^{N-1} \sum_{j=1}^{N-1} a_k b_j 2^{-k} 2^{-j} \tag{7.6}$$

Assuming now that A is the multiplier and B the multiplicand, four cases are recognized, corresponding to the four possible combinations of the signs of the two numbers.

When A and B are both positive ($a_0 = 0$, $b_0 = 0$) then the expression (7.6) reduces to

$$AB = \sum_{k=1}^{N-1} a_k 2^{-k} \sum_{j=1}^{N-1} b_j 2^{-j} = \sum_{k=1}^{N-1} a_k (2^{-k} B) \tag{7.7}$$

which corresponds to the algorithm of successive additions of the form considered earlier.

If A is positive and B is negative $(a_0 = 0, b_0 = 1)$, then the expression (7.6) can be written in the form

$$AB = -b_0 \sum_{k=1}^{N-1} a_k 2^{-k} + \sum_{k=1}^{N-1} a_k 2^{-k} \sum_{j=1}^{N-1} b_j 2^{-j} = \sum_{k=1}^{N-1} a_k (2^{-k} B) \qquad (7.8)$$

where B in this case is the 2-complement representation of a negative number. The same algorithm as described in the preceding case can therefore be used, with the added modification that the sign bit of B has to be used in the operation and the shift necessary in the algorithm are arithmetic shift, that is division by two of the negative number. This means that when the shift to the right is performed, the sign bit has to be duplicated in the next position, not merely transferred to that position.

When A is negative and B is positive $(a_0 = 1, b_0 = 0)$, the expression (7.6) can be written in the form

$$AB = -a_0 \sum_{j=1}^{N-1} b_j 2^{-j} + \sum_{k=1}^{N-1} a_k 2^{-k} \sum_{j=1}^{N-1} b_j 2^{-j} = \sum_{k=1}^{N-1} a_k (2^{-k} B) - a_0 B \qquad (7.9)$$

The above result means that, to perform the product, $N-1$ additions of the partial products obtained by multiplying B by every bit of A (sign excluded) have to be performed, but the number B has to be subtracted from the preceding result. Therefore to multiply A and B in this case, it is necessary to have an arithmetic circuit which performs not only additions, but also subtractions.

The same result is obtained in the fourth case, that is when the two numbers are both negative. In this case the product is represented by the entire expression of (7.6) which can be rewritten in the form

$$AB = -a_0 B + \sum_{k=1}^{N-1} a_k (2^{-k} B) \qquad (7.10)$$

where B is a 2-complement negative number.

A general observation on the above is that the multiplication of 2-complement numbers requires hardware more complicated than for the sign magnitude multiplication case.

However if some increase in complexity of the multiplication hardware is acceptable, it is possible to use some modifications of the multiplication algorithm which may allow a decrease of the average time necessary for carrying it out. In fact it can be observed that all methods considered for speeding up the product evaluation are based on efforts to minimize the addition time and consequently to increase the clock frequency of the system. But some improvements in the medium throughput of the multiplier can be also obtained with the simple observation that in the add and shift method an addition has really to be performed only when the bit of the multiplier is one, whilst only the shifting operation is needed when the multiplier bit is zero. The shifting operation is obviously faster than addition and if the multiplier is allowed to run asynchronously, it is possible to save time in the multiplication, except when the multiplier has all its bits equal to unity.

The number of arithmetic operations can be further reduced using an algorithm due to Booth, which requires as arithmetic units both an adder and a subtractor, but it minimizes the number of arithmetic operations in the product evaluation and can moreover be directly used with 2-complement numbers.

To illustrate the Booth algorithm it is useful to observe that the multiplier A can be written in the form

$$A=(a_1-a_0)+\sum_{k=1}^{N-1} 2^{-k}(a_{k+1}-a_k) \qquad a_N=0 \tag{7.11}$$

Thus on considering the ith step in the operation, if $a_i = a_{i+1}$ only a shift operation has to be performed. If $a_{i+1}=1$ and $a_i=0$ the difference in (7.11) is positive and the corresponding product has to be added to the partial result. The converse is true if $a_{i+1}=0$ and $a_i=1$. Therefore the algorithm consists of three different operations: shift, sum and subtraction and the choice of the operation to be performed has to be made on the basis of the value of two successive bits of the multiplier.

A serial parallel multiplier using the Booth algorithm is commercially available in one large scale circuit.[5] This circuit can perform the product of a 2-complement multiplicand of 8 bits presented to the circuit in a parallel form and a 2-complement multiplier of M bits. Several chips can be connected in cascade to allow the multiplication of numbers of any length both in the parallel and in the serial inputs. $N+M$ clock pulses are necessary to perform a multiplication of two number of N and M bits respectively and all the bits of the product are obtained sequentially starting from the least significant one.

The circuit is in fact implemented using Schottky TTL technology and can be clocked with frequencies of up to 40 MHz. A product of 8×8 bits can be obtained in 450 ns, while a 16×16 product can be obtained in 850 ns.

VIIE4 The array multiplier

Let us now consider again the first type of serial-parallel multiplier considered above. In this circuit the single additions are performed in parallel but they are organized in a serial sequence to use only one parallel adder. A conceptually simple extension of this type of multiplier is to provide a different parallel adder for every bit of the multiplier obtaining a structure (array multiplier) like the one presented in Fig. 7.11. The two numbers A and B are presented in parallel to this multiplier and the time for the product evaluation is determined only by the propagation delay of the signals inside the array.

Figure 7.11 really represents only the principle block diagram form of this type of multiplier, which is normally implemented using as building blocks not simple adders and AND circuits, but cells specific for this application.[3]

The structures obtained by such an approach are very regular and modular, that is they can be implemented by means of a regular interconnection of equal building blocks. This makes this approach very convenient from the point of view of the production of special integrated circuits.[7]

Several building blocks have been produced by several manufacturers to implement array multipliers and more recently array multipliers have been manufactured as single chip large scale integrated circuits.

As an example of the available circuits of the first kind we have the Advanced Microdevices AM 2505. It is really a 2 bit$\times$4 bit multiplier which performs an operation of the type $A \cdot B + C$. This operation can be considered as a product and addition of a partial product.

To give an idea of the complexity of an array multiplier and the obtainable speed of operation let us consider the following three cases: (a) 8 bit$\times$8 bit multiplication, (b) 12 bit$\times$12 bit multiplication and finally (c) 16 bit$\times$16 bit multiplication. In the 8 bit case, 8 chips are necessary to perform the complete product, whilst in the 12 bit case 18 chips are required and in the 16 bit case 32 chips are needed. Obviously if some form of rounding or truncation is allowed, the chip count can be reduced by not implementing the portion of the array which produces the low order bits.

As far as the speed of operation and the power consumption are concerned, three different implementations of the same chip are available as shown in Table 7.1.

	AM 25S05		AM 2505		AM 25L05	
	ns	W	ns	W	ns	W
(a)	76	4.8	115	3.6	262	1.2
(b)	115	10.8	175	8.1	396	2.7
(c)	153	19.2	235	14.4	530	4.8

TABLE 7.1

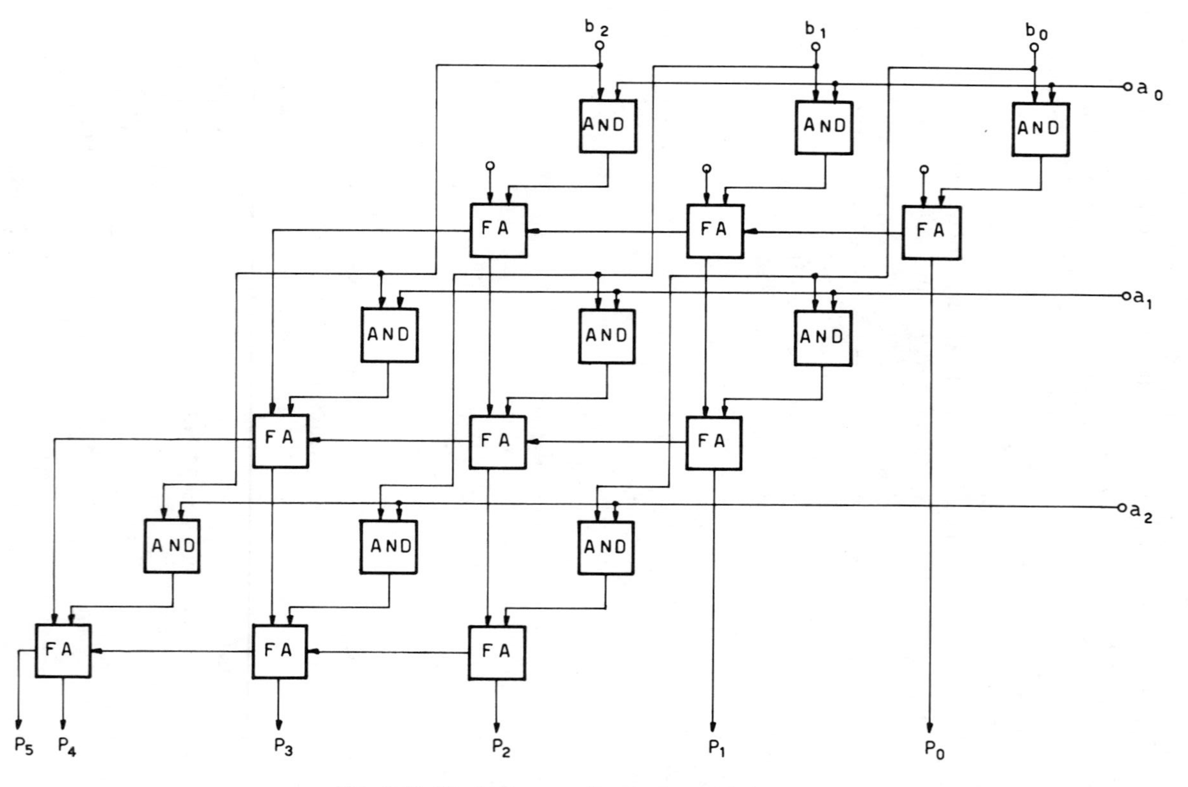

FIG. 7.11. Block diagram of a matrix multiplier (3×3).

Finally it can be observed that this chip is very flexible and can be used to implement other multiplier structures to accommodate other number representations.

As mentioned before, array multipliers are also available as single chip large scale integrated circuits.

Three different multipliers are currently available (from TRW) that is the 8 bit, the 12 bit and the 16 bit ones. They can perform a multiplication in 130, 150, 160 ns respectively while the power consumptions are 3, 5.5, 8 W.

VIIF Implementation of second order sections

Let us now briefly describe some possible implementations of the second order section which is the basic building block of some IIR digital filter structures.

A biquadratic system has in general a transfer function $H(z)$ of the form:

$$H(z)=\frac{a_0+a_1z^{-1}+a_2z^{-2}}{1+b_1z^{-1}+b_2z^{-2}} \tag{7.12}$$

The corresponding difference equation can be implemented with two registers, five multiplications and four additions in the canonic form.

However we should point out that a scaling of the signal may be necessary between such sections used to realize a larger system, so as to avoid overflow. Thus another multiplication has to be introduced between these sections. This however can in general be approximated by means of a multiplication or division by a power of 2, that is by means of an arithmetic shift.

This necessary additional multiplication suggests that (7.12) should be written in the following form:

$$H(z)=K_1\frac{1+\hat{a}_1z^{-1}+\hat{a}_2z^{-2}}{1+b_1z^{-1}+b_2z^{-2}} \tag{7.13}$$

This form reduces to four the necessary products to implement the second order section, whilst a multiplication by K_1, equal to the scaling constant, is introduced in the path between the sections.

We observe also at this point that the normalization between the sections in the structure affects the overall dynamic range of the system and its noise behaviour. Thus it is convenient to implement the sections with some extra bits with respect to the coefficients and the signal word length, so as to allow scaling to be performed without affecting the overall performance of the system. The second order section considered in (7.13) is a very convenient one to be implemented as an integrated chip, making available to the user a very flexible second order section component.

A chip implementing two biquadratic second order sections of the type (7.13) is available. This chip is implemented using PMOS (and recently NMOS) technology and it contains two biquadratic sections, which can be used separately or cascaded if necessary. The circuit is completely serial and it needs only an input for the data and an input for the coefficients. The coefficients have to be available on an external serial memory and have to be organized in sequences of 15 bits of which 12 represent the magnitude and 3 are used for the scaling necessary to avoid overflow. Thus an external memory of 60 bits per section is necessary. As far as the input data are concerned, they have to be represented as 2-complement number of 15 bits. The maximum sampling frequency of the implemented chip is about 30 kHz, and the minimum sampling frequency (the chip is dynamic) is $\simeq$85 Hz (available from PYE-TMC, England).

This circuit is a good example of the type of components which could be produced and made available commercially, covering with its maximum sampling frequency a wide range of applications (biomedical signals, speech signals, and so on) and allowing many potential users of digital filtering techniques to employ them in the applications without being involved with the difficulties of digital hardware design and construction.

Other very interesting structures have also been proposed to implement second order sections without multiplications. The simplest way is to use a memory where all possible products arising in a given biquadratic section are stored and signal samples are then used as memory address. Unfortunately, even if memory is an inexpensive component, the number of possible combinations, when the number of bits of the variables increases, becomes very soon too high to make it a viable solution. Fortunately the fact that in the second order section the coefficients are only five makes it possible to rearrange the operations and so to avoid multiplications with a very small amount of memory.[6]

The second order section operation can be written as:

$$y(n) = a_0 x(n) + a_1 x(n-1) + a_2 x(n-2) - b_1 y(n-1) - b_2 y(n-2) \quad (7.14)$$

By consider a 2-complement left justified fixed point number representation, it is possible to write for the sample x_k:

$$x_k = -x_k^0 + \sum_{j=1}^{N-1} x_k^j 2^{-j} \quad (7.15)$$

where N is the number of bits of the representation and x_k^j are binary variables. With the above notation, expression (7.14) can be written in the form

$$y(n)=a_0\left(\sum_{j=1}^{N-1} x_n^j 2^{-j}-x_n^0\right)+a_1\left(\sum_{j=1}^{N-1} x_{n-1}^j 2^{-j}-x_{n-1}^0\right)$$
$$+a_2\left(\sum_{j=1}^{N-1} x_{n-2}^j 2^{-j}-x_{n-2}^0\right)-b_1\left(\sum_{j=1}^{N-1} y_{n-1}^j 2^{-j}-y_{n-1}^0\right)$$
$$-b_2\left(\sum_{j=1}^{N-1} y_{n-2}^j 2^{-j}-y_{n-2}^0\right)$$

and defining:

$$\phi(x^1, x^2, x^3, x^4, x^5)=a_0x^1+a_1x^2+a_2x^3+a_3x^4+a_4x^5$$

it is possible to write

$$y(n)=\sum_{j=1}^{N-1} 2^{-j}\phi(x_n^j, x_{n-1}^j, x_{n-2}^j, y_{n-1}^j, y_{n-2}^j)-\phi(x_n^0, x_{n-1}^0, x_{n-2}^0, y_{n-1}^0, y_{n-2}^0)$$

It is evident therefore that the function ϕ can assume only $2^5=32$ different values. Thus implementing this function using, for example, a ROM memory or a combinational network, the filtering hardware can assume a structure like the one presented in Fig. 7.12, which uses only shifting operations and additions, equal in number to the number of bits of the input data ($\varphi=\phi$). The bits of the sample are used as addresses of the memory where the values of ϕ are stored.

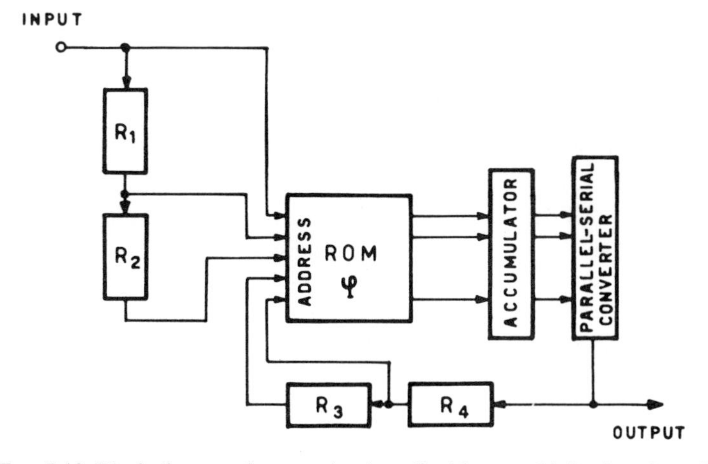

FIG. 7.12. Block diagram of a second order cell without multiplications ($\varphi=\phi$).

A complete discussion of this implementation method and also a comparison with the conventional multiplier based second order sections can be found in Peled and Liu.[6] As an example let us consider the case of a section with 12 bit signal samples implemented with TTL circuits having an addition time of $\simeq$40 ns and an access time to the memory of 50 ns. 20 ICs are typically needed, with a power consumption of 9.6 W. On overlapping the additions and the memory access and assuming a clock frequency of 20 MHz, this section has a bandwidth of 800 kHz, that is it can process up to 1.6 Msamples/s.

Higher speeds could be obtained either by using ECL technology or by using different degrees of parallelism in the structure. With a complete parallel structure, that is using a different memory for every bit in the samples and a different adder for every sum, it is possible to construct a section which uses typically 60 TTL ICs, consumes 24 W and can operate in real time on a signal with the bandwidth of 10 MHz (20 Msamples/s).

Using this approach it is possibe to construct second order sections which are faster than the conventional ones with additional savings in cost and/or power consumption. The price to pay for this increased efficiency is obviously the loss of generality that it is possible to have using the multiplier approach, where any value for the filter coefficients can be used. In fact the filter obtained by this approach is a fixed one and its form can be changed only by changing all the memory of the values of the function ϕ. If programmability is necessary and the processor may be used in connection, for example, with a computer which generates the coefficients or the value of the function ϕ, the memory for the values of ϕ could be a RAM, with the only disadvantage of having to transfer 32 words for every set of coefficients instead of 5.

VIIG New technologies

In the above considerations digital filtering implementations based on MOS or bipolar logic circuit have been examined. However new technologies have been or are being developed which could improve the situation for several applications.

In the low frequency domain magnetic bubble technology may in some cases be more convenient. Magnetic bubble memories are already available, whilst the feasibility of logical gates and also of LSI circuits useful in digital processing techniques, such as adders or pipeline serial multipliers, are under examination.[9]

Other technologies which could be very useful in signal processing are based on the surface acoustic waves (SAW) components or on charge coupled devices (CCD).[8]

In the first case an acoustic wave is generated on a substrate by a suitable transducer, when an electrical input signal is applied. This wave can be again transduced to electric signals by suitable electrodes deposited on the substrate at different distances in the wave propagation path. Therefore a tapped delay line is obtained. Because different lengths of the electrodes are equivalent to different weights of the samples, using only an additional adder for the weighted samples, a sampled structure is obtained equivalent to the FIR filters of Chapter 2. In this case however a continuous signal is appearing at the input and at the output, even if at every instant the output is obtained by summing only samples of the input signal. These filters can be designed using the same techniques of Chapter 2, even though particular care must be given to the fact that the accuracy in the coefficients realization is more important here than in the digital case.

As far as CCDs are concerned, they are conceptually analogue shift registers, where charge packets, which are proportional to the samples of the input signal, are propagated through the CCD structure. Some of the cells of the shift register can be read and made available to the output. A weighting can be obtained either by a suitable shape of the electrodes or with external analogue multipliers and the weighted outputs can be added to produce a FIR filtering operation. The CCD circuits are realized using the same technology as semiconductor circuits, and hence adders and multipliers can conceivably be built on the same CCD substrate. Thus a complete filter can be obtained on a single chip.

Finally it can be observed that the above technologies (SAW and CCD) can be used to implement Fourier transformers, using the chirp transform algorithm. This requires multiplication of the signal with suitable exponentials and a convolution operation. The convolution can be obtained using the tapped delay line structure considered before. Thus real time operations could be obtained for frequency domain transformations.

References

1. Allen, J. (1975). Computer architecture for signal processing. *I.E.E.E. Proceedings* **63**, **4**, 624–33.
2. Freeny, S. L. (1975). Special-purpose hardware for digital filtering. *I.E.E.E. Proceedings* **63**, **4**, 633–48.
3. Hoffman, J. C., Lacaze, G. and Csillag, P. (1968). Iterative logical network for parallel multiplication. *Electronics Letters* **4**, 9–178.

4. Jackson, L. B., Kaiser, J. F. and McDonald, H. S. (1968). An approach to the implementation of digital filters. *I.E.E.E. Trans. Audio Electroacoustics* **AU-16**, **3**, 413–21.
5. Mick, J. R. and Springer, J. (1976). Single-chip multiplier expands digital role in signal processing. *Electronics* **49**, 103–8.
6. Peled, A. and Liu, B. (1974). A new hardware realization of digital filters. *I.E.E.E. Trans. Acoustics, Speech and Signal Processing* **ASSP-22**, **6**, 456–62.
7. Pezaris, S. D. (1971). A 40 ns 17 × 17 bit array multiplier. *I.E.E.E. Trans. Comp.* (short notes) **C-20**, **4**, 442–7.
8. Special issue on Surface Acoustic Waves Devices and Applications (1976). *I.E.E.E. Proceedings* **64**, **5**.
9. Williams, R. P. (1977). Serial integer arithmetic with magnetic bubble. *I.E.E.E. Trans. Comp.* (short notes) **C-26**, **3**, 260–4.

Chapter 8

Applications of Digital Filters

VIIIA Introduction

The application of digital signal processing techniques in general and of digital filtering in particular has expanded in and is useful in many important areas such as speech signal processing, digital telephony and communications, facsimile and TV image processing, radar-sonar systems, biomedicine, space research and operative systems, geoscience, etc.

There are several reasons for the all-embracing nature of the subject:

(i) the use in processing and transmission systems of digital representation of signals and images (through sampling, quantization and perhaps binary coding) as a more efficient means of representation than the analogue form: this efficiency is due mainly to accuracy, stability, noise immunity and other desirable effects;

(ii) the higher efficiency with which digital filtering can fulfil any realizable frequency response requirements with great flexibility and adaptivity;

(iii) the decreasing cost of hardware and software implementation due to the production of highly reliable large scale integrated (LSI) circuits and the availability of readily realizable software on minicomputers and microprocessors.

Some of the more important areas of application mentioned above are described in some detail with typical examples in this chapter. These areas include speech signal processing, digital image processing, digital communications (digital telephony, digital telemetry, data transmission), radar-sonar systems and biomedicine.

Other examples of application of digital filtering are included in the Appendices. The purpose of the examples presented here is primarily to show the practical aspects of the use and application of digital filtering techniques.

In the following sections we consider firstly some typical 1-D and 2-D digital filtering operations which have been widely used in several specific application areas described later. Then we summarize several other areas of application with some typical performances or results wherever appropriate.[3,4,57,94]

VIIIB Some typical digital filtering operations

Digital filters in fact can be applied in many and different parts of signal and image processing-transmission systems.

Two typical examples of systems in which digital filtering can be inserted are: (i) local digital processing system (Fig. 8.1); (ii) digital communication system (Fig. 8.2).[88,94]

In the first system of Fig. 8.1 an analogue signal $x(t)$ or image $x(x_1, x_2)$ is converted to a digital form (as already mentioned through the operations

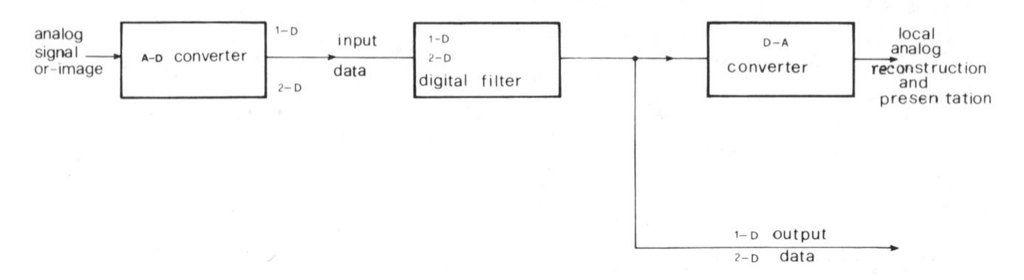

FIG. 8.1. General structure of a local digital processing system.

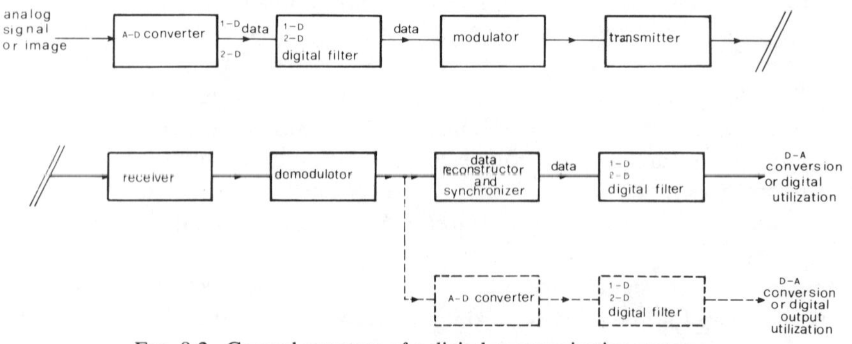

FIG. 8.2. General structure of a digital communication system.

of sampling, quantization and binary encoding) by the analogue to digital (A–D) converter; thereby the 1-D $x(n)$ or 2-D $x(n_1, n_2)$ data are obtained (see Sections IC and ID). The 1-D or 2-D digital filter in such systems performs the required operation which may correspond for example to some specific frequency requirements thus needing a suitable frequency transfer function $H(j\omega)$ or $H(j\omega_1, j\omega_2)$, and giving the output data to be immediately locally utilized (e.g. displayed) or stored.[34] These output data can be again converted back to analogue form by a D–A converter to produce a signal or indeed an image reconstruction for direct use and presentation for example on an oscilloscope, a plotter, display terminals, etc. It is interesting to observe that the same digital filter or a suitable section of it can also be used to give high accuracy data interpolation to produce a reconstructed signal or image of better quality.

In the second system of Fig. 8.2 an analogue signal or image is converted to digital form as in Fig. 8.1 and the data so obtained are processed by a 1-D or 2-D digital filter as required, to specify and delimit the bandwidths or to perform suitable corrections on the signals.

The filtered data are hence sent through a modulator and a transmitter in the communication channel: the purpose of the modulator-transmitter is to represent the binary data in terms of amplitude (AM), frequency (FM) or phase (PM) variations of a high frequency signal (carrier) suitable for transmission through the communication channel.

At the receiving terminal a receiver and demodulator reproduce the data but usually these are not identical to those transmitted, but altered due to the noise and disturbances introduced in or by the communication channel. For this reason data are recovered through a data reconstructor-synchronizer. In this way a digital filtering can then be performed on the received data so as to extract pertinent information within specific frequency bands and reducing noise and other disturbances.

Alternatively, as shown by dashed lines in Fig. 8.2, one can apply digital filtering directly to the signals as given by the demodulator converted to digital form. In this case digital filtering can have the main task of reconstructing the data by extracting them from noise.

One important difference between the two alternative systems above is related to their respective noise performances. While in the system of Fig. 8.1, in some cases at least, noise and disturbances can be negligible and we have essentially a *deterministic* process, in the system of Fig. 8.2 at the receiving terminal we have, as outlined above, the presence of noise (*random or stochastic process*). In general by considering both *multiplicative and additive noise* the received 1-D or 2-D signal can be represented as

$$x_r(t) = m(t)x(t) + n(t) \tag{8.1}$$

$$x_r(x_1, x_2) = m(x_1, x_2)x(x_1, x_2) + n(x_1, x_2) \tag{8.2}$$

where m is the multiplicative and n the additive noise part on the transmitted signal x. Some typical digital filtering operations are described below for processing 1-D and 2-D signals: some signals are considered without and some with noise. Thus for example digital filter band-pass analysis and spectral estimation can be applied to signals of both types; special kinds of filters such as matched filters, inverse filters and equalizers are needed to recover a signal from noise.

VIIIB1 Multiple filtering, band-pass analysis and spectral estimation

In general any kind of filtering (e.g. low-pass for frequency band-limiting, high-pass or band-pass for extracting some frequency band of interest, etc.) can be performed by means of FIR or IIR digital filters. If the extra facility of automatic control of the filter coefficients is taken into consideration, then adaptive digital filters are also obtainable.

There exist many practical situations in which multiple filtering and particularly band-pass analysis are required. In multiple filtering one specific filter operation (for example low-pass, band-pass, high-pass, etc.) is required to be performed on many signal channels presented as input data, e.g. in multiplex systems, while in band-pass analysis many band-pass filtering operations, which are in general adjacent to one another so as to cover the entire signal spectrum, are performed on a single signal channel presented as input data. In both of these cases several alternative approaches can be made.[7,22,26]

However by taking advantage of the multiple processing capabilities of digital techniques, efficient and economic digital filtering structures can be designed and applied. By means of appropriate organization of the operations involved and through suitable programming it is possible to reduce to a minimum the replicas of the required filtering operations and thereby make use of only one digital system realization (Section VII B).

Figure 8.3 shows a typical structure of a digital filtering processor performing multiple filtering operations on N different input data channels. It is observed that the different channel data are sent through a multiplexer to an input memory (e.g. a random-access-memory RAM) in which the data are organized in the most convenient way, subject to the actual signal processing to be performed: from the memory the data are sent to a fast arithmetical unit where they are processed using the appropriate filter coefficients (which in general may define a single digital filter) as given by another memory (e.g. a Read Only Memory: ROM). The output data can be again obtained on N separate channels through a demultiplexer. All the

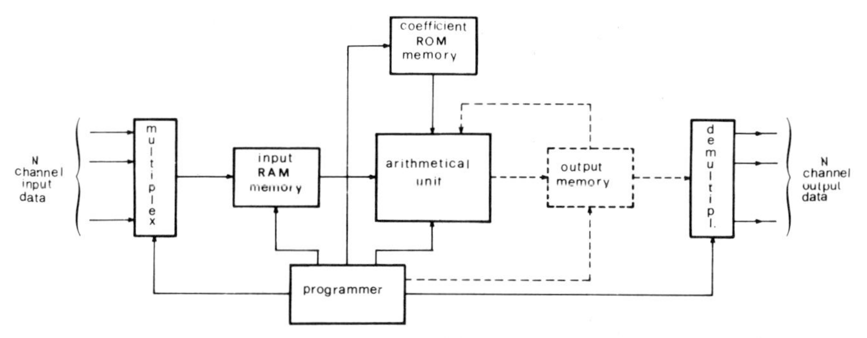

FIG. 8.3. Typical structure of a digital filtering processor performing multiple filtering operations on N input data channels.

operations are under the timing and control of a general program control. If a recursive type (IIR) digital filtering is to be performed the additional output memory, shown with dashed lines in Fig. 8.3, is required.

For band-pass analysis, Fig. 8.4(a) shows a more direct structure resulting from the use of a bank of digital filters of adjacent passbands and of frequency responses $H_i(j\omega)$, $i=1,\ldots,M$. Each passband has an output $y_i(n)$ resulting in the discrete convolution (see eq. (1.43)) of input data $x(n)$ with the impulse response $h_i(n)$ of the ith digital filter. If the filter passbands are chosen to cover the entire input signal spectrum, and in addition the filters are carefully designed, the sum $y(n)$ of all the filter outputs will be a good approximation to the original input signal (dashed lines in Fig. 8.4(a)).[19,25]

Figure 8.4(b) shows a typical structure of a digital filtering processor performing band-pass analysis (M bands) on input signal data $x(n)$, using essentially a single arithmetical unit: this structure is practically identical with that of Fig. 8.3 with the exclusion of the input multiplexer.†

There are many methods with which the structure of Fig. 8.4(b) can be used to perform the band-pass analysis.[20,21,38]

Firstly it can be observed that the different M band-pass filtering operations can be performed on the same input data set, if the coefficients $\{h_1(n)\}, \{h_2(n)\}, \ldots, \{h_M(n)\}$ are stored in the coefficient memory.

Alternatively there are other interesting methods and corresponding realization structures that make use of a single digital filter coefficient set

† The idea in Fig. 8.3 and Fig. 8.4(b) of using a unique arithmetic unit for different filtering operations has been used to extend a low order digital filter to higher order. As in the reference[89] where a variable-order digital filter has been realized, having essentially a second order recursive digital filter section multiplexed four times to achieve an eighth order filter, the data corresponding to each input sample are cycled four times through the second order filter structure and for each cycle the coefficients of this filter are changed to correspond to the appropriate filter section.

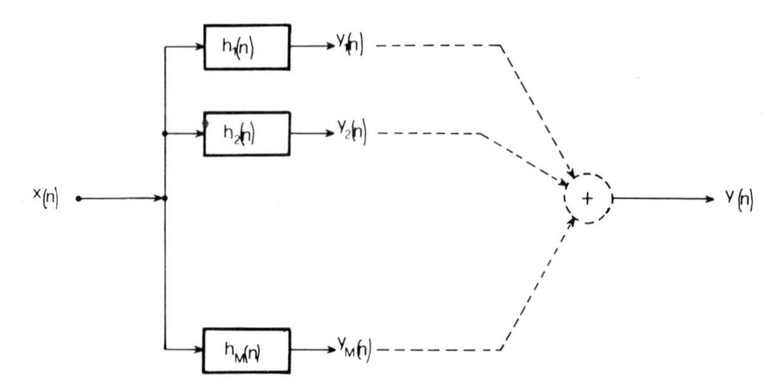

FIG. 8.4(a). Band-pass analysis using a bank of band-pass digital filters.

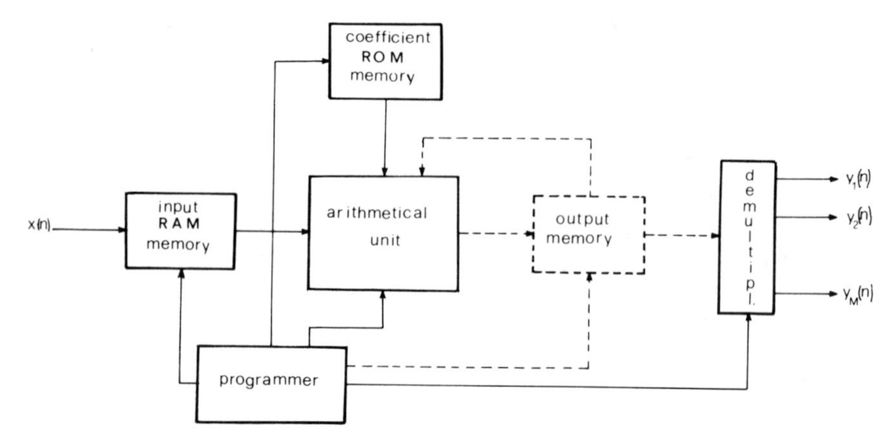

FIG. 8.4(b). Band-pass analysis using a digital filtering processor with a single arithmetic unit.

but involve modification of the input data in an appropriate way. Furthermore some of these methods have the additional interesting characteristic of producing a band-pass analysis in compressed form involving a reduction of the number of samples as is the case in the following.

The method corresponds to using a single low-pass digital filter and a frequency shift operation on the sampled signal spectrum, by multiplying the input data by an exponential factor. If an angular frequency shift ω_i is performed on the sampled signal spectrum $X_p(j\omega)$ (see relation (1.12)), to produce a spectrum $X_p(j\omega - j\omega_i)$, then this corresponds to a new sampled signal of the form (with T sampling interval)

$$x_i(n) = x(n)\, e^{jn\omega_i T} \tag{8.3}$$

If ω_M is the signal maximum angular frequency (for example, the analogue signal $x(t)$ is band limited to ω_M) then angular frequency shifts $\omega_i = i(\omega_M/M)$, $i = 1, \ldots, M-1$, are required to be operated on the signal so as to have M-band analysis by using a single low-pass filter of angular cutoff frequency $\omega_c = \omega_M/M$. It is of course apparent from this that all the original spectrum frequency bands will be shifted to the low frequency band $(0, \omega_M/M)$. In this case the arithmetic unit of Fig. 8.4(b) has to perform complex multiplications on $x_i(n)$ using the coefficients of the single low-pass digital filter.

On this line a particular method has been proposed by Cappellini[27] using the values $\omega_i = \omega_s/2$ with $\omega_s = 2\pi/T = 2\pi f_s$ as the angular sampling frequency. It is quite straightforward to verify that now the relation (8.3) becomes very simple indeed, as shown below

$$x_i(n) = x(n)(-1)^n$$

simply by using an alternative sign change.

If moreover $\omega_s = 2\omega_M$ and $\omega_c = \omega_M/2$ the situation of Fig. 8.5 is produced. It can be seen from this that low-pass filtering of the original signal $x(n)$ produces the sampled signal $x_1(n)$ which corresponds to the original spectrum band $(0, \omega_M/2)$ (*Fig.* 8.5(*a*)), whereas low-pass filtering of $x(n)$ $(-1)^n$ produces the sampled signal $x_2(n)$ corresponding to the original spectrum band $(\omega_M/2, \omega_M)$, shifted and inverted within the frequency band $(0, \omega_M/2)$, as shown in Fig. 8.5(b).

The procedure described above can be applied again to the signals $x_1(n)$ and $x_2(n)$ by considering only one of two consecutive samples (i.e. involving sample *decimation* or *reduction*) so that the preceding relationship can be still valid if an angular frequency shift equal to $\omega_M/2 = \omega_s/4$ is used. Then four bands can be so obtained. Therefore this method can separate, in a tree structure as shown in Fig. 8.6 for $f_M = \omega_M/2\pi$, the signal spectrum in a rapidly increasing number of bands. In general if r successive operations are performed, then 2^r bands are obtained.

The band-pass analysis resulting from the procedure described presents the following advantages with respect to the normal band-pass filtering:

(1) the band-pass analysis can be obtained by using a unique digital filter without changing its coefficients (i.e. it depends only on the number of processed samples and on the ratio ω_s/ω_c);

(2) high and constant efficiency of digital filtering;

(3) the speed of analysis can, in some cases, increase because we can stop the analysis in those parts of the frequency spectrum having zero value or which are of no interest;

(4) due to the sample reduction procedure a compressed representation

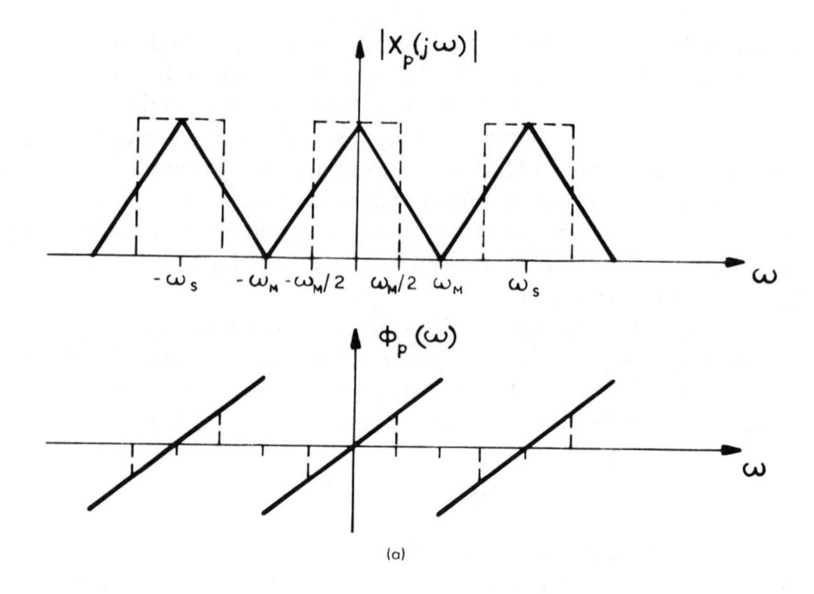

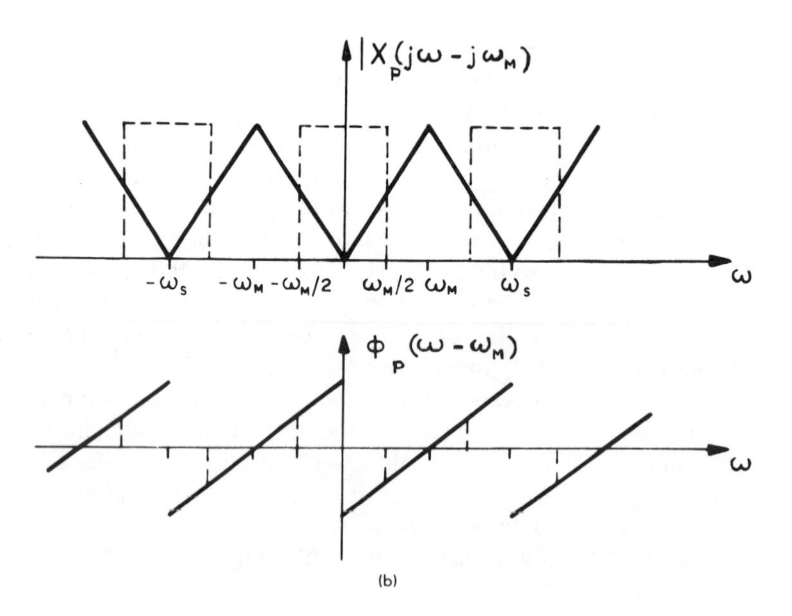

FIG. 8.5(a). Low-pass digital filtering of a sampled signal with $\omega_c = \omega_M/2 = \omega_s/4$. (b). Low-pass digital filtering of a sampled signal after a spectrum frequency shift.

- $0 \div f_M$
  - $0 \div \frac{1}{2}f_M$
    - $0 \div \frac{1}{4}f_M$
      - $0 \div \frac{1}{8}f_M$
      - $\frac{1}{8}f_M \div \frac{1}{4}f_M$
    - $\frac{1}{4}f_M \div \frac{1}{2}f_M$
      - $\frac{1}{4}f_M \div \frac{3}{8}f_M$
      - $\frac{3}{8}f_M \div \frac{1}{2}f_M$
  - $\frac{1}{2}f_M \div f_M$
    - $\frac{1}{2}f_M \div \frac{3}{4}f_M$
      - $\frac{1}{2}f_M \div \frac{5}{8}f_M$
      - $\frac{5}{8}f_M \div \frac{3}{4}f_M$
    - $\frac{3}{4}f_M \div f_M$
      - $\frac{3}{4}f_M \div \frac{7}{8}f_M$
      - $\frac{7}{8}f_M \div f_M$

FIG. 8.6. Procedure of band-pass analysis using frequency shift algorithm of the sampled signal spectrum and a single low-pass digital filter.

of each analysed passband is obtained with fewer samples than normally needed;

(5) the practical implementation can be relatively simple because the coefficient memory contains only one coefficient set and the operation $x(n)(-1)^n$ is very simple.

For a complete description of this method see Cappellini and Emiliani.[27]

Other methods have been proposed and applied, by employing again a single band-pass digital filter and involving compressing or expanding the sampled signal spectrum in the band of the digital filter. The advantage of digital signal representation is used in this technique in a way that the required time for expansion or compression is immediately obtainable simply by selecting the suitable sampled values in the input data set $x(n)$.

An alternative technique for band-pass analysis involving compression of the data relies on digital Hilbert filtering[13] with the evaluation of the in-phase and quadrature components. In general, a band-pass signal $s(t)$ with frequency spectrum extending over a bandwidth B centred on the frequency f_0 may be represented in the form

$$s(t) = a(t)\cos(2\pi f_0 t) - b(t)\sin(2\pi f_0 t) \tag{8.4}$$

where $a(t)$ and $b(t)$ are its in-phase and quadrature components. The spectra of both $a(t)$ and $b(t)$ extend over the frequency range $(-B/2, B/2)$. The two low-pass signals $a(t)$ and $b(t)$ may then be sampled at their Nyquist rate B, and the band-pass signal $s(t)$ can be recovered from the knowledge of both $a(t)$ and $b(t)$ according to relation (8.4). Therefore a complete identification of the band-pass signal $s(t)$ requires only the minimum sampling rate sequences $a(n/B)$ and $b(n/B)$. If the band-pass signal is originally sampled at a higher frequency f_s, these sequences can be obtained as shown in Fig. 8.7 from the samples $s(n/B)$ of the band-pass signal and the samples $\hat{s}(n/B)$ of its Hilbert transform.[15]

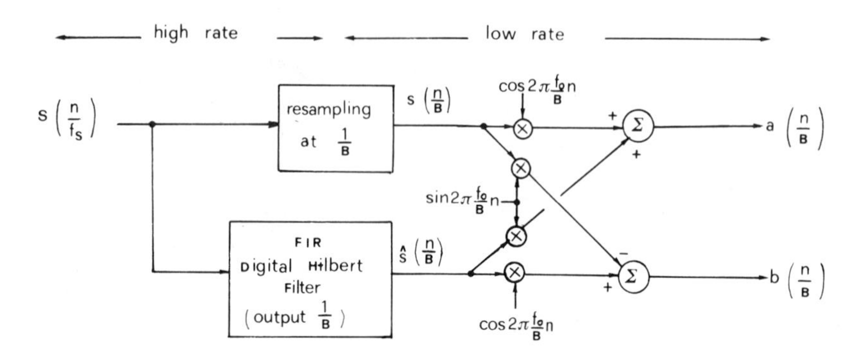

FIG. 8.7. Block diagram of a band-pass filtering using a FIR digital Hilbert filter.

The use of FIR digital filters is known to be highly desirable when, as in this case, the output samples are required at a lower sampling rate than the input samples. In fact a FIR digital filter allows filtering operations to be performed only at the required sampling instants and avoids evaluation of all the unnecessary output samples. The two sequences of samples $a(n/B)$ and $b(n/B)$—or equivalently $s(n/B)$ and $\hat{s}(n/B)$—may then be stored or transmitted for the complete characterization of the band-pass signal $s(t)$.

The above technique can be applied to obtain a complete band-pass analysis of a wide spectrum signal $x(t)$ sampled at the frequency f_s, by producing first the band-pass signal samples $s(n/f_s)$ and then processing these samples as shown above. Another solution of complete band-pass analysis corresponds to obtain the samples $s(n/B)$ and $\hat{s}(n/B)$ directly from the original wide-spectrum signal samples $s(n/f_s)$ through a FIR digital band-pass Hilbert transformer.

In summary, the advantage of this signal processing technique is that full information about a band-pass signal $s(t)$ can be obtained by a reduced number of samples—taken at the minimum rate B—of its in-phase and quadrature components, and therefore a compressed form band-pass analysis can be performed.

It is, of course, evident from the above discussion that digital filtering can also be used to produce spectral estimations of given signals. Indeed, as described in Chapter 1, the DFT and particularly FFT (see Appendices 1, 2) is a very useful and widely applied technique for performing short-time spectral estimation. The FFT can in fact give the amplitude and phase components of the spectrum. From these values the power spectrum can be estimated. Alternatively one can always apply the FFT to the autocorrelation function of a signal so as to estimate its power spectrum. Digital filtering can also be used to produce spectral estimations, particularly by performing band-pass analysis as described earlier.

If we consider a sufficiently high number M of bands, accurate spectral estimation is obtained. A particularly useful method with compact results, for many applications, is based on the short-time power spectral estimation obtained by evaluating the r.m.s. (root mean square) value $\overline{y_i(n)}$ in each output band $\{y_i(n)\}$ of the band-pass analysis, $\overline{y_i(n)}$ is given by

$$\overline{y_i(n)} = \sqrt{\frac{1}{L} \sum_{k=0}^{L-1} |y_i(n-k)|^2} \tag{8.5}$$

where L is the number of processed samples. The set of M values $\overline{y_i(n)}$, $i = 1, \ldots, M$ or of $\overline{y_i(n)^2}$, $i = 1, \ldots, M$, represent or are related to power spectral estimation in the *short-time* LT, T being the sampling interval.

In general we can observe that while FFT techniques are conveniently performed by means of computers or minicomputers and can give, if properly applied, detailed spectral information (by processing a sufficiently high number of data), spectral estimation obtained through band-pass digital filtering can be more economic for some hardware implementations, especially when some signal frequency bands are at the same time required in output.

VIIIB2 Matched filtering

A *matched filter* on a signal $x(t)$ with spectrum $X(j\omega)$ is a filter having impulse response $h(t)$ and transfer function $H(j\omega)$ defined by the following relations[91]

$$\begin{aligned} h(t) &= Kx(t_0 - t) \\ H(j\omega) &= KX^*(j\omega)\, e^{-j\omega t_0} \end{aligned} \tag{8.6}$$

where K is a constant, and t_0 is the time delay. Therefore the transfer function of a matched filter is, except for a possible amplitude and delay factor of the form $e^{-j\omega t_0}$, the *complex conjugate* of the spectrum of the signal to which it is matched. For this reason, a matched filter is often called a *conjugate filter.*

It is well known that a matched filter on a signal $x(t)$ when processing the signal $x(t)$ altered by additive white Gaussian noise (see relation (8.1)) as

$$s(t) = x(t) + n(t) \tag{8.7}$$

gives an output signal having the maximum signal-to-noise ratio (SNR) (maximum ratio of peak signal amplitude to r.m.s. noise). Hence this kind of filter is very important for many practical situations.

The output of the matched filter is in fact the cross correlation between $x(t)$ and $s(t)$.

The response at the output of a matched filter, to $x(t)$ applied at the input, is given by

$$y(t)=\frac{1}{2\pi}\int_{-\infty}^{\infty} X(j\omega)H(j\omega)\,\mathrm{e}^{j\omega t}\,\mathrm{d}\omega=\frac{K}{2\pi}\int_{-\infty}^{\infty}|X(j\omega)|^2\,\mathrm{e}^{j\omega(t-t_0)}\,\mathrm{d}\omega \tag{8.8}$$

In particular, when $t=t_0$, $y(t)$ will have an amplitude A given by

$$|y(t_0)|=\frac{K}{2\pi}\int_{-\infty}^{\infty}|X(j\omega)|^2\,\mathrm{d}\omega=A \tag{8.9}$$

where the amplitude A is thus proportional to the signal energy E.

Further if the input signal $x(t)$ is symmetrical in time, that is $x(t)=x(-t)$, $X(j\omega)$ will be a real function of the frequency. From eq. (8.6) we have for this case

$$h(t)=Kx(t-t_0)$$

$$H(j\omega)=KX(j\omega)\,\mathrm{e}^{-j\omega t_0} \tag{8.10}$$

Therefore the impulse response of a matched filter to a symmetrical input signal is a *delayed replica* of such an input signal.

In practice the main problem is to approximate the matched filter characteristics to those of an actual filter. It is important to point out that by using digital filters this problem can be solved in general in a better way than with analogue filters, due to the fact that digital filters can approximate any time function more directly and effectively.

In particular for the finite time duration signal $x(t)$ any matched filter defined by (8.6) or (8.10) can be implemented by using a FIR digital filter.

VIIIB3 Inverse filtering and restoration problems

Let us consider a system having a frequency transfer function $S(j\omega)$, then its *inverse filter* is defined as a filter having a frequency response $H(j\omega)$[91]

$$H(j\omega)=\frac{1}{S(j\omega)}=S^{-1}(j\omega) \tag{8.11}$$

This situation is illustrated in Fig. 8.8, from which it is clear that the output $y(t)$ of the inverse filter will be theoretically equal to $x(t)$. In practice it will be an approximation to $x(t)$ due to the fact that $H(j\omega)$ will have to approximate $1/S(j\omega)$ mainly for stability reasons.

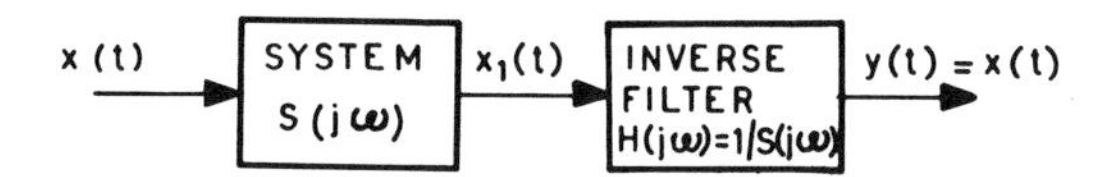

FIG. 8.8. Block diagram of inverse filtering.

Analogous considerations can be developed for two-dimensional signals or images $x(x_1, x_2)$: considering a 2-D system having a transfer function $S(j\omega_1, j\omega_2)$, a 2-D *inverse filter* is defined as a filter having a transfer function $H(j\omega_1, j\omega_2)$

$$H(j\omega_1, j\omega_2) = \frac{1}{S(j\omega_1, j\omega_2)} = S^{-1}(j\omega_1, j\omega_2) \tag{8.12}$$

Let us consider in more detail, for 1-D and 2-D cases, two situations: time or space variant linear systems and time or space invariant linear systems and the corresponding inverse filters.

A *time variant linear system* processing an input signal $x(t)$ can be defined by the equation

$$y(t) = \int_{-\infty}^{\infty} h(t, \tau)x(\tau)\, d\tau \tag{8.13}$$

where $h(t, \tau)$ is the impulse response, while a *time invariant linear system* is defined by the well known convolution

$$y(t) = \int_{-\infty}^{\infty} h(t-\tau)x(\tau)\, d\tau \tag{8.14}$$

For 2-D systems we have in an analogous way for a 2-D *space variable linear system* the following[78]

$$y(x_1, x_2) = \int_{-\infty}^{\infty}\int_{-\infty}^{\infty} h(x_1, x_2, \eta_1, \eta_2)x(\eta_1, \eta_2)\, d\eta_1\, d\eta_2 \tag{8.15}$$

where $x(x_1, x_2)$ is the input image and $h(x_1, x_2, \eta_1, \eta_2)$ is the point-spread function (2-D impulse response). For a 2-D *space invariant linear system* we have

$$y(x_1, x_2) = \int_{-\infty}^{\infty}\int_{-\infty}^{\infty} h(x_1-\eta_1, x_2-\eta_2)x(\eta_1, \eta_2)\, d\eta_1\, d\eta_2 \tag{8.16}$$

This last relation is in fact the well known 2-D convolution.

We can now define the so-called *restoration problem.*

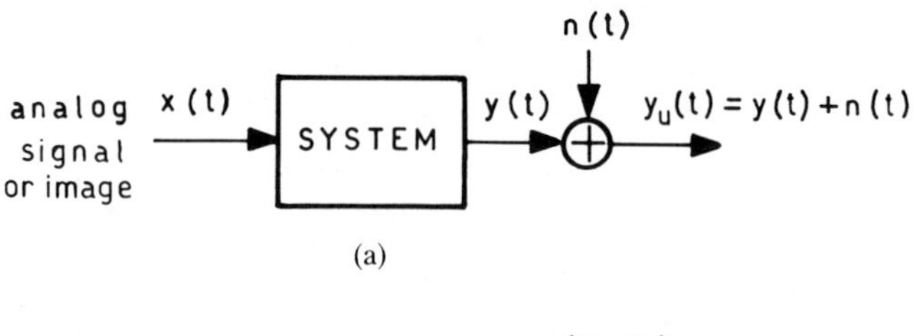

(a)

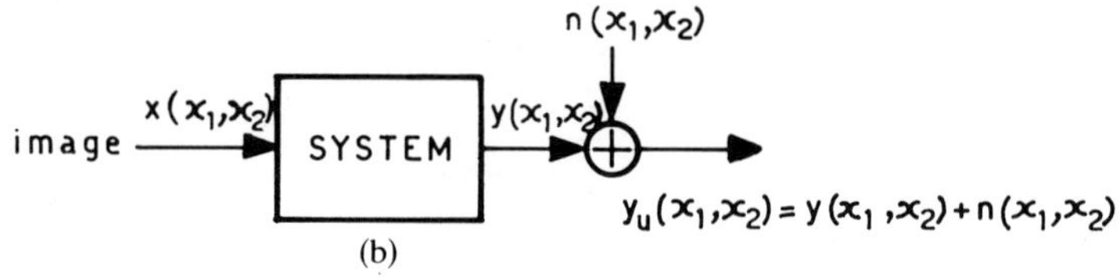

(b)

FIG. 8.9(a). Block diagram of a system adding noise to 1-D signal or scanned image. (b) Block diagram of a system adding noise to 2-D signal.

Let us consider 1-D or 2-D signal passing through a linear system and altered by additive noise (see relations (8.1) and (8.2)), as shown in Fig. 8.9; we have

$$\begin{aligned} y_u(t) &= y(t) + n(t) \\ y_u(x_1, x_2) &= y(x_1, x_2) + n(x_1, x_2) \end{aligned} \tag{8.17}$$

The process by means of which we recover the signal $x(t)$ from $y_u(t)$ or the image $x(x_1, x_2)$ from $y_u(x_1, x_2)$ is often referred to as *restoration process.* In general an *estimate* or approximation of $x(t)$ or $x(x_1, x_2)$ is in practical terms obtainable.

If the noise term $n(t)$ or $n(x_1, x_2)$ is zero or neglected, the signal or image restoration problem reduces to solving an integral equation. For $x(t)$ for example, we solve eqs (8.13), (8.14) whereas for $x(x_1, x_2)$ we solve eqs (8.15) (8.16), depending on whether the systems are variant or invariant. Even in this quite ideal case the problem is difficult, because it can be shown by the Rieman–Lebesque lemma that, for instance, in (8.13) or (8.15) an arbitrarily small perturbation in $y(t)$ or $y(x_1, x_2)$ can correspond to arbitrarily large perturbations in $x(t)$ or $x(x_1, x_2)$.[78]

The use of digital filtering in these problems is by employing a *digital inverse filter.* This is defined as the filter satisfying the inverse conditions of eqs (8.11) and (8.12) and when used for restoration we then have a *1-D or 2-D restoration digital filter.* Let us now consider in some detail the case of image restoration, which is of interest in many image processing areas.

Firstly it can be observed that, by a simple rectangular approximation of the integral relations (8.15) and (8.16), we can obtain the following relationship[78]

$$y(n_1, n_2) \simeq \sum_{m_1=0}^{N_1-1} \sum_{m_2=0}^{N_2-1} h(n_1, n_2, m_1, m_2)x(m_1, m_2) \tag{8.18}$$

$$y(n_1, n_2) \simeq \sum_{m_1=0}^{N_1-1} \sum_{m_2=0}^{N_2-1} h(n_1 - m_1, n_2 - m_2)x(m_1, m_2) \tag{8.19}$$

Equations (8.18) and (8.19) can also be represented in matrix form[78]

$$y = H_V x \tag{8.20}$$

$$y = H_I x \tag{8.21}$$

where H_V is a space variant restoration matrix and H_I is a space invariant restoration matrix.

By neglecting the noise contribution in (8.17), eq. (8.18) or (8.20) define a *linear space-variant digital restoration* while eq. (8.19) or (8.21) define a *linear space-invariant digital restoration.*

Further eq. (8.16), (8.19) or (8.21) describe a linear spatial filter, and the restoration problem can be described as the choice of the *filter inverse* to the system, defined by $h(x_1, x_2)$ or $h(n_1, n_2)$ (as described at the beginning of the section), so as to yield the original image when the degraded image is processed by the filter:† 2-D FIR and IIR digital filters can be used (see relations (1.104), (1.105)).

Indeed if the noise contribution on the original image can be neglected, the inverse filter can give good results.[94]

For the more general case including the noise contribution, some image restoration methods have been proposed and used. Let us consider the structure (8.21) with noise contribution n

$$y = H_I x + n \tag{8.22}$$

One method is based on the minimum-mean-square-error (MMSE) criterion which corresponds to minimizing the mathematical expectation E of the difference $x - \hat{x}$, i.e.[78]

$$\min E(x - \hat{x}) \tag{8.23}$$

where $\hat{x}$ is the restored image estimate. Since H_I in (8.22) is block Toeplitz, the block circulant approximation may be employed. The resulting

† Analogously in 1-D case eq. (8.14) describes a linear filter, and the restoration problem can be described as the choice of the *filter inverse* to the system defined by $h(t)$, as outlined at the beginning of the section.

restoration algorithm is a Wiener linear digital spatial filter described in DFT form by[78]

$$H_w(k_1, k_2) = \frac{H^*(k_1, k_2)}{|H(k_1, k_2)|^2 + \dfrac{G_n(k_1, k_2)}{G_x(k_1, k_2)}} \tag{8.24}$$

where $H(k_1, k_2)$ is the 2-D DFT of $h(n_1, n_2)$ (see relation (1.84)) and $G_n(k_1, k_2)$, $G_x(k_1, k_2)$ are power spectra of noise and image, respectively. The restored-image Wiener estimate results in DFT form as

$$\hat{X}(k_1, k_2) = H_w(k_1, k_2)Y(k_1, k_2) \tag{8.25}$$

where $Y(k_1, k_2)$ is the 2-D DFT of $y(n_1, n_2)$.

Wiener filter restoration requires *a priori* information, that is the point-spread function $h(n_1, n_2)$ and knowledge of image and noise autocovariance functions to evaluate the power spectra G_x and G_n.

Other methods do not have all these requirements.

Constrained least squares estimation eliminates the requirement of covariance knowledge. The resulting digital spatial filter has DFT definition as[78]

$$H_c(k_1, k_2) = \frac{H^*(k_1, k_2)}{|H(k_1, k_2)|^2 + \lambda|C(k_1, k_2)|^2} \tag{8.26}$$

where λ is a parameter determined by iteration and $C(k_1, k_2)$ is the 2-D DFT of a constraint matrix $c(n_1, n_2)$. The relation (8.26) is similar to (8.24): a family of filters can be generated of which (8.24) is a special case.

Homomorphic techniques of Stockham, Cole and Cannon[94] assume the point-spread function as unknown. This is estimated from the degraded image by taking averages of image segments in the log-spectral domain. The homomorphic filter is described in the DFT frequency domain by[78]

$$H_H(k_1, k_2) = \sqrt{\frac{1}{|H(k_1, k_2)|^2 + [G_n(k_1, k_2)/G_x(k_1, k_2)]}} \tag{8.27}$$

Considering a point-spread function with zero phase, the homomorphic filter is the geometric mean between the Wiener filter (8.24) and the inverse filter

$$H_I(k_1, k_2) = \frac{1}{H(k_1, k_2)} \tag{8.28}$$

An interesting statistical approach to restoration of noisy images has been developed by Lebedev: it is based on a statistical description with probability density evaluation of structural properties of images.[80]

Some of the above restoration techniques have been compared. Some of the results are as follows:[78]

(1) if the signal to noise ratio is great enough no particular preference results, the techniques converging however to the inverse filter;

(2) as regards the noise, due to the fact that images are basically low-pass in behaviour, high-pass noise presents the simplest restoration problem, while low-pass and white noise present practically comparable difficulties;

(3) as far as the *a priori* information is concerned, the Wiener filter is the worst, whilst the homomorphic filter is the best (it can construct the restoration from the degraded image itself);

(4) for the visual quality of restored images, the situation is as in (3) above, namely that the homomorphic filter is the best and the Wiener filter the worst.

VIIIB4 Equalization

Equalizer is in general called a filter designed to compensate for the normal distortion that a signal has suffered by travelling through a transmission medium or a communication channel (see relation (8.1) and Fig. 8.2).

Theoretically it reduces essentially to an inverse filter. In general it is the introduction of the noise that poses a significant problem: the signal distortion could at least in theory be equalized to as fine a degree as required, while the noise ultimately provides a limit on signal detectability. Restoration techniques as described in Section VIIIB3 can be applied in this case.

For digital signal processing and transmission, with which we are concerned here, the typical structure of an equalizer is a *transversal filter* as shown in Fig. 8.10, where T is a constant time delay and $a_0, a_1, \ldots a_{N-1}$ are the weights. The output $y(t)$ is given by

$$y(t) = \sum_{k=0}^{N-1} a_k x(t-kT) \tag{8.29}$$

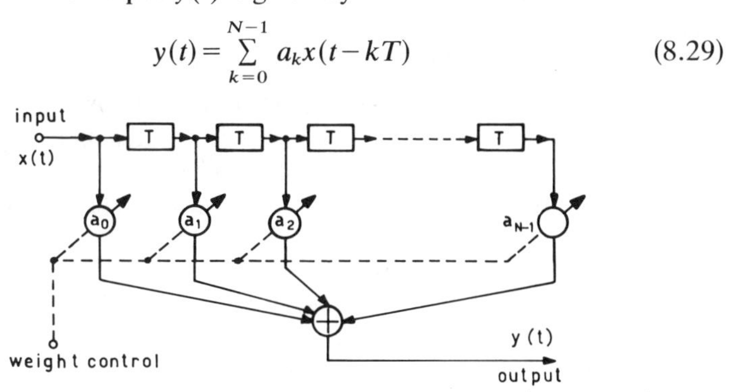

FIG. 8.10. Block diagram of an equalizer, with adaptive control (dashed lines).

To obtain efficient equalization, *automatic equalizers* and *adaptive equalizers* have been developed and used.[47,91]

The main idea in automatic or adaptive equalizers is to adjust automatically the filter weights according to some control criterion or algorithm (shown by dashed lines in Fig. 8.10).

Two main approaches to this are used:

(i) a test signal (in general pulse train) is sent through the distorting noisy system prior to the actual information signal so as to enable adjustment of the weights (tap gains) of the equalizer to optimum positions to take place; after the training period the weights remain fixed and the normal information signal is sent;

(ii) the equalizer derives the control signals for weight adjustment directly from the signals received in input: the equalizer tracks continually the distorting noisy system (communication channel) and readjusts itself when required so as to provide optimum equalization.

The second solution of adaptive equalization is in general more efficient than the first one because it is done automatically and does not have the disadvantage of requiring test signals and long training periods.

In the analogue implementation of the structure in Fig. 8.10 tapped delay lines have been used.

It is quite evident from the above structures that there exists an immediate connection with digital filters. Indeed eq. (8.29) corresponds to that of a FIR digital filter (see eqs (1.95) (1.166)) and the structure of Fig. 8.10 to a nonrecursive digital filter (see Fig. 1.15). The adaptive digital equalizer is therefore essentially an adaptive nonrecursive (FIR) digital filter.

However, digital adaptive equalizers of the recursive type (corresponding to IIR digital filtering) have also been designed and applied.

VIIIC Digital speech processing

Speech signals are for obvious reasons one of the most important signals and therefore it is quite natural that their processing by means of digital techniques has received so much attention.[62]

Indeed the representation and transmission of speech signals by digital means can result in more efficient and often more economic techniques than in the corresponding analogue case.

Let us consider firstly in general terms the main aspects of speech generation and the characteristics of the principal sounds and then we shall describe the application of digital filtering so as to obtain a digital model for speech production.[86]

Speech is produced by the *human vocal apparatus*, in particular by excitation of an acoustic tube, called the *vocal tract* which is terminated at one end by the *vocal cords* (glottis) and at the other end by the lips. An ancillary tube, the *nasal tract*, can be connected or disconnected to the vocal tract by the movement of the velum. The actual shape of the vocal tract is determined by the position of the lips, jaw, tongue, and velum.

Speech is generated in three basic ways:[86]

(i) *voiced sounds* are produced by exciting the vocal tract with quasi-periodic pulses of air flow caused by the vibration of the vocal cords;

(ii) *fricative sounds* are produced by forming a suitable constriction in the vocal tract and pushing air through the constriction, thereby creating turbulence which represents a noise-like source exciting the vocal tract;

(iii) *plosive sounds* are produced by completely closing off the vocal tract, increasing pressure and then abruptly releasing it.

All these sources act as a wide-band excitation of the vocal tract. The vocal tract is characterized by its natural frequencies called formants, which correspond to resonances in the sound-transmission characteristics of the vocal tract. The vocal tract can in general be modelled as a *linear time-variant system* (see relation (8.13)), while in short time intervals (less than 10 ms) it can be approximated to a *linear time-invariant system* (see relation (8.14)). In this last case the output is the convolution of the impulse response of the vocal tract with the excitation waveform.

The application of digital filtering to speech processing and representation was made possible many years ago by the relatively low frequency range of speech and consequent availability of low cost analogue to digital (A–D) and digital-to-analogue (D–A) converters. Sampling frequencies in the range 8 KHz–100 KHz are used with word-length of 7–12 bits, depending on the nature of the processing.

Three main areas of application of digital filtering to speech processing are:

(i) digital model for speech production;

(ii) filtering, band-pass analysis and spectral estimation;

(iii) vocoders (analysis and synthesis).

Other aspects regarding speech transmission, e.g. digital telephony, are examined in Section VIIIE.

VIIIC1 Digital model for speech production

According to the mechanism of speech production in the human vocal apparatus, described above, some digital models have been proposed and used for the production or simulation of speech signals.

One of the most important digital models is shown, as a block diagram, in Fig. 8.11. The time varying digital filter represents the vocal tract as a linear time-variant system: as already observed, in short time intervals (less than 10 ms) the digital filter can be characterized by a fixed impulse response or by a precise set of coefficients.[85] To represent voiced sounds, the digital filter is excited by an impulse train generator which gives a quasi-periodic impulse train in which the spacing between impulses corresponds to the fundamental period of the glottal excitation. To represent unvoiced sounds (fricative and plosive sounds), the digital filter is excited by a random number generator corresponding to a flat spectrum noise. In both cases an amplitude control, inserted before the digital filter, regulates the intensity of the signals to reproduce actual speech intensity.)[86]

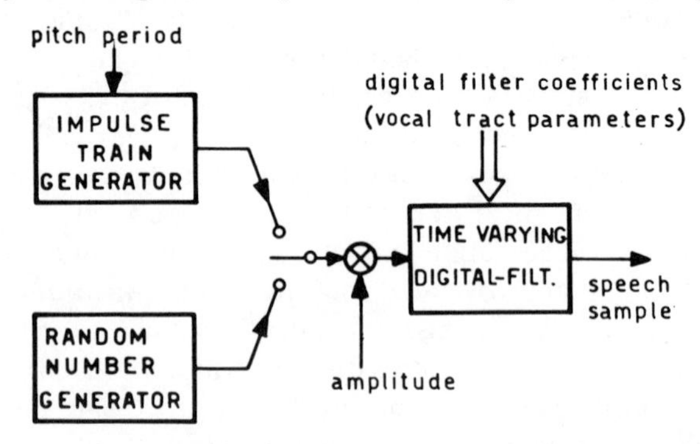

FIG. 8.11. Block diagram of a digital model for speech signal production.

For voiced sounds, except nasals, the transfer function of the digital filter excited by a train of impulses $p(n)$ (in which the spacing between impulses corresponds to the fundamental or pitch period of the voice) can be expressed as[85,86]

$$H_V(z) = V(z)G(z) \tag{8.30}$$

where $V(z)$ represents the vocal tract component and has an expression of the following type[86]

$$V(z) = \frac{A}{\prod_{k=1}^{L} (1 - C_k z^{-1})(1 - C_k^* z^{-1})}, \quad |C_k| < 1 \tag{8.31}$$

with C_k corresponding to the natural frequencies of the vocal tract, and $G(z)$, taking into account the characteristics of finite-duration glottal pulses, is of the form (with a_k, b_k coefficients)[86]

$$G(z) = B \prod_{k=1}^{M} (1 - a_k z^{-1}) \prod_{k=1}^{N} (1 - b_k z) \tag{8.32}$$

For unvoiced sounds the transfer function of the digital filter excited by a random-noise sequence $r(n)$ can be expressed as (with d_k coefficients)

$$H_u(z)=\frac{A\prod_{k=1}^{P}(1-d_k z^{-1})(1-d_k^* z^{-1})}{\prod_{k=1}^{L}(1-C_k z^{-1})(1-C_k^* z^{-1})}, \quad |C_k|<1 \tag{8.33}$$

According to the preceding relations and by using discrete convolution notation (see eq. (1.42)), we can represent the voiced sound sequences $\{s_v(n)\}$ and the unvoiced sound sequences $\{s_u(n)\}$ as[86]

$$\{s_v(n)\}=\{p(n)*g(n)*v(n)\} \tag{8.34}$$

$$\{s_u(n)\}=\{r(n)*h_u(n)\} \tag{8.35}$$

Homomorphic deconvolution has been applied by Oppenheim to estimate the short-time parameters of the speech model.[85]†

The interest in the digital speech model above lies in the fact that it can give speech production with the required accuracy, by increasing the complexity of the time-varying digital filter or by changing the pulse generation. This is possible because of the high degree of flexibility and adaptivity that the digital representation affords. These valuable characteristics are not easily obtainable, if at all, at the same levels of flexibility and adaptability in models of the analogue type.

VIIIC2 Digital filtering, band-pass analysis and spectral estimation of speech signals

As far as single filtering is concerned (e.g. low-pass for frequency band limiting, high-pass or band-pass for extraction of some frequency band of interest), FIR and IIR digital filters can be easily applied. One example of low-pass FIR digital filter frequency response, using 17 coefficients with the window technique (Lanczos-type window, $m=1.8$, see relation (2.66)) used for normal speech filtering, is shown in Fig. 8.12.

For speech band-pass analysis, the different solutions using digital filtering described in Section VIIIB1 can be applied. In particular the methods using a single digital filter and modifying in the meantime the input data in a suitable way can be attractive for hardware implementations.

A complete system, called ANPAD, using the technique described in Section VIIIB1 (see Figs. 8.5 and 8.6) with a single digital filter, frequency shift of the sampled signal spectrum and sample reduction criteria, was

† Further methods for the direct estimation of the vocal tract shape by inverse filtering of acoustic speech waveforms have been proposed and applied.[86]

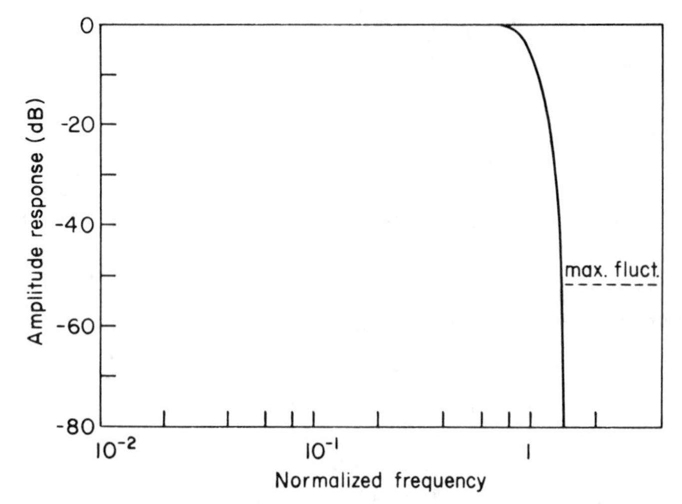

FIG. 8.12. Example of low-pass FIR digital filter frequency response, used for normal speech filtering in the ANPAD system.

constructed in hardware at IROE, CNR in Florence. The block diagram of the band-pass analyser, giving up to 16 bands (4 steps in Fig. 8.6) and processing input signals with a maximum frequency of 4 KHz (8000 samples/s, each sample represented with 8 bit accuracy) is shown in Fig. 8.13. The usual coefficients are those giving the frequency response in Fig. 8.12. Figure 8.13 shows clearly the organization of the input sample memory with multiplications by $(-1)^n$ (simple alternative change of the sign bit) as required by the algorithm described in Section VIIIB1.

Figure 8.14 shows the constructed digital analyser, including the A–D converter (on the left) operating at a maximum rate of 8000 samples/s, 8 bits/sample; the simple programming of the required number of bands, i.e. 2, 4, 8 or 16 bands is shown in the central part and the band-pass analysis outputs on the right reconstructed in analogue form (digital outputs are available at the back of the analyser).

It is interesting to observe that the analogue reconstructed outputs after D–A conversion are available in two forms: the more usual one with zero order holding and a special form with linear interpolation. This last analogue reconstruction is based on the insertion of a staircase waveform with a constant high number of steps connecting two consecutive samples. With reference to Fig. 8.15(a) two consecutive samples x_n and x_{n+1} are processed by the interpolator to obtain the data in the form[30]

$$A_i = x_n + \frac{i}{N}(x_{n+1} - x_n), \qquad i = 1, \ldots, N \tag{8.36}$$

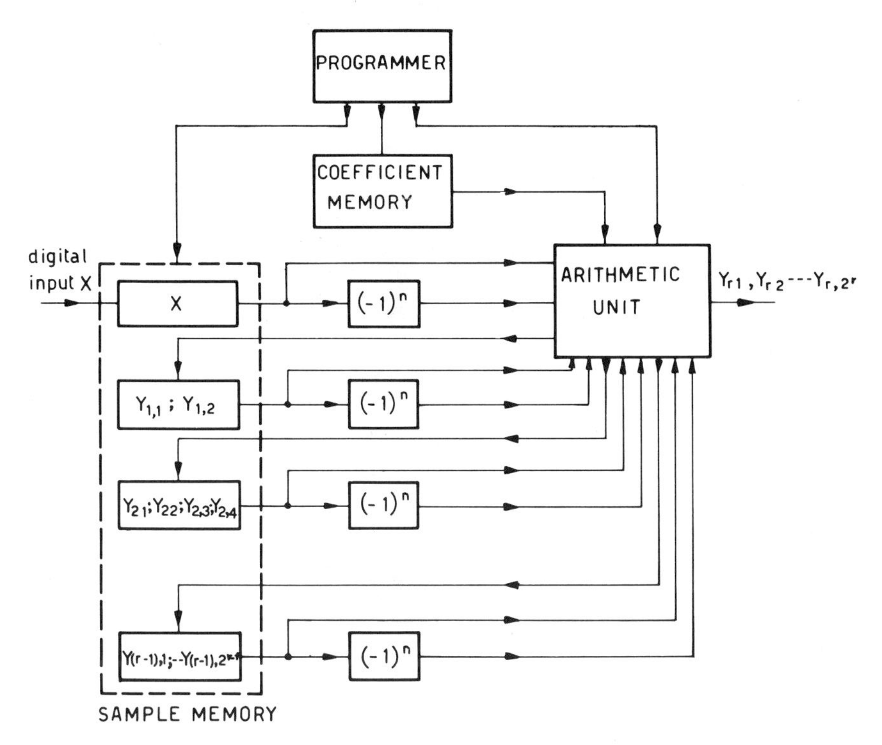

FIG. 8.13. Block diagram of the band-pass analyser of the ANPAD system.

where N is a constant sufficiently large integer ($N = 16$ in the band-pass analyser). The relation (8.36) defines a very simple FIR digital filter. The A_i values are sent to a standard D–A converter. An example of output in comparison with a standard zero-order interpolator is shown in Fig. 8.15(b); an example of speech band-pass analysis is shown in Fig. 8.16: 13 bands (250 Hz each) of the spectrum part 0–3250 Hz are shown corresponding to the words *digital processor* (it should be remembered that the amplitudes—waveform envelope—are maintained by the algorithm used in the processor, while some frequency components are shifted in the frequency domain).

As far as speech spectral estimation is concerned, the methods described in Section VIIIB1 can be applied (FFT, autocorrelation with FFT, band-pass analysis with power evaluation in each band). In particular, by evaluating the r.m.s. value in each output-band according to the relation (8.5) the ANPAD system includes a digital spectrum analyser (at top in Fig. 8.14): the value L in (8.5) (LT, with T sampling period, defines the *short time* or *epoch* of spectral estimation) can be changed from 16 to 256 (from

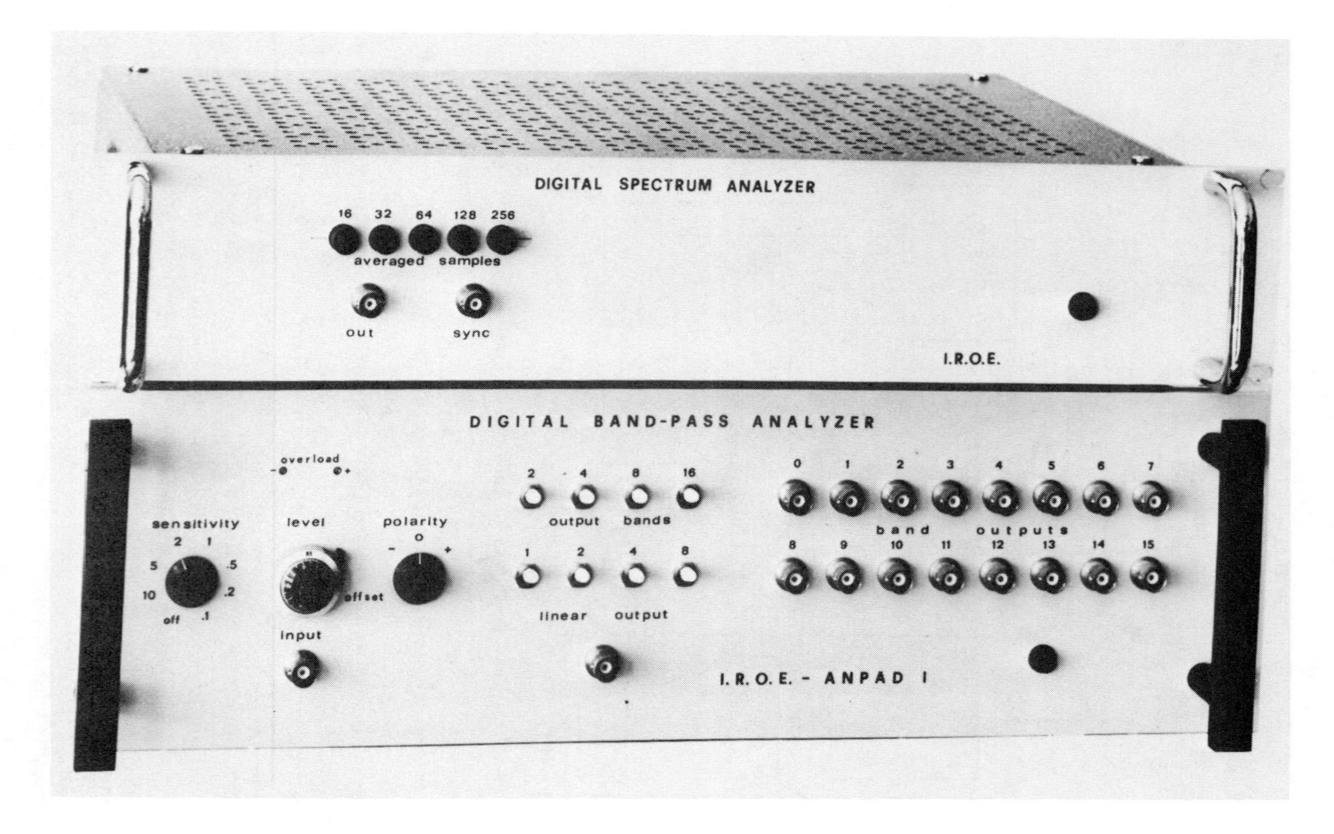

FIG. 8.14. Photo of the constructed digital analyser (ANPAD) at IROE, CNR, of Florence.

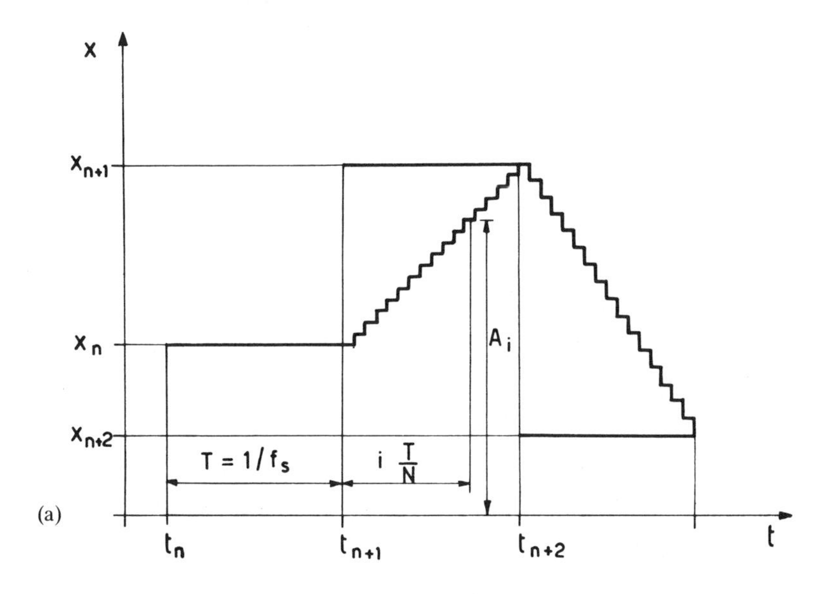

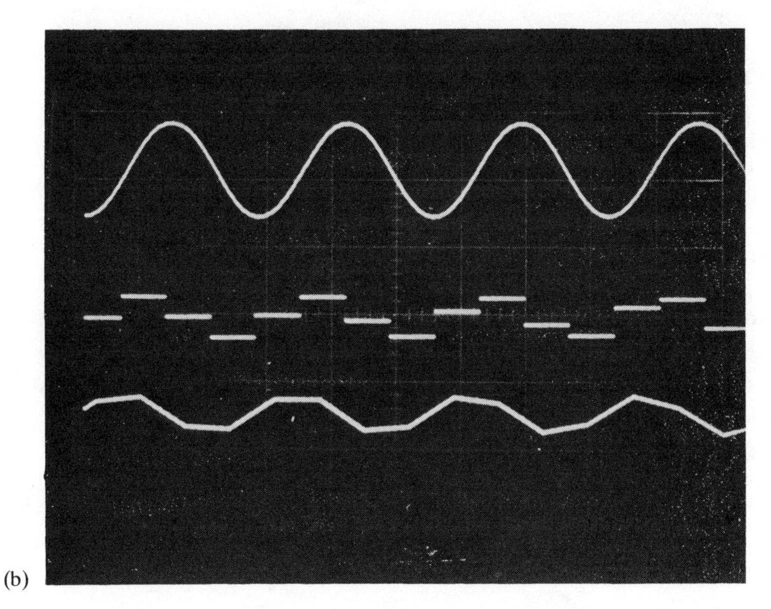

FIG. 8.15(a). Digital linear interpolation with a constant and high number of steps. (b) Example of signal reconstruction by means of the digital linear interpolator and a standard zero-hold device.

FIG. 8.16. Example of speech band-pass analysis by means of the ANPAD system: 13 bands (250 Hz each) of the spectrum part (0, 3250 Hz) are shown from the bottom to the top corresponding to the words "digital processor" (lowest line).

128 to 2048 input samples). An example of spectral estimation (16 band data) performed by means of this system is shown in Fig. 8.17.†

The particular technique used by the ANPAD system was applied also in software by using minicomputers: in this case a low-pass digital filter with higher efficiency (higher cutoff slope and lower level of fluctuations out of band) than that shown in Fig. 8.12 was used (in particular filters using the Cappellini window were applied[1]).

Some examples of comparison of processing speech units between this spectral estimation technique (which is particularly useful for giving power estimation) applied with 32 bands and standard FFT are shown in Fig.

† The ANPAD system was also used for automatic speech identification and recovery, by comparing the incoming short-time spectral evaluation (16 data) with a memorized *a priori* information on the short-time spectrum (other 16 data).

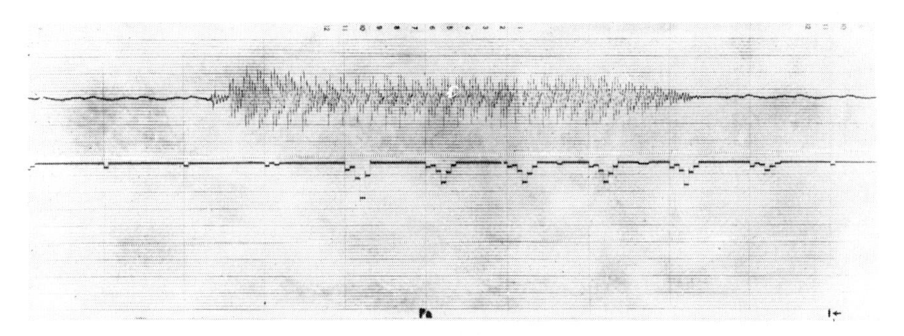

FIG. 8.17. Example of spectral estimation on a phonetic unit (SSRU):pa.

8.18.[1] The following values are used: $L=32$ (512 input samples), FFT with 512 points. We can see the good agreement between the two estimations.

VIIIC3 Vocoders

Vocoders are speech processors studied and developed for the analysis of signals for the purpose of transforming and representing in highly compressed form speech signals, and also in signal synthesis for recovering from the compressed representation the original speech signal. Vocoders are therefore an economic form of transmission or storage of such signals.

Digital filtering can be used in many and different ways in vocoder systems as proposed and implemented so far.[70]

A first and immediate application of digital filtering is found in the *channel vocoder* (Fig. 8.19): the band-pass analysis in transmission (*analyser*) and that in reception (*synthesizer*) can be performed through a digital filtering band-pass analysis as considered previously (ANPAD could, for example, be used). In Fig. 8.19 the information in each band in compressed form is extracted through a bank of band-pass digital filters, subsequent rectification and low-pass digital filtering. In parallel a voice sound detector (*buss-hiss detector*) recognizes voiced and unvoiced sounds, while a *pitch detector* measures the frequency or period of the glottal excitation in voiced sounds (see also Fig. 8.11). The short-time representation of the m-band analysis with outputs from voice sound detector and pitch detector can be economically stored locally or transmitted at a distance through a communication channel. In the synthesis (i.e. at the receiving terminal) the m-band data are used to modulate the output of a pitch pulse generator (for voiced sounds) or a noise generator (for unvoiced sounds), the switching between the two generators being controlled by the received buss-hiss detector binary information and the fundamental period

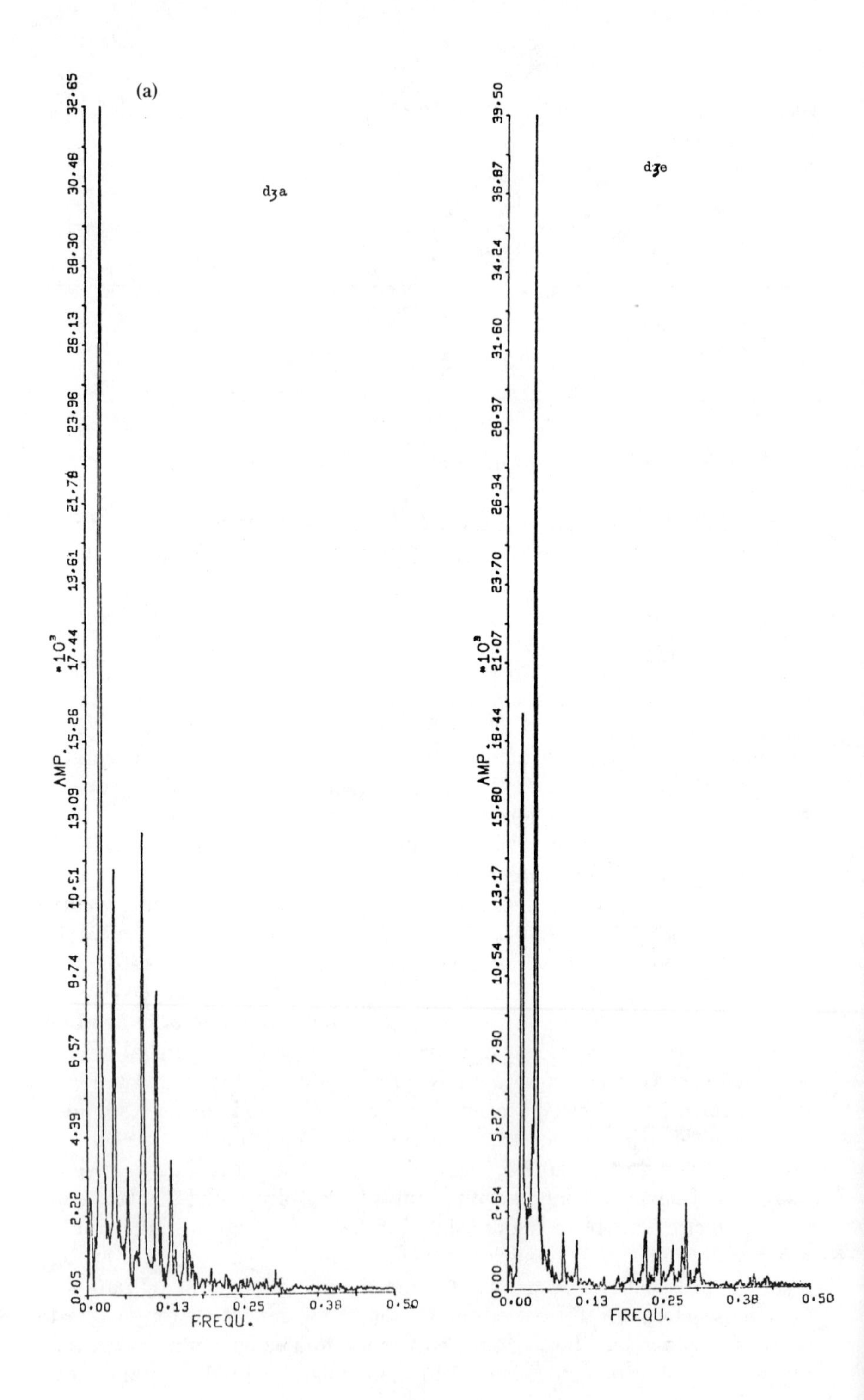
(a)
dʒa
AMP. *10³
32.65
30.48
28.30
26.13
23.96
21.78
19.61
17.44
15.26
13.09
10.91
8.74
6.57
4.39
2.22
0.05
0.00
0.13
0.25
0.38
0.50
FREQU.
dʒe
AMP. *10³
39.50
36.87
34.24
31.60
28.97
26.34
23.70
21.07
18.44
15.80
13.17
10.54
7.90
5.27
2.64
0.00
0.00
0.13
0.25
0.38
0.50
FREQU.

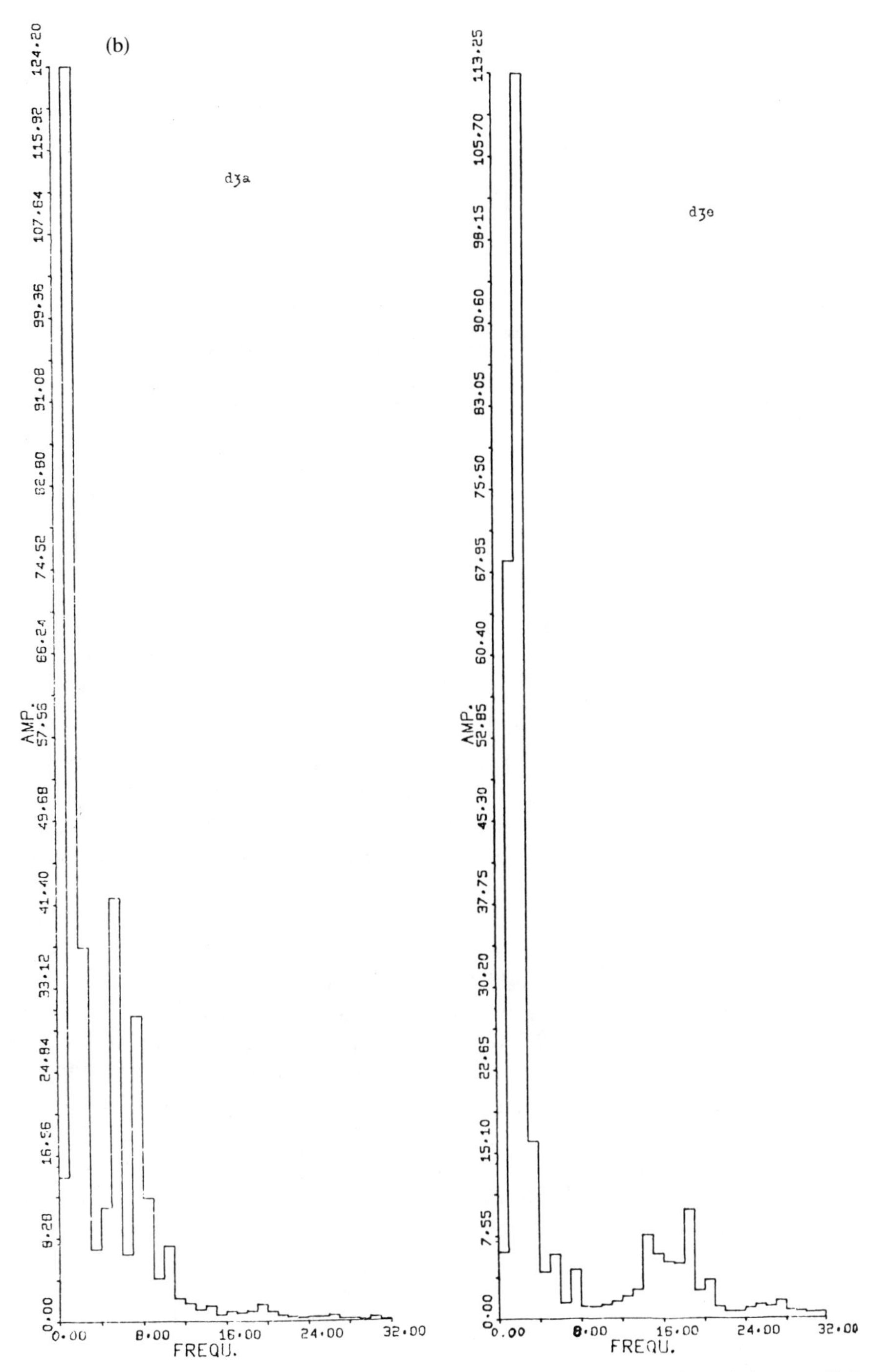

FIG. 8.18. Comparison of processing speech units between frequency shift technique applied with 32 bands (b) and standard FFT (a).

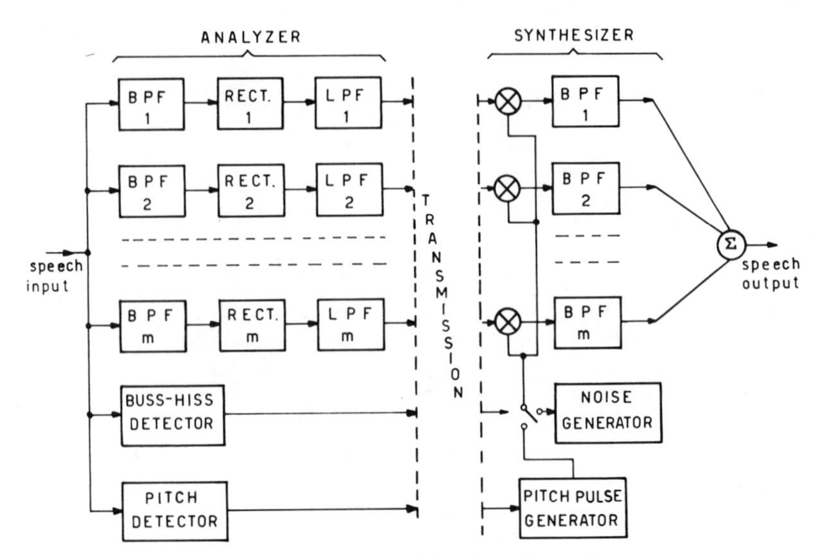

FIG. 8.19. Block diagram of the channel vocoder.

of pitch pulse generator being determined by the received pitch detector data.

A second type of vocoder, using digital filtering, is that resulting directly from the digital speech model of Fig. 8.11. In the analyser a pitch detector[69] and a voice sound detector can be used as above, while the *vocal tract parameters* defining the short-time performance of the digital filter (see also relations (8.30)–(8.35)) are in parallel estimated with different techniques. In the synthesis the actual system is practically very close to the structure of Fig. 8.11.†

Many other types of vocoder systems have been proposed and applied, but in principal characteristics they are similar to the two types considered above: phase, homorphic and linear prediction vocoders.

By considering the transmission rate, from the standard rate of 50 000 –100 000 bits/s (i.e. 64 000 bits/s, by considering 8000 samples/s and 8 bits/sample), vocoder systems of the above types can reduce the rate to few thousands of bits/s. Compression ratios of 10–50 can be obtained.

Interest has also been focused on systems capable of synthesizing speech or *standard voice* from a compressed coded representation of it. Some of these systems are based on speech segmentation (division in time). They employ the so-called standard speech reproducing units (SSRU) of a

† An interesting digital synthesizer has been implemented:[17] it is a microprogrammed processor which, operating as a minicomputer peripheral, can execute speech synthesis algorithms in real time.

language. There are approximately 140 such units in the Italian language including vowels. The situation of the English language is more complex.

By using SSRU any message can be analysed and divided into units (Fig. 8.20) as it enters a computer through a teletype. Then by taking from the computer memory the suitable SSRU (of a typical speaker) in sequence and sending it to a D–A output converter, a computer voice can be obtained.

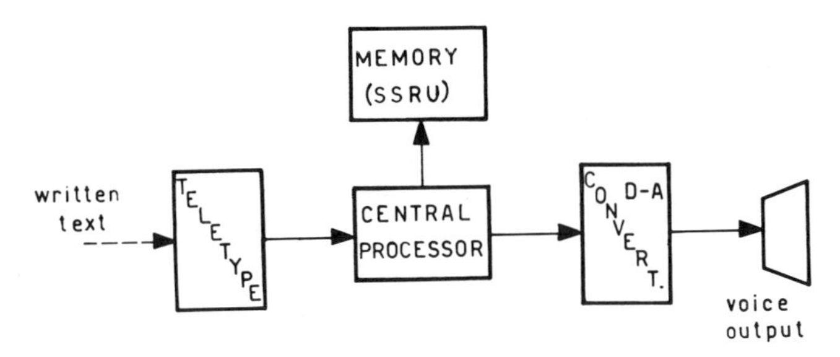

FIG. 8.20. Block diagram of a computer voice system.

A practical system of this type was defined by Debiasi and Francini at Padua University and applied more recently at Florence University.[1] Digital filtering, in particular for spectral estimation, was applied to the purpose of defining the frequency structure of each SSRU (see Fig. 8.13) and finding a more compact language representation (reduction of the number of SSRU). If these SSRU units are transmitted, to represent speech, bit rates of 200–500 bits/s can be obtained.

Another interesting speaking computer output peripheral was designed and implemented by Madams and Witten at the University of Essex.[82] This system, called parametric artificial talker (PAT), uses the model of speech with formant filters (of the type in Fig. 8.11) implemented with a special purpose processor with a unique arithmetic unit multiplexed for different operations and controlled by a stored program.

VIIID Digital image processing

Vision is the most powerful of the human senses and it is quite natural to find wide interest in image processing systems.

In the vision process the image is formed by the human eye upon the retina. Much effort has been directed towards explaining the vision process and in particular towards devising an acceptable model for it, as is the case for speech production (see Section VIIIC) and the hearing process. The physical processes in vision however are somewhat more complicated than the corresponding case of speech.

In some proposals of vision process interpretation, frequency-spatial models are considered with many parallel processors (suitable filters) working in the low frequency range of eye response to external stimuli, carrying the information over the retina.[72] In particular Sakrison[84] has proposed a model for the visual system represented as a collection of parallel channels with detection occurring whenever activity in any one of these channels reaches a certain level; each channel is an incoherent detector consisting of a band-pass filter, followed by a power law, and then spatial summation.

From the technological point of view there are different methods of image formation which can be classified in three main categories, depending on the wavelength of the electromagnetic wave or radiation used for the purpose:

(i) images obtained by means of electromagnetic waves at wavelengths longer than that of light, such as microwave maps (as given for example by radar);

(ii) optical images (as in the eye) and images obtained by means of electromagnetic waves of wavelengths proximate to those of visible light such as in the infra-red and ultraviolet regions. Infra-red waves are related to thermal effects and for this reason infra-red images are also called *thermic images*;

(iii) images obtained by means of penetrating radiation such as X-ray, gamma rays, neutrons, etc.

In all these cases the planar image formation process can be schematized as shown in Fig. 8.21.[78] The *image formation system* (intermediate box)

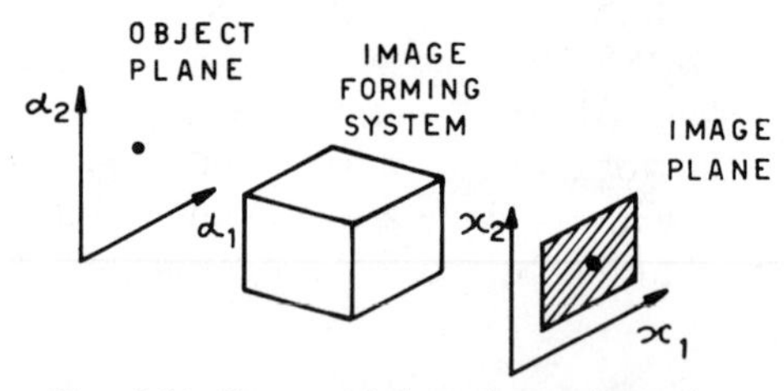

FIG. 8.21. Process of planar image formation.

generates the image by acting upon an electromagnetic or radiant energy component of the object. In optical systems the radiant light intensity reflected or emitted by the object is transformed by a set of lenses and apertures (as in the eye) to form the actual final image on a given plane.

Images of present day importance include the following:†

facsimile and television images;

biomedical images;[61]

† Emphasis is given to images in general due to the large quantity of information they contain.

aerial photographs and images from satellites for earth resource evaluation, territorial planning and archaeology;[60]
astrophysical maps;
meteorological maps;
seismic maps;[2,9]
oceanography maps;
artistic paintings.

In all the above areas there are some common processing problems such as:[77]

linear filtering
nonlinear filtering
redundancy reduction (data compression)
feature extraction
image interpretation (segmentation).

It is interesting to compare analogue (optical systems) and digital systems to solve the above problems.[90]

The analogue (optical) systems such as lenses, apertures, etc. are characterized by a very high processing speed, by the capability of giving immediately the 2-D Fourier transform of an image (it is well known that this is simply obtained at the output of a lens), by the practical impossibility (few sophisticated attempts) of performing nonlinear (inhomogeneous) operations and by very low flexibility and adaptivity.[95]

Digital systems (e.g. computers, minicomputers, microprocessors or digital hardware circuits) are characterized by a relatively low processing speed, especially in comparison with optical systems, high processing efficiency (see also the Section VIIIA) with high accuracy and stability, great flexibility and adaptivity, and the possibility of performing nonlinear (inhomogeneous) operations.[8,52,73,92]

From the above general and highly compressed considerations in addition to the comments made in the previous chapters regarding the evolution of digital circuit technology and the discovery of fast digital processing algorithms such as the FFT and fast digital convolution, the increasing interest in 2-D digital processing and the expansion of digital image processors for many applications become evident.

However this does not mean that optical systems are not of interest. Indeed they are very useful image processing systems and, due to their characteristics of extremely fast processing speed and immediate production of 2-D Fourier transforms, hybrid optical digital systems of the type sketched in Fig. 8.22, may be the kind of processing advisable for the near future.[71]

As far as a completely digital image processing system is concerned, a typical organization of the system pertinent to the above indicated

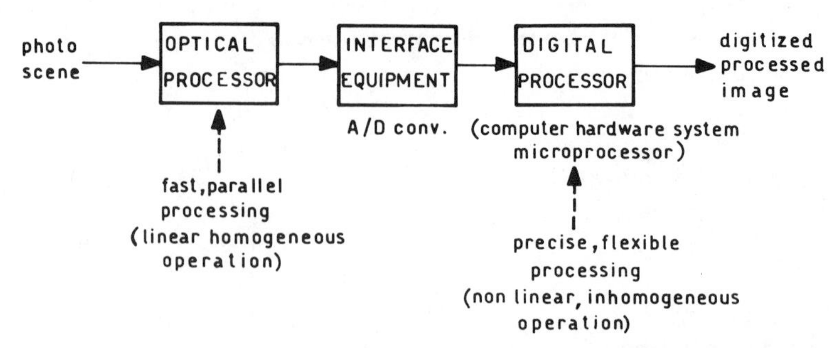

FIG. 8.22. Block diagram of an hybrid optical digital system.

problems is shown in Fig. 8.23. At first we have the A–D conversion, which is in general performed by means of a scanner (vidicon, multispectral scanner, synthetic aperture radar, scanning radiation detector, etc). Once 2-D data are obtained, we have the following operations to perform:

preprocessing, in which very simple operations can be performed, such as grey level adjustment, or digital linear and nonlinear filtering for smoothing, space bandwidth extraction, enhancement, correction and data compression (reduction of the amount of data such as redundancy reduction);

processing and pattern recognition, in which final processing operations are performed in 2-D, on already preprocessed data, such as *feature extraction, image interpretation* and *image description*;

output presentation and display, in which a suitable criterion of 2–D processed data presentation is selected and the actual output 2-D data are sent in different forms to a display; depending on the requirements of display we can distinguish: (a) binary outputs, in which only two grey levels (white-black) are used; (b) use of binary display for multi-level grey presentation (as with teletype, plotter, alphanumeric display, etc.) (c) multi-level output on oscilloscope or z-display (grey scale data can be also presented in many colours, *pseudocolour form*).

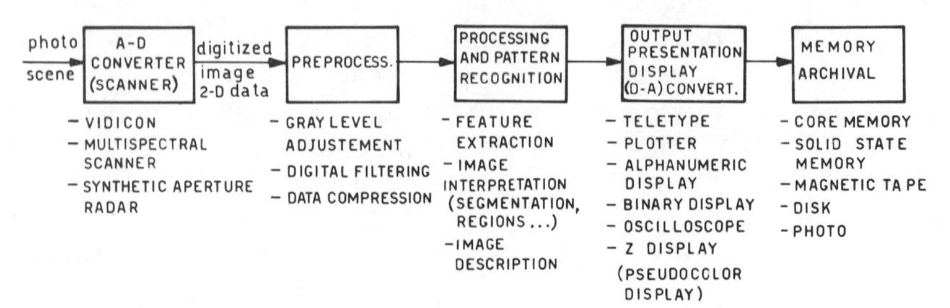

FIG. 8.23. Typical organization of a digital image processing system.

Finally the processed and presented 2-D data, eventually converted in analog form, can be stored in digital form or optical form (photo).

Figure 8.24 shows an example of implementation of a digital image processing system (computer, hardware processor) performing the above operations, with the required input–output terminals.

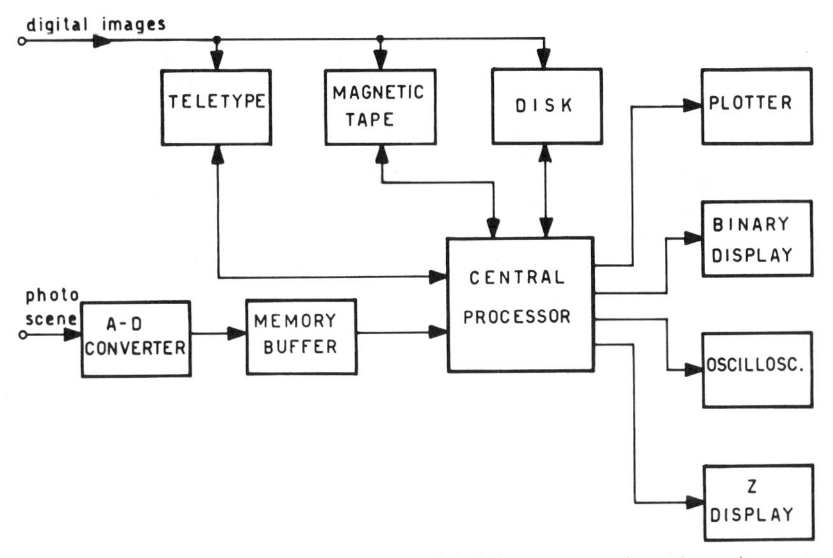

FIG. 8.24. Example of implementation of a digital image processing system (computer, hardware processor).

From the filtering point of view different kinds of 2-D digital filtering can be performed by means of FIR filters and IIR filters; for example low-pass filtering, band-pass filtering, high-pass filtering, matched filtering, equalization, inverse filtering and restoration are all possible digital operations (see Section VIIIB).

It is important to point out that the phase characteristics of the digital operations involved in a specific application are very significant indeed. Phase linearity response is almost always requested.

A practical demonstration of the importance of phase information in the image Fourier transform $X(j\omega_1, j\omega_2)$ is shown in Fig. 8.25. In this figure the inverse transform of the phase spectrum only is performed, the amplitude of each frequency component of the image transform is in fact ignored by putting $|X(j\omega_1, j\omega_2)| = 1$, and the result is illustrated in Fig. 8.25(b) in comparison with the original image in Fig. 8.25(a).[67]

For the above reasons FIR 2-D digital filters can be directly used because many design methods are available assuring linear phase characteristics. IIR 2-D digital filters must be applied with some care not only

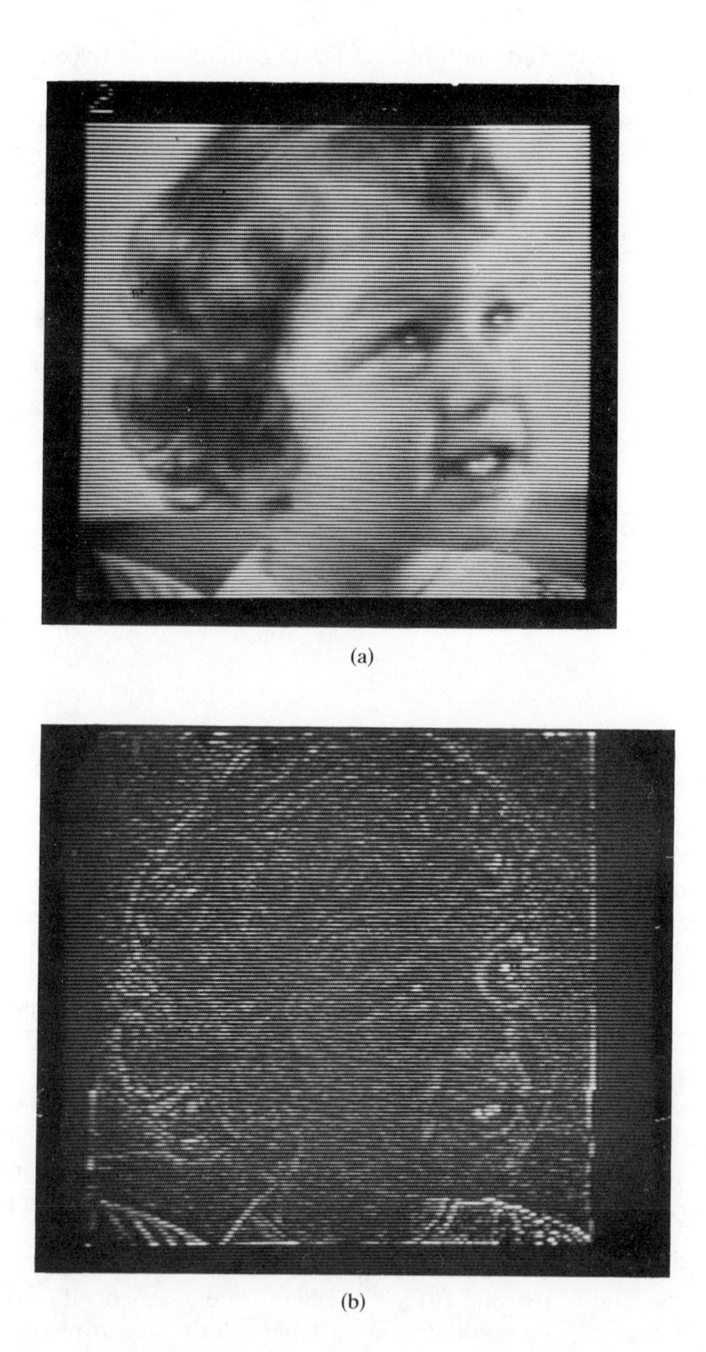

(a)

(b)

FIG. 8.25. Practical demonstration of the importance of phase information: (a) original image; (b) inverse transform of the phase spectrum only.

because of the stability problems but also because a phase linearity is not always possible.

Another general observation that can be made concerns the processing speed in the applications of 2-D digital filters to image processing. As it has been observed already this is quite limited and therefore applications are limited to static images and not to real time applications. Fast special purpose hardware digital parallel processors can however be designed for simple digital filtering in real time.

Let us now show a few typical digital processing examples on test-simulation images regarding band-pass filtering, restoration and data compression, while specific digital filtering applications to two important areas, biomedicine and airphoto-images from satellites, are described in Section VIIIG and Appendix 6, respectively.

A 2-D FIR band-pass digital filter with circular symmetry has been designed for pre-processing in terms of enhancing the 2-D data. The design method is based on the window technique (see Section IIE) where a Cappellini window extended to the 2-D case[34,55] with $m=6$ has been employed. The 2-D frequency response has an expression as given by eq. (2.32).[55] Figure 8.26(a)(b) shows the result of applying this filter to a simple test image (64×64 points) presented on an oscilloscope in perspective form (a HP-2100A minicomputer was used).[55]

Another example is included which is a 2-D IIR recursive band-pass digital filter. This filter designed by Garibotto at CSELT, Italy, approximates in amplitude a band-pass function in accordance with the selective frequency-response of the human eye, as shown in Fig. 8.27(a).[67] The phase characteristic is equalized. The digital filter consists of three second order sections connected in cascade followed by all-pass phase equalizers. The filter designed by using a nonlinear optimization technique has the transfer function given below

$$H(z_1, z_2)=G\prod_{i=1}^{3}\frac{\sum\limits_{n_1=0}^{2}\sum\limits_{n_2=0}^{2} a(n_1, n_2)z_1^{-n_1}z_2^{-n_2}}{\sum\limits_{n_1=0}^{2}\sum\limits_{n_2=0}^{2} b(n_1, n_2)z_1^{-n_1}z_2^{-n_2}}$$

with the constraints: $a(n_1, n_2)=a(n_2, n_1)$, $b(n_1, n_2)=b(n_2, n_1)$, $a(0,0)=b(0,0)=1$. G is the gain. The phase is equalized for each one of the three second order sections by means of a filter having the following expression

$$H_e(z_1, z_2)=\frac{z_1^{-2}z_2^{-2}D(z_1^{-1}, z_2^{-1})}{D(z_1, z_2)}$$

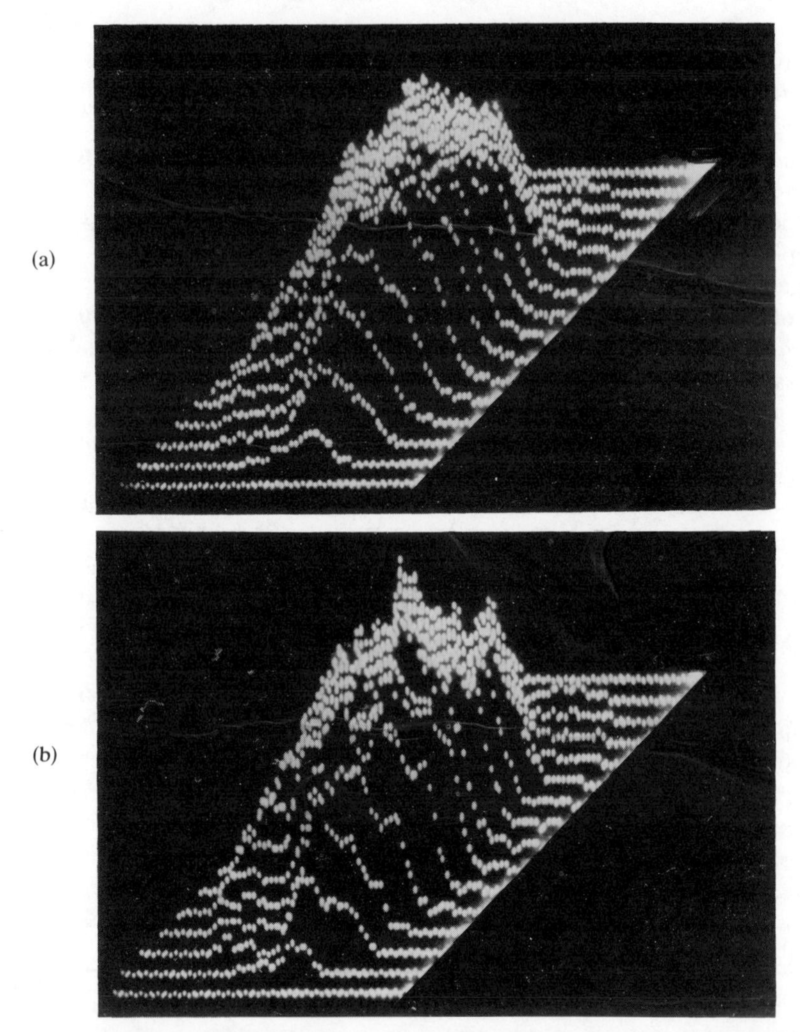

FIG. 8.26. Example of 2-D FIR band-pass filtering with window technique a test image with perspective presentation: (a) original image; (b) processed image.

with

$$D(z_1, z_2) = 1 + \sum_{n_1=0}^{2} \sum_{n_2=0}^{2} d(n_1, n_2) z_1^{-n_1} z_2^{-n_2}$$

The amplitude response is shown in Fig. 8.27(b) while the phase response is shown in Fig. 8.27(c). The result of application of the filter to the image of Fig. 8.28(a) is illustrated in Fig. 8.28(b): the processed image is filtered by the three second order sections of digital filters connected in cascade and is shifted backwards along rows and columns in order to compensate for the overall group delay (all-pass phase equalizers). It can be seen that linear phase preserves the information content in the original image. Both the original and processed images are originally represented with 8 bits samples but reduced to 64 equi-probable levels via histogram equalization to obtain enhanced line printer plots which are then photographed.

A third example is a 2-D digital filter designed by Mottola at the University of Trieste to perform image restoration.[51,54] A degrade-defocused image, 256×256 points, shown in Fig. 8.29(a) is considered as a test of processing. The restoration filter is of the Wiener type (see relation (8.24)) and has a transfer function

$$H_{WM}(j\omega_1, j\omega_2) = \frac{2k \cdot |H(j\omega_1, j\omega_2)|^2}{H(j\omega_1, j\omega_2)[|H(j\omega_1, j\omega_2)|^2 + k^2]}$$

The actual used filter has a transfer function

$$H_t(j\omega_1, j\omega_2) = H_{WM}(j\omega_1, j\omega_2) \cdot H_c(j\omega_1, j\omega_2) \tag{8.37}$$

where $H_c(j\omega_1, j\omega_2)$ is a frequency window of cosine type to improve the performance at high spatial frequencies.

The test defocused image is transformed through 2-D FFT and the power spectrum is displayed to extract the zeros of the distortion function to define the filter; then the transformed image is multiplied by $H_t(j\omega_1, j\omega_2)$. An example of restored image is shown in Fig. 8.29(b).

The experience and tests developed by Mottola have shown the relative facility of determining the impulse response $h(n_1, n_2)$ and that by means of digital filters of the type (8.37) it is possible to restore defocusing and displacements of image parts up to 7% of the linear size of the image, the dynamic range k of the restoration filter varying from 40 to 20 dB, depending on the characteristics of the original image and on the degradation level (a HP-21MX minicomputer was used).

Finally let us underline the interest of 2-D digital filtering for data compression. This last problem is becoming of primary importance in many applications due to the large amount of data contained in some images (i.e. earth resource images) and to the great quantity of images to be processed, stored or archived. We can observe:

(i) 2-D digital filtering (low-pass, band-pass) represents by itself a sort of data compression, because a limited part of the space frequency spectrum is extracted, requiring a lower number of data to be represented;
(ii) 2-D low-pass digital filtering is in general useful as a preprocessing, before the application of particular data compression algorithms,

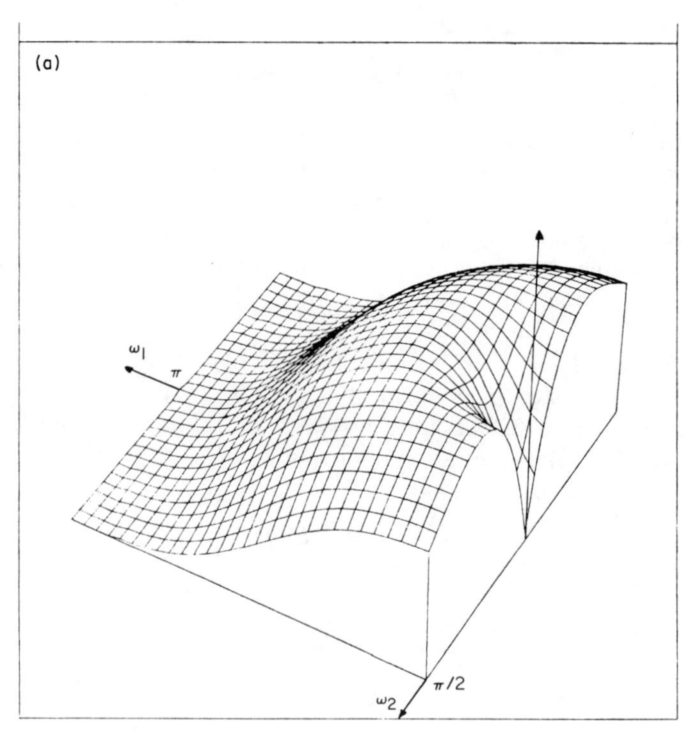
(a)
ω1
π
π/2
ω2

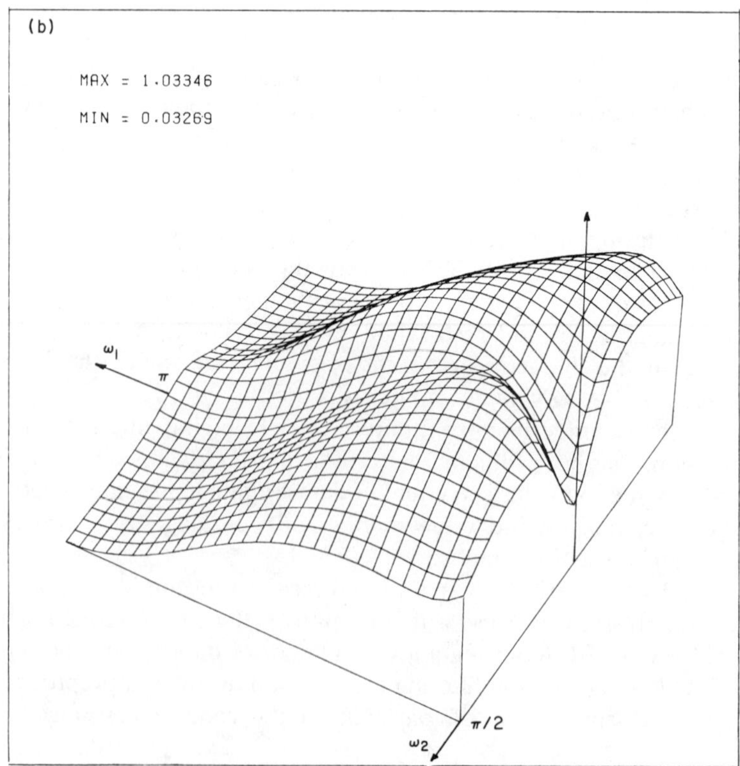
(b)
MAX = 1.03346
MIN = 0.03269
ω1
π
π/2
ω2

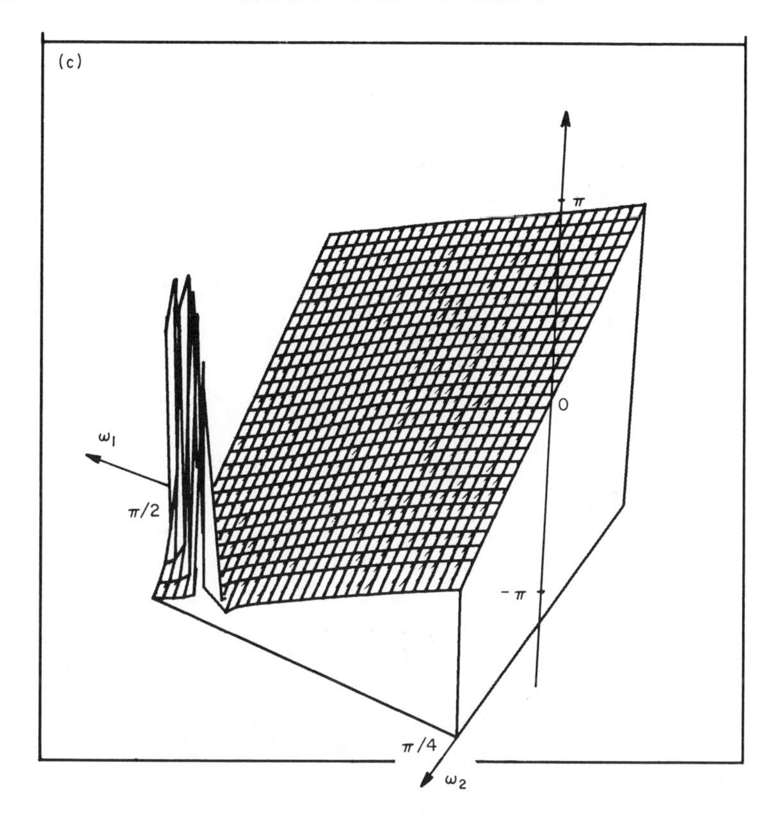

FIG. 8.27. Example of 2-D IIR recursive digital filtering (from Garibotto): (a) required response; (b) amplitude response of the used filter; (c) phase response of the used filter.

because the *smoothed data*[10] can be more efficiently compressed by specific compression algorithms.

Many 2-D data compression techniques have been developed for static images and TV images, with prediction-interpolation, differential pulse code modulation, coding, etc.[45,63]† Some data compression transformations (Fourier, Hadamard, Walsh, Haar, Slant, Karhunen, Loeve) are moreover very near to actual digital filtering operations.[48,74,75] Indeed with this type of compression technique the 2-D transform is performed on the image and then some data reduction is performed in the transformed

† An interesting approach to reduce redundancy in TV images, by means of inter-intra frame coding, which exploits movement compensation was developed by Rocca at Politecnico di Milano: the coding consists essentially in determining for each picture element (pixel) the prediction mode (spatial, temporal) and transmitting, when necessary, the quantized differences and the prediction mode changes.[11]

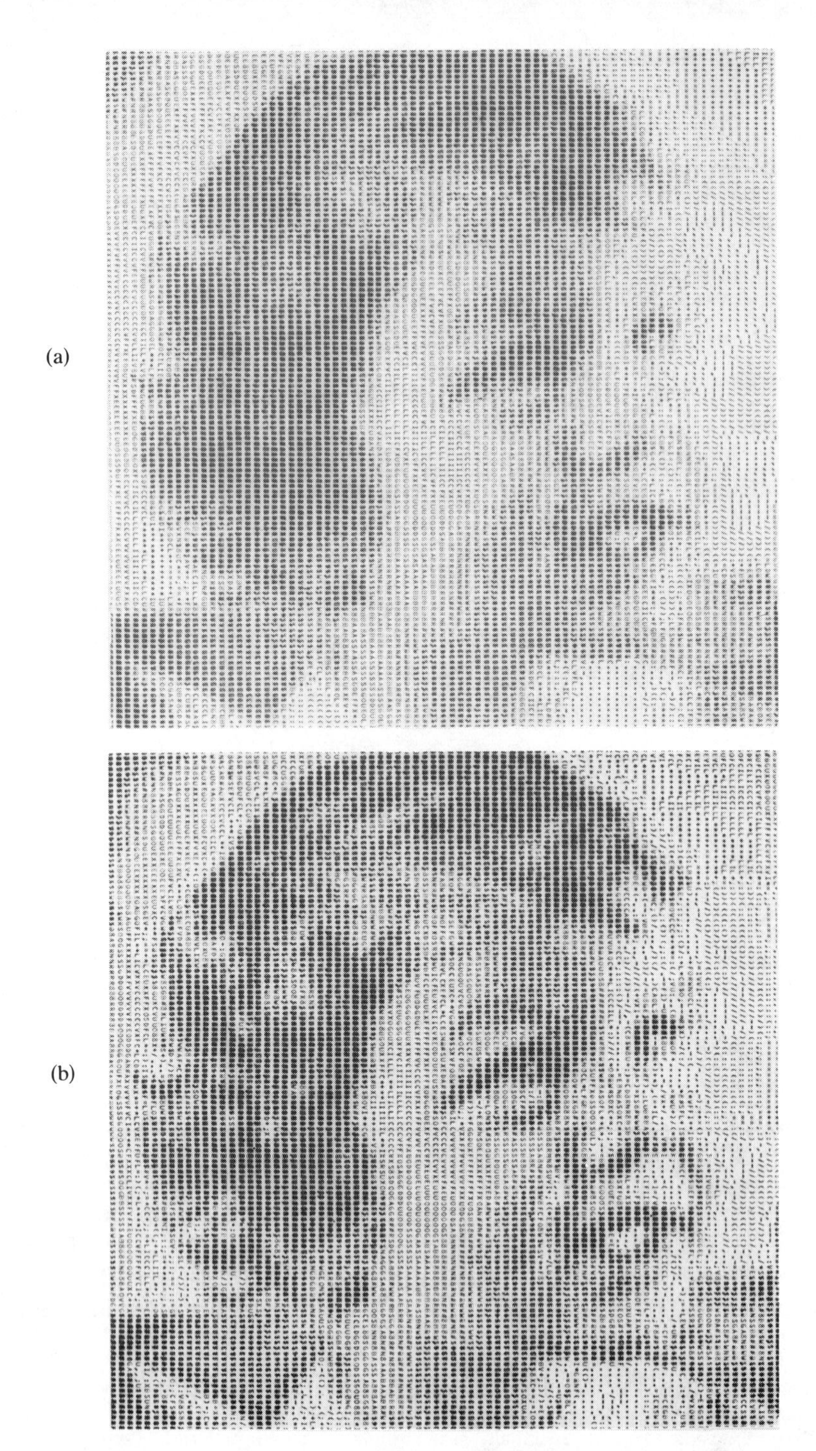

FIG. 8.28. Example of image processing by means of the filter in Fig. 8.27: (a) original image; (b) processed image.

(a)

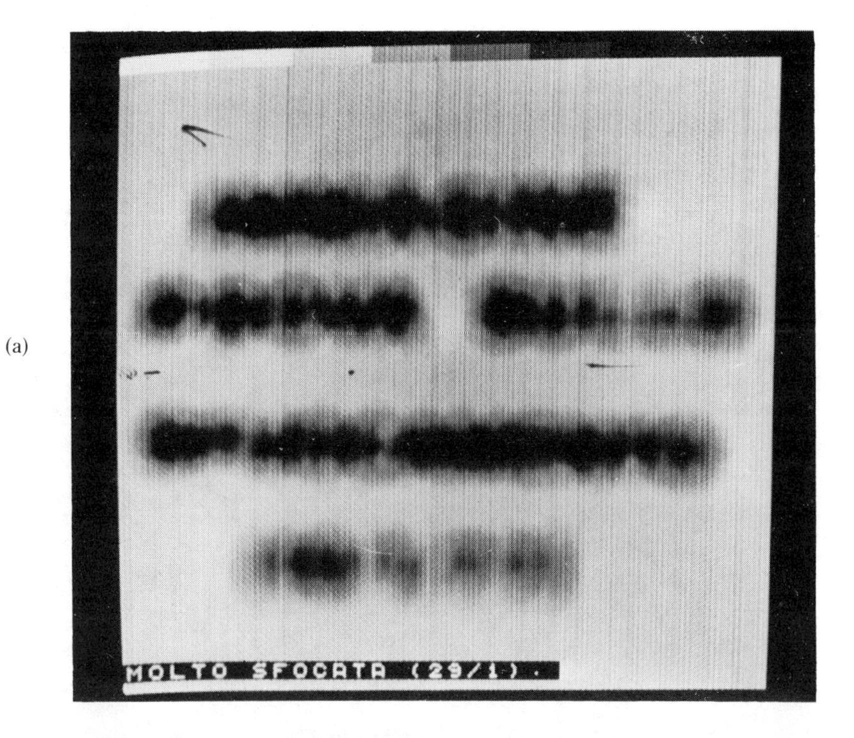

(b)

FIG. 8.29. Example of image restoration by means of a 2-D FIR digital filter (from Mottola): (a) original image; (b) processed image.

domain, with thresholding or variable-length encoding of groups of transformed data. An example of 2-D data compression on a test image, performed through 2-D fast Walsh transform (FWT) and variable length encoding of small squares (8×8) of the transformed image is shown in Fig. 8.30.[36,37,40]

FIG. 8.30. Example of 2-D data compression by using the 2-D FWT; on left original image, and two reconstructed images (L_m = mean word-length).

VIIIE Digital communications

In a communication system a message is to be transmitted from one place to another through different physical communication channels (lines, cables, radio-links, atmosphere laser links, optical fibres, etc.).

Often many messages of the same type or different type have to be sent in parallel to utilize the communication medium in a more efficient way (cable, satellite, etc.). *Multiplex communication* systems are used for this purpose. Two important types of multiplex systems are represented by frequency division-multiplex (FDM) and time division-multiplex (TDM):

in the first the single message signals are set in adjacent frequency bands in the frequency domain, while in the second the signal messages are organized in subsequent time intervals, in general by sending one sample of each message after the other in a cycle or *frame* and sending one frame after the other. By representing each sample in the TDM system in digital form (word of a given number of bits) a pulse-code-modulation (PCM) multiplex is obtained (Fig. 8.31).

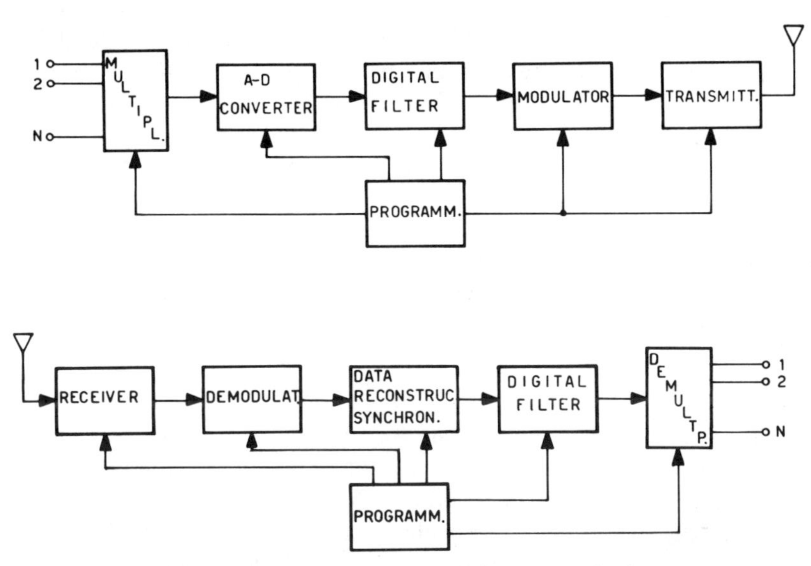

FIG. 8.31. General block diagram of a digital communication system.

Communication systems, in which digital (binary) data corresponding to a single channel or multiple channels are transmitted from one place to another through different communication channels, have expanded recently in an exponential manner. This expansion is due of course to the factors already mentioned in the Introduction (Section VIIIA). Essentially these communication systems can assure higher efficiency than most of the classical analogue systems. A decreasing cost of implementation due to the technological evolution in solid state digital integrated circuits makes such communication systems all the more attractive.

The general block diagram of a digital communication system, as results also in the case of PCM in Fig. 8.31, corresponds substantially to that already described in Fig. 8.2 (see Section VIIIB), where it also shows the application of 1-D and 2-D digital filtering both in the transmission part before the modulator and in the receiving terminal after the demodulator or data reconstructor-synchronizer. In Section VIIIB the useful operations which digital filters can carry out in this digital communication system have been also pointed out.

A special problem in digital communication systems, as indeed in any communication system, is represented by the noise and disturbances, especially at the receiving terminal, which alter in a multiplicative and additive way the useful information data. The relations (8.1) and (8.2) represent this situation for 1-D and 2-D signals. There are however, as described in Section VIIIB, several filtering operations, such as matched filtering, inverse filtering, restoration and equalization, that can solve, at least in part, this crucial problem. Further, as also pointed out in Section VIIIB, digital filtering can be more efficient in general than analogue filtering for this purpose because of its better performance, flexibility and adaptivity.

In the following sub-sections some aspects of other important digital communication systems as digital telephony, digital telemetry and data transmission are considered giving examples of digital filtering application.†

VIIIE1 Digital telephony

Telephone communication systems are continuously expanding in the world to satisfy the human communication problems: telephone systems represent one of the most important branches of communication.

Because of the need to transmit many speech channels simultaneously from one place to another, multiplex communication systems as mentioned earlier are required; digital multiplex systems of the PCM type are continuously gaining importance in comparison with the more classical FDM systems.[44]

Digital filtering can be applied in the transmission path and the reception path.

In normal telephone systems the channel signals are band-limited to 4 KHz approximately and hence in their digital representation a sampling frequency of at least 8 KHz is required. For this relatively low sampling frequency, digital filtering can be performed on multiple speech channels and hence the general criteria described in Section VIIIB are applicable. In particular a single fast arithmetical unit, as shown in Fig. 8.3, can be used with suitable memory organization and control to perform the required operation.

† An interesting case of digital filtering technique for digital communications is represented by the *simulation* of the communication channel (digital channel simulation). For this purpose a *time varying digital filter* as described for speech production (see Section VIIIC) is in general used: by varying the coefficients of the digital filter the dynamic evolution of the channel can be simulated.

In addition to the normal filtering operations of low-pass, high-pass, band-pass, etc., special filtering operations of matched filtering, equalization, inverse filtering and restoration can be performed on the communication system data in much the same way as already mentioned in a different context. These operations are very useful especially at the receiving terminal, where noisy signals are usually required to be processed.

A specific interesting application of digital filtering in telephone systems is represented by the implementation of high efficiency *tone receivers* for *signalling* and data transmission. In these systems a bank of band-pass digital filters, in general implemented with a single arithmetical unit, is used in the receiving terminal of the system to extract from the received signal the individual frequency components containing the transmitted information. This is shown in block diagram in Fig. 8.32.

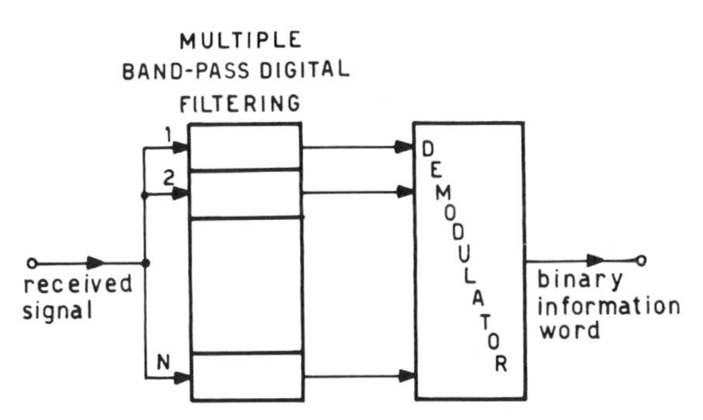

FIG. 8.32. Structure of a digital tone receiver.

A further interesting application of digital filters recently developed in this area is represented by the *interfacing system* between PCM multiplex and FDM multiplex, to make both systems compatible and easily connectable: indeed digital filters are the ideal coupling devices due to their capability of processing sampled data as in PCM multiplex and their *spectral relocation* within any particular frequency band.[64,65] For this purpose a digital phase splitting network for digital FDM applications was recently designed by Constantinides *et al.*:[49] the block diagram of a digital "A" channel bank FDM system for a single channel PCM input is shown in Fig. 8.33.

A crucial point in digital telephony expansion, which is of paramount importance as far as the digital filtering application is concerned, is represented by the replacement of *traditional electro-mechanical switching by digital electronic switching* in telephone switching centres. Indeed many PTT organizations and firms in the world are considering the development

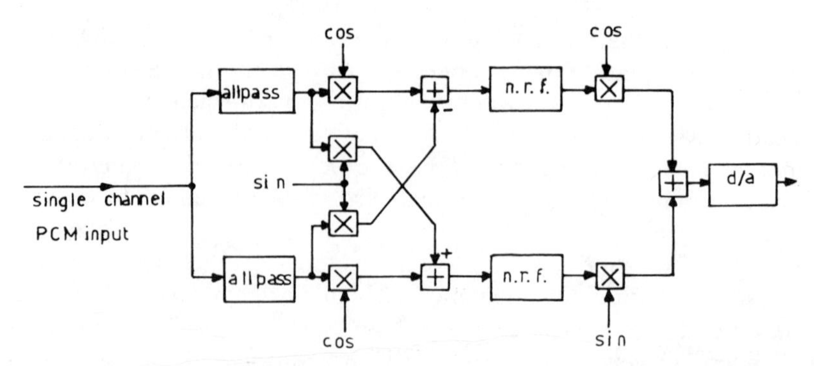

FIG. 8.33. Block diagram of a digital "A" channel bank FDM system for a single channel PCM input (n.r.f. = nonrecursive filter).

of prototypes for this kind of switching application, in the implementation of all digital integrated telephone systems.

A solution for the above problem has been proposed which satisfies the CCITT specification[12] for signal frequency components, shown in Fig. 8.34. It uses as illustrated in Fig. 8.35 an analogue prefilter followed, after quantization-encoding, by a two-section digital filter (the first section specifically designed to reject the quantization noise, the second section providing the narrow transition bandwidth and stopband rejection specified in Fig. 8.34).[43]

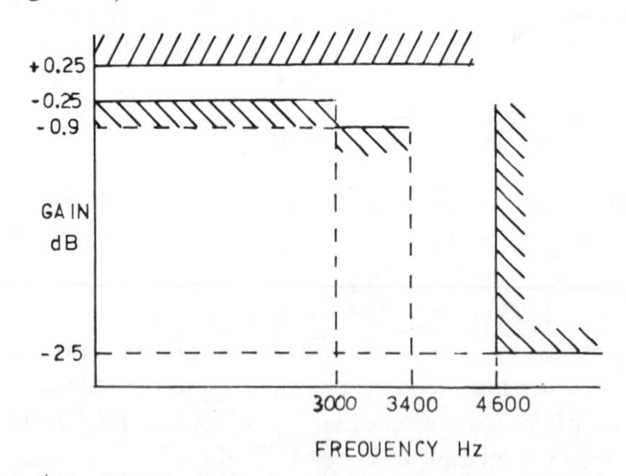

FIG. 8.34. CCITT specifications for signal frequency components.

VIIIE2 Digital telemetry

A *telemetry system* is a communication system in which many physical measurements in analogue or digital form, given by transducers or equip-

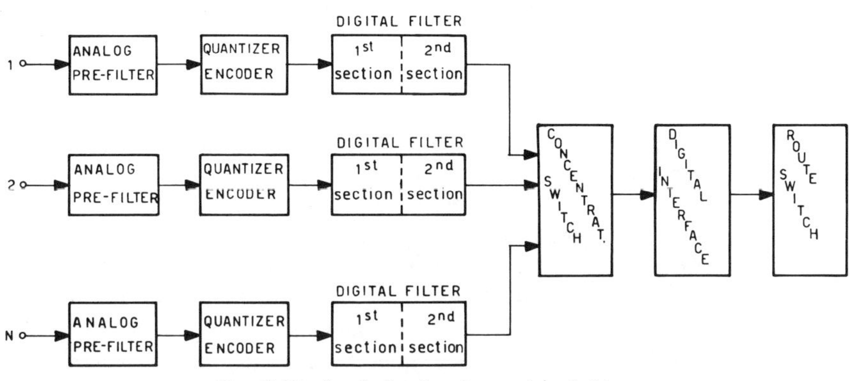

FIG. 8.35. A solution for electronic switching.

ments, are transmitted from one place to another through different communication channels.

As described already multiplex communication systems (such as FDM, TDM) can be used for this purpose.† Digital PCM systems present the advantages outlined earlier and indeed telemetry systems, especially space telemetry systems for telemetering of physical measurements collected by on-board satellite instrumentation, were the first to employ digital communication techniques in PCM.

A PCM digital telemetry system has substantially the same structure as that shown in Fig. 8.31, according also to the general criteria of Fig. 8.2.

However, due to the particularly difficult communication channels encountered in some telemetry problems, for example, the telemetering of a large amount of data by a long distance space vehicle, sophisticated digital processing and transmission techniques have to be designed and applied. In these advanced systems digital filtering plays a very important role and some examples are briefly described below.

In one system a particular class of encoded signals, the so-called M-ary orthogonal signals, have been employed.[91] By orthogonal signals $x_i(t)$ and $x_j(t)$ we mean signals satisfying the following conditions

$$\int_0^T x_i(t)x_j(t)\,\mathrm{d}t = \delta_{ij}, \qquad \delta_{ij}=0,\ i\neq j \qquad \delta_{ij}=1,\ i=j \tag{8.38}$$

One particular example is the class of signals, each one of which corresponds to a sine-wave signal with a particular frequency value, called frequency-shift-keyed (FSK) signals. Such signals can be denoted as

$$x_i(t) = A\cos\omega_i t \qquad 0\leq t\leq T \qquad i=1,\ldots,M \tag{8.39}$$

† In particular, for many signal telemetry systems, standard IRIG frequency bands and data formats have been defined for FDM and TDM systems.

with T time interval of definition and frequency $f_i = \omega_i/2\pi$ and

$$f_{i+1} - f_i = \frac{1}{T} \tag{8.40}$$

These signals are in practice equiamplitude sinusoidal pulses, of duration T, and of progressively increasing carrier frequency, with the spacing between two adjacent frequencies just $1/T$. The binary-M-ary conversion is indeed very easy: if, for example, we have binary words of $RT = 4$ bits, the binary-M-ary converter or encoder, based on the particular 4-bit sequence read, sends out one of $M = 2^{RT} = 16$ possible frequencies. As the number M increases, increasing the transmission time T, the bandwidth required for the transmission of the M possible signals increases too and can be expressed by[91]

$$B = \frac{M}{T} = \frac{2^{RT}}{T} \tag{8.41}$$

Another practical example of M-ary signals is represented by orthogonal pulse sequences: there are 2^M possible sequences utilizing the on–off pulses of duration T, each single pulse being a rectangular pulse of duration $T/2M$. Such orthogonal pulse sequences, requiring the same bandwidth given by (8.41) have the interesting characteristic of being easily generated by using shift-register techniques.

An interesting aspect is that the above indicated M orthogonal signals are examples of *optimum encoded signals*, in the sense that the probability of error in reception goes to zero as T (and hence $M = 2^{RT}$) goes to infinity.

Another very interesting aspect, regarding digital filtering, is the *optimum receiver* for the M orthogonal signals with additive band-limited white Gaussian noise (see also relations (8.1) and (8.7)) which is represented by a bank of M parallel matched filters (see Section VIIIB2) followed by a sampling and decision circuit (sampling is done every T s). As shown in Fig. 8.36(a) the M filters are each matched to one of the $x_i(t)$ signal, $i = 1, \ldots, M$.

The all-digital implementation of this receiver, as shown in Fig. 8.36(b), is quite direct and offers all the advantages outlined in previous sections and in particular the possibility of implementing the M digital matched filters by the multiple filtering techniques of Section VIIIB1 by using a single fast arithmetic unit. The last operation of the receiver of sampling and decision involves in the digital implementation simply a digital comparison of some samples which can be done very simply and extremely accurately.

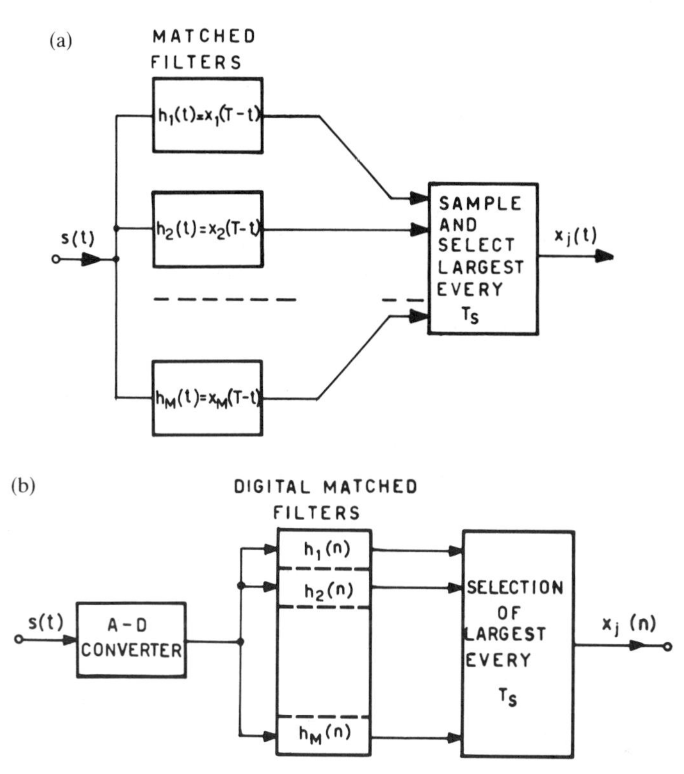

FIG. 8.36. (a). Receiver for orthogonal signals; (b) all digital implementation of the receiver.

Practical telemetry systems of the above type have been actually realized.[79]

Other digital telemetry systems have been designed, tested and applied, at first for space applications and after also for ground applications, in which the two fundamental coding transformations of Information Theory, *source coding* and *channel coding*, were used practically. By means of the first transformation, called in general *data compression*, a reduction of the amount of data to be transmitted (redundancy reduction, variable-length encoding, parameter extraction, etc.) is performed: if the remaining important data are reorganized at constant time intervals, a *bandwidth compression* is also achieved.[32]† By means of the second transformation, called in general *error control coding*, a protection of the data going into the communication channel is performed by adding suitable control data (a given number of bits in an information word or control bits interleaved

† Some examples of this operation have also been considered for digital speech processing (in particular relating to vocoder systems).

with blocks of information bits), which are utilized at the receiving terminal to detect (*error detection*) or correct (*error correction*) the errors introduced by the communication channel or other disturbances and interferences.[6,87]

A general block diagram of a digital telemetry system using these two transformations is shown in Fig. 8.37. In this figure digital filtering is also inserted both in the transmission and reception paths because digital filters

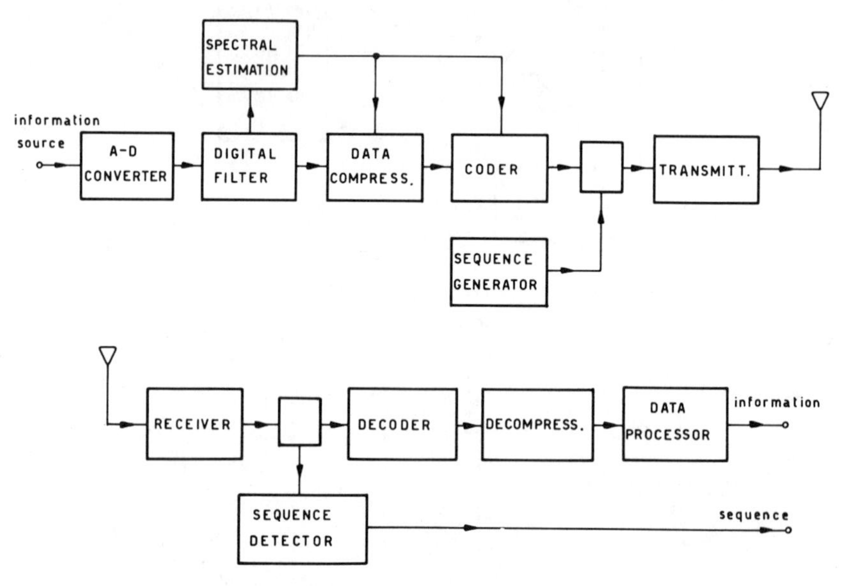

FIG. 8.37. General block diagram of a digital telemetry system using digital filtering, data compression and error control coding.

are very useful in systems of this type, in that they increase the efficiency (particularly that of data compression) of the overall system. Digital filtering can in fact perform the following interesting operations:[24]

(i) define the bandwidth of each telemetry information source thereby controlling the amount of data (a bandwidth reduction corresponds to data reduction). This bandwidth can be varied in an adaptive way so as to ensure high efficiency of the data compressors following the filter;

(ii) perform short-time spectral estimations, through one of the several techniques described in Section VIIIB1† and thereby adapting at medium or long time intervals the sampling frequency (adaptive sampling) to a more appropriate value (in general the sampling frequency is kept at the minimum possible value consistent with the signal frequency content in order to reduce the amount of data sampled).

† Compressed short-time spectral estimation as described in Sections VIIIB1 and VIIIC2 can be used as compressed representation of the signal.

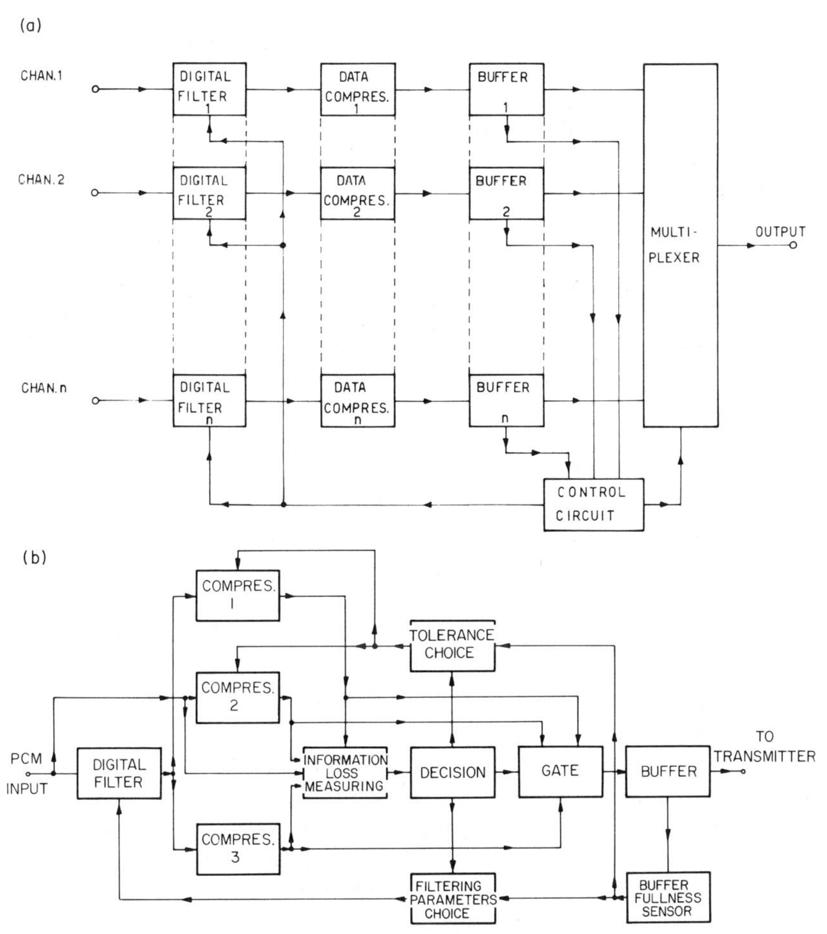

FIG. 8.38. Two examples of implementation of adaptive data compression by using digital filtering.

Figure 8.38 shows two examples of implementation of operation (i). In Fig. 8.38(a) many telemetry signals are digitally prefiltered (in general low-pass filtering is applied) in such a way that the following data compressors, which may use for example prediction-interpolation algorithms, can operate in the most suitable way. The correct amount of data is sent to the output buffers, where the remaining important data are reorganized at constant time intervals to achieve bandwidth compression: the bandwidth of the digital filters is varied (in general at constant time intervals) by a *control circuit* which continuously measures the state of fullness of the output buffers, avoiding the undesirable situations of *overflow* and *underflow*. In Fig. 8.38(b) a more sophisticated adaptive data compressor is shown, for a single telemetry signal in which three compressors are working in parallel: one (3) working as above, another (2) with a control of tolerance parameters (such as amplitude tolerance in prediction-interpolation algorithms), the last one (1) with both the preceding controls (bandwidth and amplitude tolerance). A decision circuit selects, in accordance with measurements of an information loss circuit, the most efficient data compressors among these three.

Finally in some digital telemetering systems, digital filters are used as *monitoring devices*, to detect the correct performance of important operative systems, plants and industrial processes. An interesting example of this type, using a digital processor to detect the presence of a second harmonic after a fault occurs on power lines, is described in the Appendix 5. In this Appendix we also include an attractive power system protection technique based on DFT and FFT interpretation of digital filtering where faults on power lines can be recognized to within one quarter of the period of the fundamental under certain conditions and in any case not more than one half of the fundamental period.

VIIIE3 Data transmission

A *data transmission* system is usually a digital communication system, in which low-rate or medium-rate data corresponding to written messages (letters, numbers) or sampled data (also scanned digital image) are exchanged between two terminals.[66] A typical data transmission system is that connecting two computers or one computer with a peripheral terminal set at a given distance. One example of this last structure is the so-called *time-sharing system*, in which a unique computer or computer centre is connected to many peripheral terminals, in different places, in a suitably organized time division organization. Conversely we can have the connection of different data acquisition-processing terminals to a single central computer: here the normal flow of information is from terminal to computer, where all the received data are processed, used in some way and stored. (Some examples of this structure can be an interconnected bank system or a biomedical data collection and control centre (see Section VIIIG).)

Typical transmission media can be represented by telephone channels (lines, cables), but also radiolinks and more recently optical beams (by laser in atmosphere or by LED or laser in optical fibres) are used for carrying data at different rates.

Standard bit rates, especially on telephone lines and cables, are: 600, 1200, 2400, 4800, 9600 bits/s.

Let us consider, as a typical example, the data transmission on telephone channels. The main problem here is represented by two effects (along the lines of relation (8.1)):

(i) noise presence in the received signal, which can be approximately schematized as additive noise;

(ii) spectral distortion of the received signal.

For the noise effect (i) a digital matched filter can be used (see Section VIIIB2) while for the distortion (ii) a digital equalizing filter can be applied (see Section VIIIB3). Indeed both digital filtering operations can be used together to solve the above problem.

Often, especially when additive noise is of low value, a digital equalizer is able to recover the useful information data from the received signals. In practice, however, due to the time evolution of the characteristics of distorting transmission medium, a variable equalization must be performed on received signals, that is an adaptive equalizer (as described in Section VIIIB3) is required.

An example of a digital adaptive equalizer (FIR adaptive digital filter) is shown in Fig. 8.39: the coefficients are here automatically changed by means of a continuous or step-by-step process, that minimizes r.m.s. difference between the received signal and the desired signal.

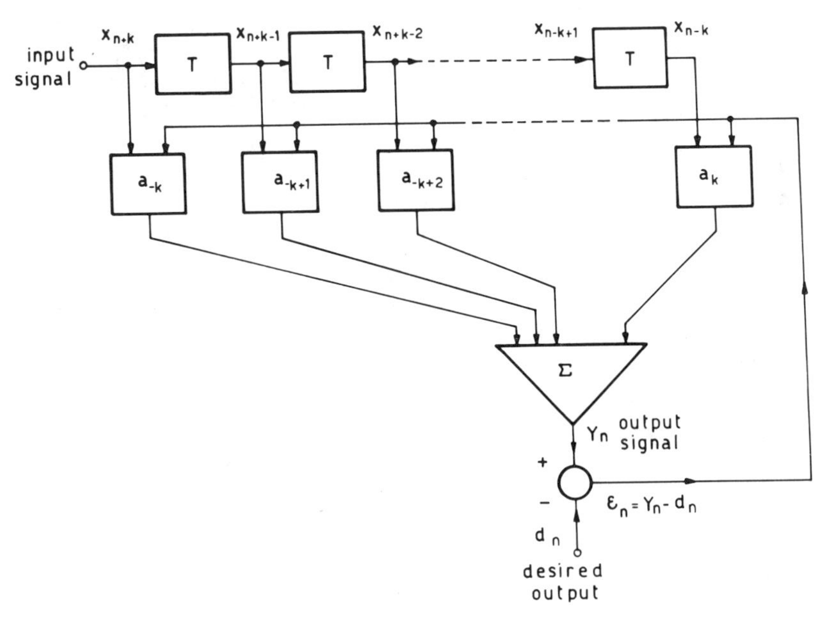

FIG. 8.39. Block diagram of a digital adaptive equalizer.

As far as the hardware implementation of this particular kind of system operation is concerned, in addition to large scale integration digital circuits (LSI) we have the new development of charge-coupled-devices (CCD) which are of immense practical interest (Section VIIG). Indeed CCD FIR filters are very attractive and easily implementable in a very compact form, by using the CCD as analogue sampled data shift register, employing either on-chip or external weighting coefficients. An actual approach employs these CCD registers with different sets of weights available and controlled by a microprocessor so as to obtain an adaptive transversal filter and consequently an adaptive equalizer. The block diagram of a system of this type is shown in Fig. 8.40.[14,56]

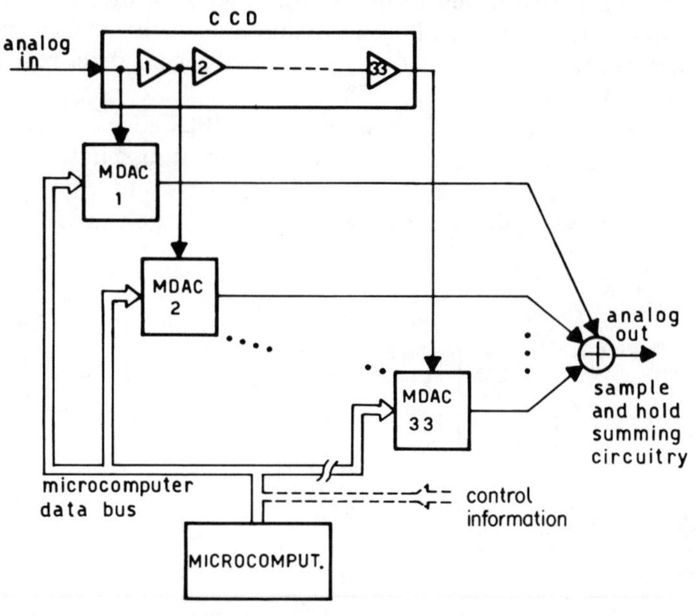

FIG. 8.40. Block diagram of a microprocessor controlled adaptive filter (MDAC = Multiplying A–D converter).

VIIIF Applications in radar–sonar systems

Radar systems are of great importance not only for military applications, but also for civil applications.[50,58,93] A typical example is the radar for air traffic control (ATC).

The same consideration is true for sonar systems for ship navigation and security.

The principal characteristics of a classic radar system are sketched in Fig. 8.41(a): it includes a pulse generator, a modulator and an oscillator-amplifier in transmission, while in reception, after the common antenna,

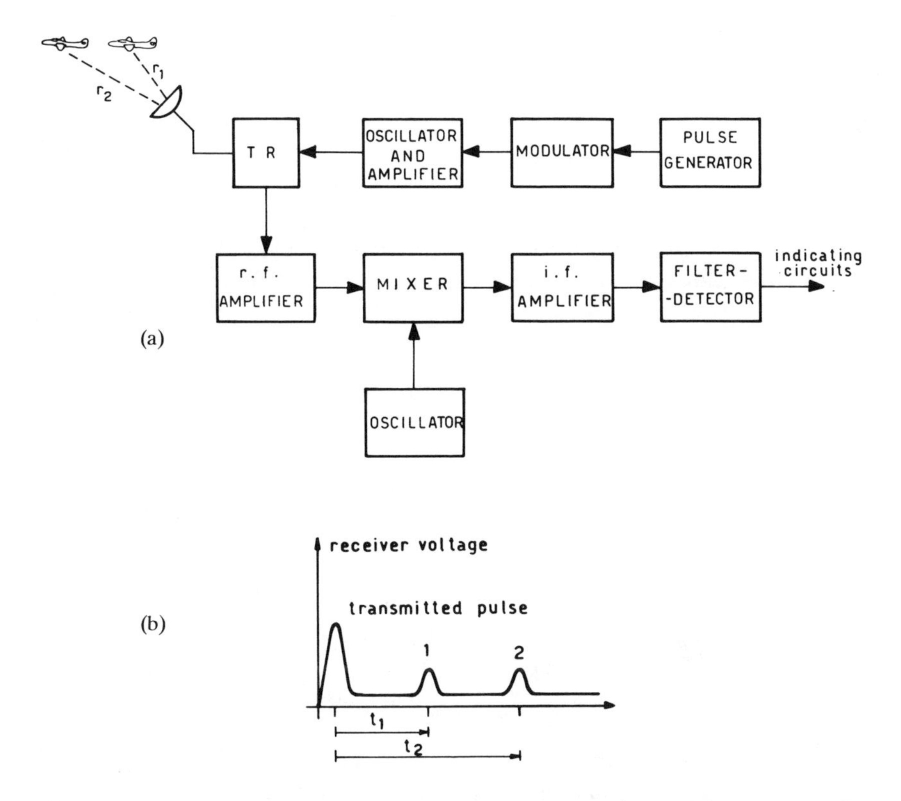

FIG. 8.41. Principal characteristics of a classical radar system (a) and of signal presentation (b).

we have a radio frequency (r.f.) amplifier followed by a mixer (with a local oscillator), an intermediate frequency (i.f.) amplifier and a filter detector.[91]

The radar transmitter provides a periodic series of pulsed sine waves, normally generated at microwave frequencies to obtain a well-defined narrow beam of electromagnetic energy with antennas of reasonable size (for sonar systems acoustic oscillations are used to propagate in the water medium). Targets (fixed or moving) in the path of the beam reflect a part of the pulsed high-frequency energy. A portion of this reflected energy is picked up at the antenna and through an heterodyne circuit, i.f. amplifier and filter detector is sent to indicating circuits. In the most simple indication, in the situation of two targets as in Fig. 8.41(a), the presentation as in Fig. 8.41(b) is obtained, where two return pulses appear t_1 and t_2 s, respectively, after the transmitted pulse. The information carried by the received pulses is primarily *range or distance information.* Since electromagnetic waves travel at light speed $c = 3 \times 10^8$ m/s in air (actually vacuum), the time taken for the transmitted pulse to travel a distance of r m, be reflected, and return to the antenna is

$$t=\frac{2r}{c} \tag{8.42}$$

The distance of two targets considered in Fig. 8.41(a) can be thus obtained. If T is the pulse repetition interval, the maximum unambiguous range which can be obtained by the radar is given by

$$r_{\max}=\frac{cT}{2} \tag{8.43}$$

The *range resolution* Δr is a measure of how well two targets that are near each other can be resolved by the radar.

Other useful information, regarding the target location (azimuth and elevation angle, for example) is obtained from the known direction in which the antenna is pointing at the precise time of transmission and reception (this is one reason why narrow beams of energy are required).

The main problem of a typical radar described above (as in sonar†) is the *detection* of return pulses: substantially it corresponds to determining the presence or absence of a pulse in noise (pulse shape or other pulse properties are in this respect secondary).

For such radars systems, the bandwidth is usually chosen as the reciprocal of the pulse width to maximize received signal-to-noise ratio (SNR). The matched filter strategy thus applies in this case (see Section VIIIB2).

Digital processing of radar signals and in particular digital matched filtering can therefore be performed for the advantages already outlined (see in particular Section VIIIB2). The block diagram of a modern radar system is therefore as illustrated in Fig. 8.42. We see that in the receiver we

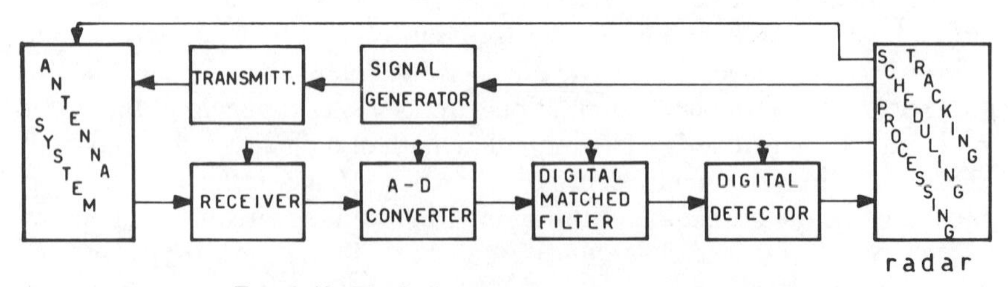

FIG. 8.42. Block diagram of a modern radar system.

† Sonar problems are in general easier to deal with than those of radar, because of the relatively low frequencies involved. Digital filtering in such cases is, for example, easier from the implementation point of view.

have an A–D converter to permit the processing of the return signals in a digital way:† the digital data are sent to the digital matched filter, followed by a digital detector. It is interesting to observe that subsequent digital processing and interpretation can be done by a *radar computer* (software or hardware implemented), which also carries out the supervision and control of the whole radar system (in particular with procedures of tracking-scheduling it controls and adjusts the performance of the antenna, receiver, A–D conversion, digital matched filtering and digital detection): the radar system is thus becoming a sophisticated *digital adaptive system.*

Let us now describe in more detail the design of two typical digital filters, useful for these modern radar sets: a digital matched filter for chirp signals and a digital MTI filter. Later some considerations about radar (sonar) maps are developed.

VIIIF1 Digital matched filter for chirp signals

The rectangular-envelope signals, having a frequency modulated (FM) carrier, have a great weight in modern radar-techniques.[23] The most popular signal is the linearly frequency-modulated signal, called *chirp signal,*

$$x(t) = \mathrm{rect}\left(\frac{t}{T}\right) \cos\left(\omega_0 t + \frac{\mu}{2} t^2\right) \tag{8.44}$$

rect (t/T) being an operator which is equal to 0 everywhere, except for $|t| \leq T/2$, where it is equal to 1. The chirp-matched-filter has a transfer function which is approximated[23] by an amplitude shape

$$|H(\omega)| = \begin{cases} 1 & \omega_1 = (\omega_0 - \mu T/2) \leq \omega_0 \leq (\omega_0 + \mu T/2) = \omega_2 \\ 0 & \text{for } \omega < \omega_1 \quad \text{and} \quad \omega > \omega_2 \end{cases} \tag{8.45}$$

and a phase shape:

$$\Phi(\omega) = \frac{(\omega - \omega_0)^2}{2\mu} = a(\omega - \omega_0)^2 \tag{8.46}$$

To perform digital matched filtering (see Section VIIIB2), a FIR digital filter can be used, symbolized by (see relation (1.95) where $N = 2M + 1$)

$$y(n) = \sum_{k=-M}^{M} a(k) s(n-k) \tag{8.47}$$

† Fast A–D converters have been developed for this task and are capable of rates more than 10 Msamples/s. It is not uncommon to have A–D conversion say at words of 8 bits with a rate of 30 MHz.[16,53]

where $s(n) = s(nT)$ are the samples of the input signal (the received signal $s(t)$ of relation (8.7)), $T = 1/f_s$ is the sampling period and the coefficients $a(k)$ are given by (window design method)

$$a(k) = h(k) \cdot w_L(k) \cdot T \tag{8.48}$$

In the relation (8.48) $w_L(k)$ are the Lanczos correction factors and $h(k)$ are the samples of the impulse response $h(t)$ of the matched filter, the frequency characteristics of which are as in (8.45) and (8.46). In the time domain the impulse response becomes

$$h(t) = \frac{1}{\pi} \int_{\omega_1}^{\omega_2} \cos[\omega t + \Phi(\omega)] \, d\omega \tag{8.49}$$

The samples of this impulse response of course will give the digital version of it. On using eq. (8.48) in conjunction with eq. (8.49) and that the frequency response is given by the expression

$$H(e^{j\omega}) = \sum_{k=-M}^{M} a(k) \, e^{-j\omega k T} \tag{8.50}$$

then for

$$a = 0.01\pi \, s^{-2}$$

$$\omega_0 = 0 \, \text{MHz}$$

$$m = 2$$

$$M = 50$$

$$f_s / f_c = 4$$

we obtain the amplitude response of Fig. 8.43(a) and the phase response of the digital matched filter shown in Fig. 8.43(b). This is in fact a parabolic response as expected from eq. (8.46). The parameter m above is the exponent of Lanczos window (see relation (2.26)), and f_c is the cutoff frequency.

The design example above does, of course, meet the required chirp signal conditions exactly† with the order of the filter of 101. If such complexity is unacceptable to the designer the stringency of the requirements can naturally be relaxed with the subsequent reduction of the order of the filter using the same design procedure. With modern technology, however, an order of filter of 100 is not considered unduly large.

† The phase constraint in particular would be difficult to meet by means of analogue filters.

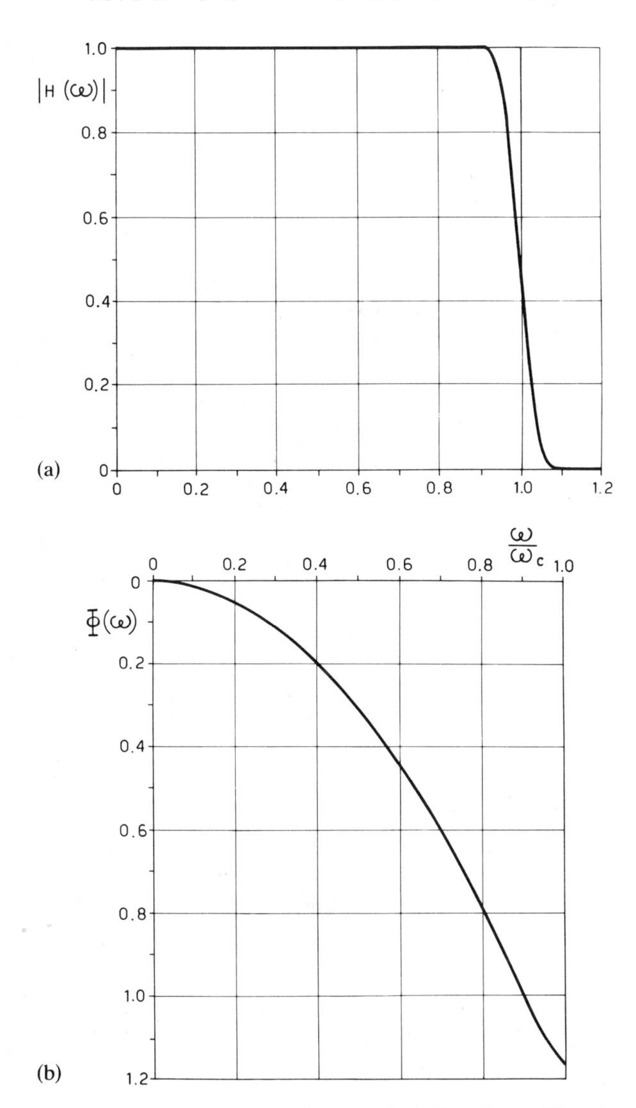

FIG. 8.43(a). Amplitude response of a digital matched filter for a chirp signal; (b) phase response of a digital matched filter for a chirp signal.

VIIIF2 Digital MTI filter

In many practical radar systems both range and velocity of a target are required. To obtain velocity information a moving target indicator (MTI) radar is required, where information is extracted about moving targets from the clutter background.

In MTI radars the Doppler shift in frequency is used to discern moving targets even when the echo signal from fixed targets is many orders of magnitude greater than the Doppler shift. Fixed target echoes or clutter are included within the same radar pulse-packet as the target information but the signals from fixed targets are not shifted in frequency and the vital process extracting the required information is that of filtering which in the digital form is of course a digital filter.[96]

Basically MTI radar can be mathematically described as follows. In the pulse Doppler radar, a small portion of carrier-wave power is diverted to the receiver to take the place of the local oscillator (see Figs 8.41 and 8.42). In addition, it also acts as the *coherent reference*, needed to detect the Doppler frequency shift (coherent means that the phase of the transmitted signal is preserved also in the reference signal).

We can indicate the oscillator signal as[96]

$$x_0(t) = X_0 \sin \omega_0 t \tag{8.51}$$

and the reference signal as

$$x_1(t) = X_1 \sin \omega_0 t \tag{8.52}$$

The Doppler shifted echo can be expressed as

$$s(t) = S \sin\left[2\pi(f_0 - f_d)t - \frac{2\omega_0 r_0}{c}\right] \tag{8.53}$$

where c is the velocity of propagation (relation (8.42)) and r_0 is the distance to target at time $t = 0$.

Now the reference signal and the target echo signal are mixed in the mixer stage of the receiver (Fig. 8.41). Only the difference frequency component from the mixer is of interest

$$s_d(t) = S_d \sin\left[\omega_d t - \frac{\omega_0 r_0}{c}\right] \tag{8.54}$$

By the measure of the difference frequency f_d, the target velocity can be obtained by using the Doppler equation

$$v = \frac{c f_d}{2 f_0} \tag{8.55}$$

where $f_0 = \omega_0 / 2\pi$ is the carrier frequency. Equation (8.55) refers to the case of a constant velocity target. For stationary targets we have $f_d = v = 0$.

Two cases are of interest:

(i) if the Doppler frequency is large in comparison with the reciprocal of the pulse width, the Doppler information can be extracted from the single pulse;

(ii) if the Doppler frequency is small, many pulses will be needed to extract the Doppler information.

The first case is, for example, typical for satellites and ballistic missiles, while the second one is typical for aircraft.

On a pulse position indicator (PPI) the display of Doppler information is usually done with a *delay-line canceller* as shown in Fig. 8.44. The delay has the task of eliminating the d.c. component of fixed targets and passing the a.c. component of moving targets.

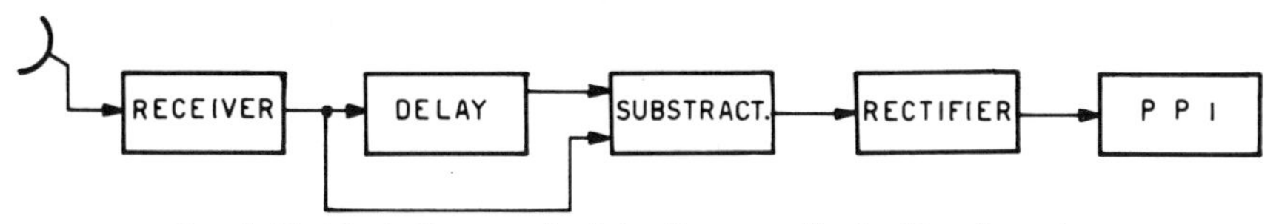

FIG. 8.44. Block diagram of a delay-line canceller for Doppler radar.

The equivalent circuit of the delay-line canceller is a transversal filter, already described in previous sections (see Figs. 1.15 and 8.10): the taps are spaced exactly as the radar interpulse period. The simple $h(t)$ (with $a_0 = 1$, $a_1 = -1$, all other $a_i = 0$ in Fig. 1.15) results in

$$h(t) = \delta(t) - \delta(t - T) \tag{8.56}$$

and the frequency response

$$H(j\omega) = 1 - e^{j\omega T} \tag{8.57}$$

More powerful filters are obtained by cascading two or more single delay filters.

Most MTI radars were implemented in the past by using analogue techniques, in particular based on *ultrasonic delay lines* for storing the phase information for one repetition period. However more efficient digital techniques, *digital MTI filters*, have recently been employed. In fact such. digital filters and techniques are essential in order to meet present day exacting constraints. Analogue implementations have the following important disadvantages:

(i) difficulty in controlling the delay;

(ii) critical gains must be controlled for the proper weighting coefficients;

(iii) good operation when well tuned, but deterioration in performance with time;

(iv) effective use in radar limited to two to three pulse cancellers, one to four interpulse periods.

The digital MTI filter (DMTI) indeed eliminates all the above disadvantages, with the additional advantage of inherent flexibility and adaptivity. The design of DMTI must however be carried out with care to consider all aspects of the digital implementation. In particular attention is to be given to the so-called *blind speeds*: DMTI not only eliminates the d.c. component caused by clutter, but also rejects any moving target whose Doppler frequency happens to be the same as the pulse repetition frequency (PRF) or its multiple. Blind speeds are those relative target velocities that result in zero MTI output (see Fig. 8.45). The solution to this effect is in general found by using *variable interpulse periods* (*staggered PRF*) FIR and IIR digital filters can be used for this application.†

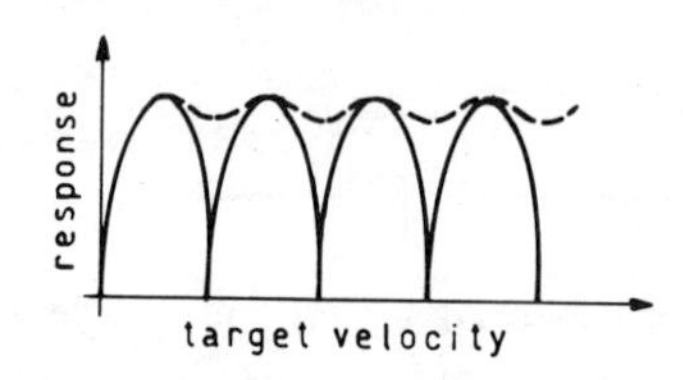

FIG. 8.45. Frequency response of a digital MTI filter without (continuous line) and with (dashed lines) variable interpulse periods (staggered PRF).

A particular case of velocity measurement is that related to low velocity moving targets (as ships, for example). In this case, as in the analysis of fluctuations of static targets with normal radar, of particular interest is the application of band-pass digital filter analysis (as described in Section VIIIB1) to extract in a definite way the velocity information.

In Fig. 8.46 we show a real radar signal (courtesy of SMA of Florence), regarding a fluctuating target and the resulting band-pass analysis using a software bank of FIR digital filters designed with the window technique (Cappellini window, see Table 2.2).

For the purposes of comparison we have included the alternative analysis based on a 512 point FFT: it is apparent that the band-pass analysis processing (which can also be done with smaller bandwidth than that applied here) can give very useful information in difficult cases like the one considered here.

VIIIF3 Digital processing of radar (sonar) maps

An important area of application now emerging is concerned with the use of 2-D digital filters for radar, and indeed sonar, map processing. The following problems are currently under examination:

(i) 2-D noise reduction of the kind of noise which is located at specific areas in the 2-D frequency plane;

† FIR filters are often preferred to avoid long transient phenomena.

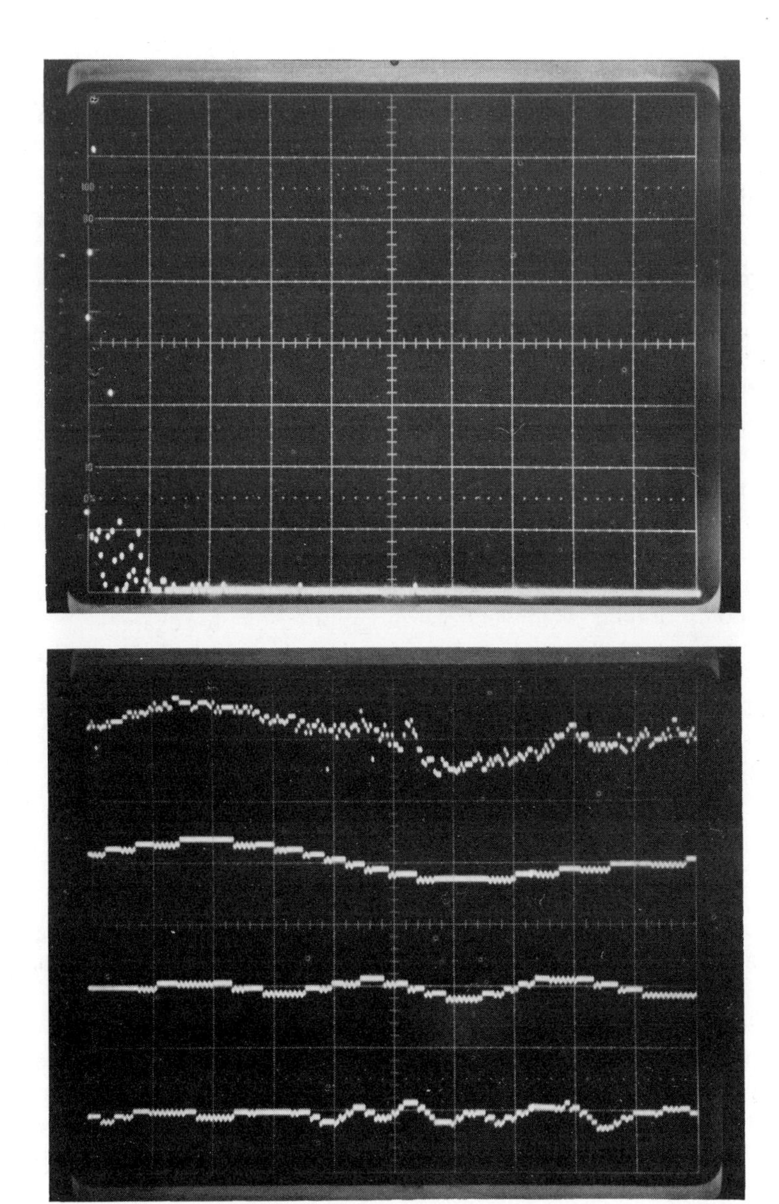

FIG. 8.46. Example of digital filtering a real radar signal for detecting low frequency fluctuations (courtesy of SMA of Florence), in comparison also with FFT (at top, 5 Hz per division). Starting from the bottom we have: 4.5–15.5 Hz band, 2.5–4.5 band, 0.5–2.5 band and the input signal.

(ii) edge enhancement and extraction of boundaries for military and civil applications and also for pattern recognition purposes;
(iii) equalization of the smearing effects of finite aperture of the radar antenna where in general an inverse filtering process is required in two dimensions (see Section VIIIB4).

The very nature of the above problems makes them eminently suitable for digital processing in 2-D.

VIIIG Applications in biomedicine

Biomedicine is an application area, in which the use of digital signal processing techniques has had great impact.

There are several reasons for this, some of a general nature (advantages of digital signal processing) as already pointed out in the preceding sections and some more specific ones:

(i) the demand from a clinical viewpoint to have better quality signals and images (e.g. no phase distortion inherent in analogue systems), from which a better diagnosis can be carried out;
(ii) the desire and necessity for automatic fast processing of large amount of data (as in electrocardiography ECG for example).

Many biomedical signals and images can be very efficiently processed with digital techniques and in particular with digital filtering.

Here we examine in general terms three main areas of digital signal processing applications: (i) in biomedical signals (such as ECG, EEG, etc.); (ii) in the processing of biomedical images or maps (for nuclear medicine, for example) and (iii) in computer tomography.

VIIIG1 Digital processing of biomedical signals

There are many biomedical signals of clinical interest such as: electrocardiogram (ECG), electroencephalogram (EEG), phonocardiogram, temperature, pressure, blood flow velocity, and others.[5,29,33,41,68]

All these signals $x(t)$ have in general a frequency spectrum $X(j\omega)$ limited to relatively low frequency bands (ECG and EEG for example to 100 Hz, phonocardiograms to 1–2 KHz, etc.)

Specific external signals, used as sounding signals through the body are however extended to higher frequency bands: an example is ultrasonic signals covering some MHz of bandwidth.

Two very useful processing operations which are required to be performed on these signals are:

(i) filtering (in particular extraction of signal from noise);
(ii) frequency spectrum analysis.

If we consider a filter with impulse response $h(t)$ and frequency response $H(j\omega)$, we will have for the filtered signal

$$y(t)=\frac{1}{2\pi}\int_{-\infty}^{\infty} X(j\omega)H(j\omega)\,e^{j\omega t}\,d\omega \tag{8.58}$$

One specific problem in this filtering of biomedical signals is the constraint to have *no distortion* of processed signal. As it is well known in linear system theory this implies that

$$H(j\omega)=|H(j\omega)|\,e^{-j\beta(\omega)}=K\,e^{-j\omega t_0} \tag{8.59}$$

that is the gain of the filter must be exactly constant with respect to the frequency in the passband, while the phase $\beta(\omega)$ must be exactly linear (this corresponds to the output signal having a constant time delay t_0 with respect to the original input). Further if multiple filtering is performed, each output band must have exactly the same delay, in such a way that theoretically with $K=1$ the sum of the outputs must give the original signal.

The difficulty of designing linear phase analogue filters is well known but FIR digital filtering has this property of exact linear phase which is essential for the processing of biomedical signals. In addition to the above, for biomedical signals in the low frequency range, economic hardware processors can be easily built.

Some interesting digital filtering applications to ECG, EEG and phonocardiograms with experimental results are described in the following:[31,39]

(1) Through low-pass filtering (cutoff frequency at 40–50 Hz) a reduction of disturbance and noise is obtained (muscular tremor, power supply, etc); an example of low-pass digital filtering applied to an ECG signal by means of the ANPAD system described in Section VIIIC is shown in Fig. 8.47; the digital filter used in ANPAD has the transfer function shown in Fig. 8.12.

(2) Through band-pass filtering, with bandwidths of a few Hz, it is possible to extract some particularly important frequency components from the signal such as the α or β components in EEG; an example of EEG processing is shown in Fig. 8.48, where in Fig. 8.48(a) we show the frequency response of the two band-pass digital filters used here, designed with the window-technique (Cappellini window, see Table 2.2), the first centred on 10 Hz, the second on 20 Hz, each one of bandwidth 2 Hz (sampling frequency = 96 Hz). In Fig. 8.48(b) these two filters are applied (along with a low-pass filter) to process a typical EEG waveform to extract the α and β components. In Fig. 8.48(c) a complete band-pass analysis with 12 filters, 2 Hz each, from 0 to 24 Hz is shown.[42] An alternative solution, which has the advantage of giving compressed-data band output (minimum data from the sampling view point) is represented by the

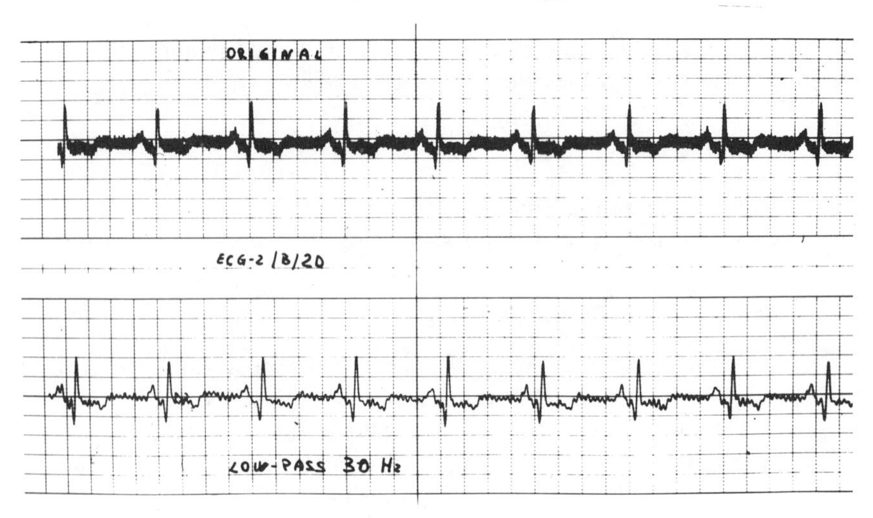

FIG. 8.47. Example of low-pass digital filtering an ECG signal.

complex demodulation described in Section VIIIB1. A clear example of this aspect is shown in Fig. 8.49 on EEG signal: we see the low number of samples representing the bandwidth 6–14 Hz and the perfect reconstruction of EEG by the samples.

(3) Through multiple band-pass filtering, covering the frequency range of interest, it is possible to perform a spectral estimation by evaluating the r.m.s. values in each output band (as described in Section VIIIB1). This is very useful for a better physical understanding of the phenomena observed and for a more efficient overall clinical diagnosis. The same filters of item (2) above were applied to the same EEG signal for spectral estimation based on r.m.s. values: results with 16 filters are shown in Fig. 8.50. An alternative example (8 bands), using the ANPAD system applied to phonocardiogram for spectral estimation, is shown in Fig. 8.51. A particular real case of comparing a normal phonocardiogram with a pathological case is shown in Fig. 8.51: the differences are immediately apparent.[28]

VIIIG2 Digital processing of biomedical images or maps

It is the nature of many biomedical signals to be in the form of 2-D data, such as nuclear medicine machine signals, X-ray pictures, ultrasonic images, thermal detector images, etc.[35]

These 2-D signals require in general filtering operations for such purposes as smoothing, noise reduction or restoration operations. As in the 1-D case, if we consider an image $x(x_1, x_2)$ with spectrum $X(j\omega_1, j\omega_2)$ and a

2-D system filter having impulse response $h(x_1, x_2)$ and frequency transfer function $H(j\omega_1, j\omega_2)$, we have for the filtered processed image

$$y(x_1, x_2) = \frac{1}{(2\pi)^2} \int_{-\infty}^{\infty} \int_{-\infty}^{\infty} X(j\omega_1, j\omega_2) H(j\omega_1, j\omega_2)\, e^{j\omega_1 x_1}\, e^{j\omega_2 x_2}\, d\omega_1\, d\omega_2 \tag{8.60}$$

Again, a specific problem in this filtering of biomedical images is the constraint to have no distortion. This implies that

$$H(j\omega_1, j\omega_2) = |H(j\omega_1, j\omega_2)|\, e^{-j\beta(\omega_1, \omega_2)} = K\, e^{-j\omega_1 x_{1a}}\, e^{-j\omega_2 x_{2b}} \tag{8.61}$$

that is the gain of the filter must be exactly constant with respect to ω_1, ω_2 in the passband, while the phase $\beta(\omega_1, \omega_2)$ must be exactly linear (x_{1a} and x_{2b} are constant).

The advantages already outlined in the 1-D case are also applicable to 2-D digital signal processing.

In the following we describe two examples of 2-D digital filtering in this area, one relating to nuclear medicine the other to X-ray analysis.

In nuclear medicine a penetrating radiation (see Section VIIID) is in general sent through the body and a nuclear detector scanning the body collects the radiation intensity passed through different body parts. A *map* is in this way obtained.

These maps are easily converted into digital form because the data are merely readings of radiant energy intensity. In addition the normally low number of points, typically 64×64, used in the image representation makes such 2-D signals eminently suitable for 2-D digital signal processing.

The problems of noise reduction, edge-boundary extraction and inverse filtering appear in this area in the same way as they do for radar maps.

Two types of widely used filters are: (i) low-pass filters reducing fluctuations and noise in high spatial frequency components; (ii) inverse filtering or restoration filters to correct the image.[18]

In the case of noise reduction through low-pass filtering we show in Fig. 8.52(a) the raw signal of a liver image and in Fig. 8.53(b) we show the low-pass filtered version of it. The filtering operation has been achieved by means of a FIR filter of type[46]

$$H(j\omega_1, j\omega_2) = \sum_{k_1=0}^{N-1} \sum_{k_2=0}^{N-1} h(k_1, k_2)\, e^{-j\omega_1 k_1}\, e^{-j\omega_2 k_2} \tag{8.62}$$

whose frequency response is shown in Fig. 8.53.

From the same processing we can obtain the very useful result from the diagnostic point view of *iso-contour displays* of Fig. 8.54.

For edge-boundary extraction and area identification for pattern recognition purposes we can apply a suitably selected band-pass 2-D digital filter

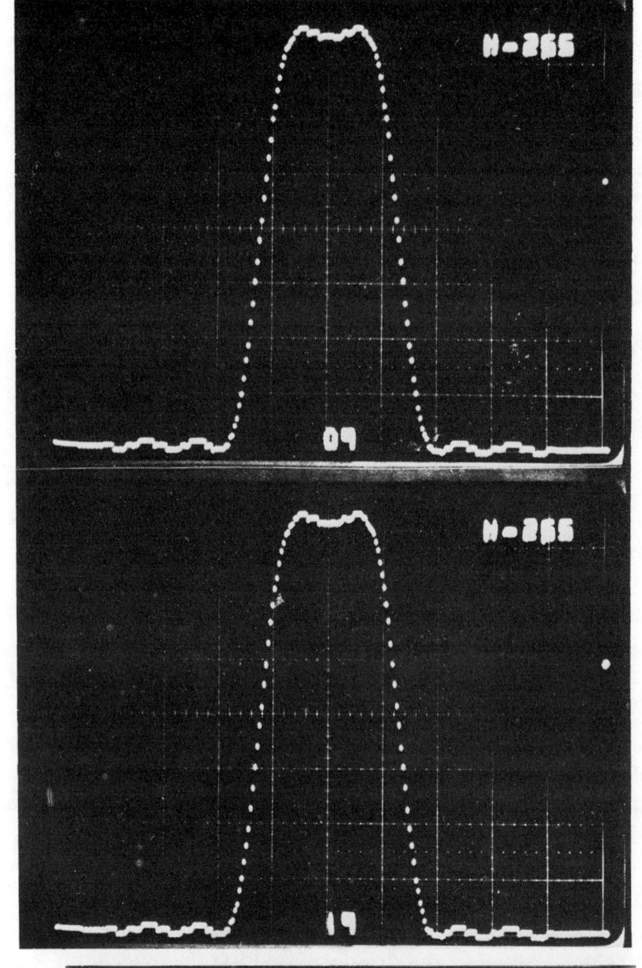

$F_S = 96$ Hz

$B_W = 2$ Hz

(a)

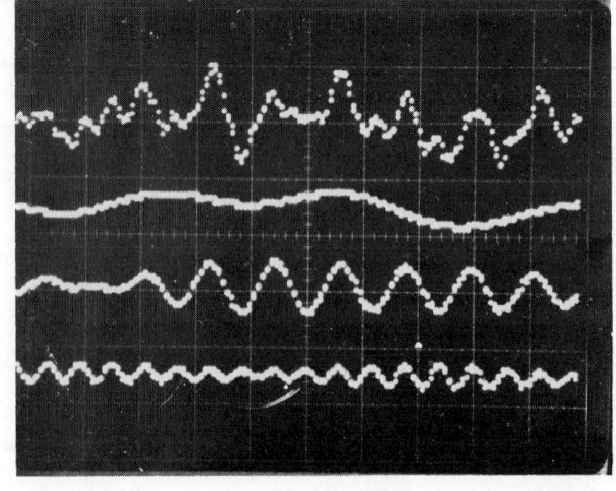

EEG

0 - 4 Hz

8 - 12 Hz

18 - 22 Hz

(b)

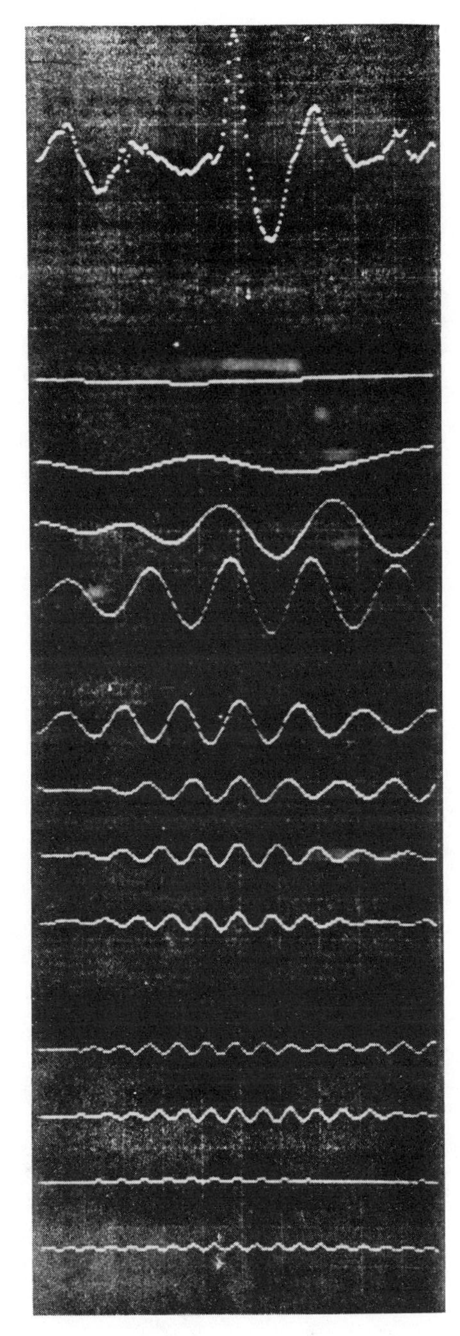

(c)

FIG. 8.48(a). Frequency responses of two digital filters for EEG processing; (b) example of EEG digital filtering; (c) example of complete band-pass analysis on EEG signal (with permission from Laboratorio di Elettroencefalografia, Ospedali Neuropsichiatrici and OTE Biomedica, Florence).

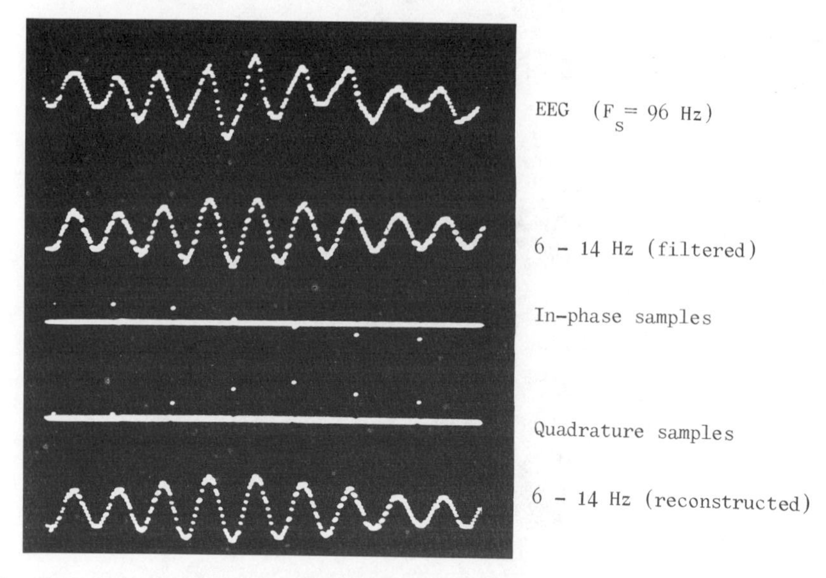

FIG. 8.49. Example of compressed representation of an EEG band by using Hilbert filtering method.

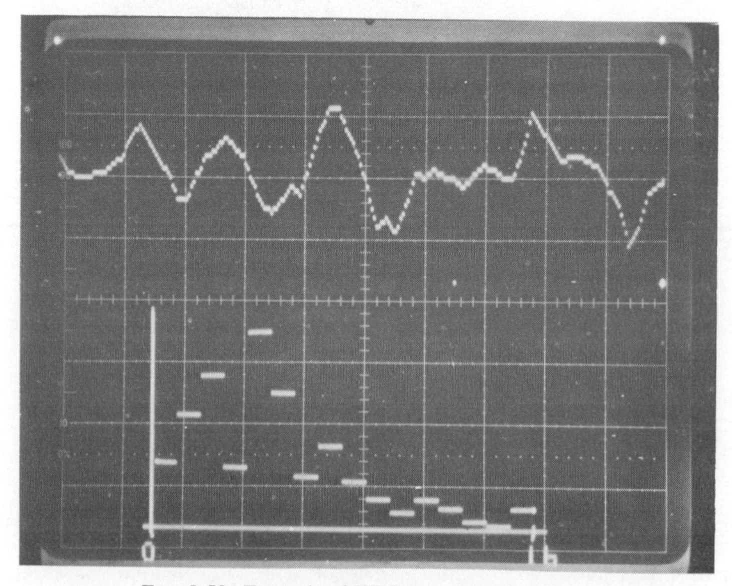

FIG. 8.50. Example of EEG spectral estimation.

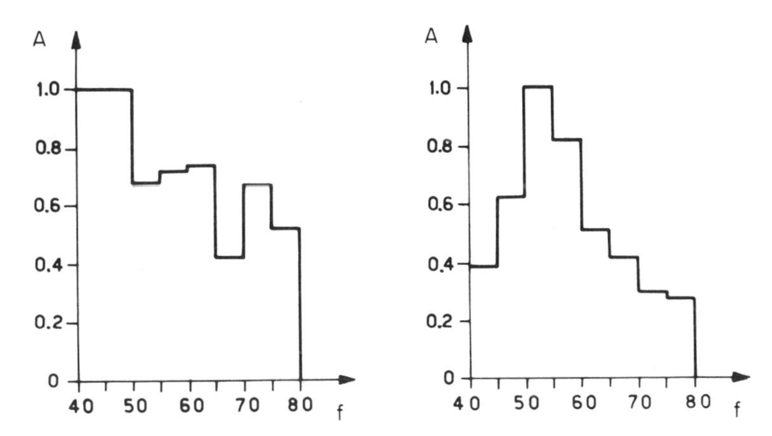

FIG. 8.51. Example of band-pass analysis on phonocardiograms.

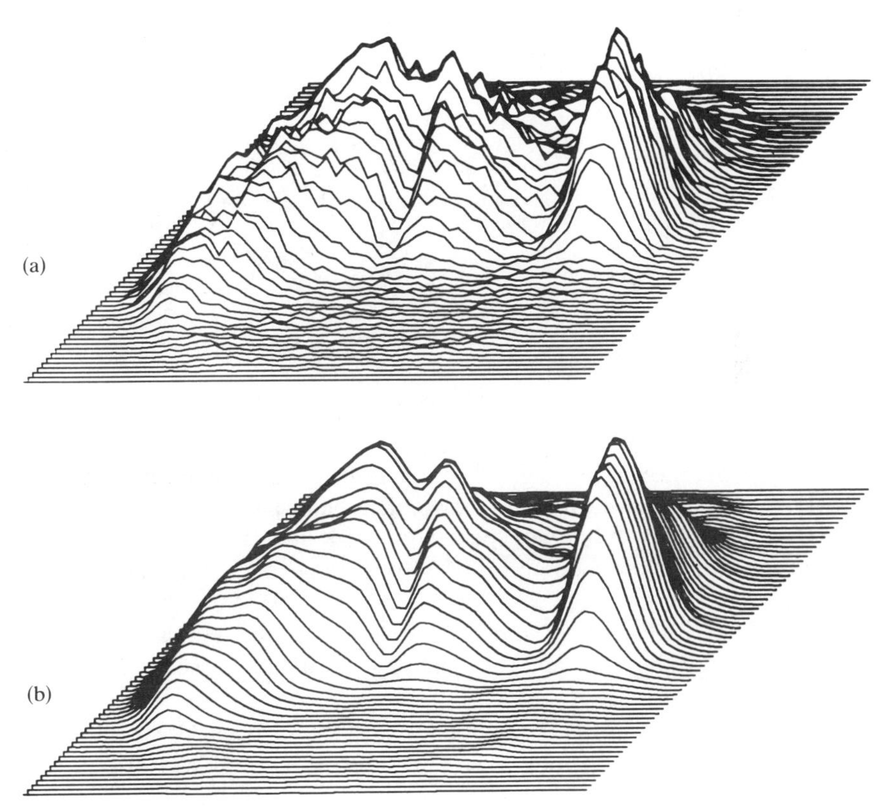

FIG. 8.52. Example of 2-D FIR digital filtering a liver scintigraphy: (a) original image; (b) processed image (with permission from Istituto di Medicina Nucleare of Florence).

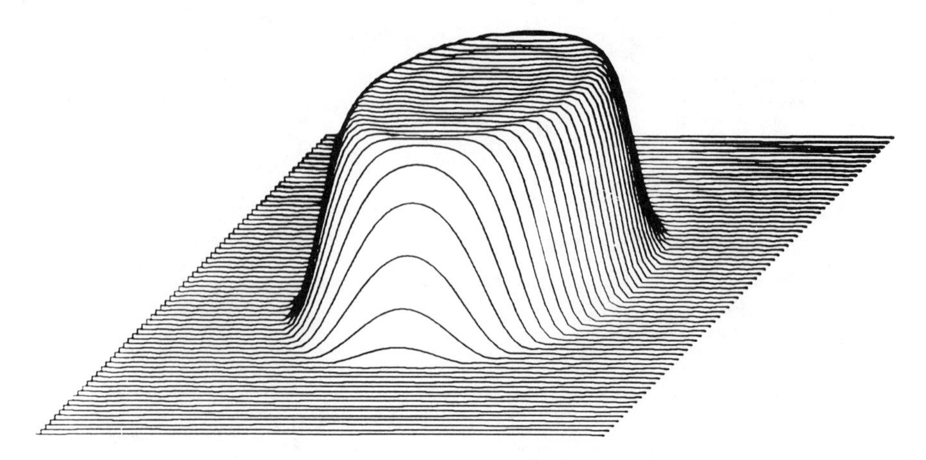

FIG. 8.53. Response of the FIR digital filter used for the processing of Fig. 8.52.

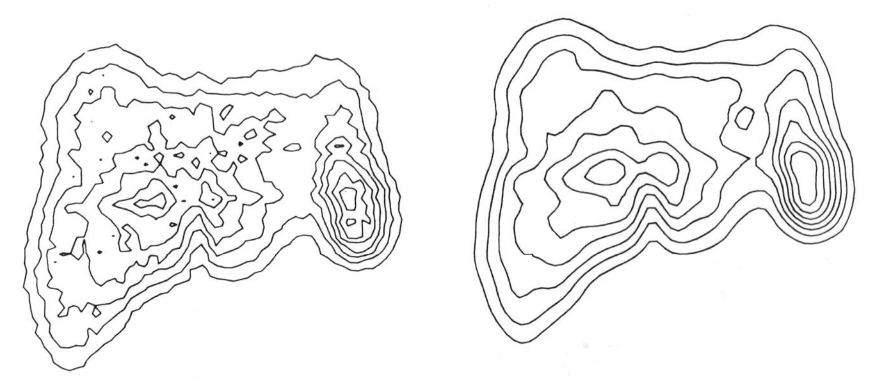

FIG. 8.54. Iso-contour displays of the image in Fig. 8.52.

as in Fig. 8.55(a),(b). The original image (a liver image different from the previous case) is shown in Fig. 8.55(a). The band-pass filter, of the type (8.63) with a window (Lanczos type window with $m = 10$) is used to process this image with the consequent result of extracting and enhancing the area around a cyst.[55]†

In the X-ray analysis the X-ray radiation is sent from one side through to the other side of the body and the fraction of radiation passed is recorded on a sensitive film.

The operations of interest are similar to nuclear maps. A typical processing required here is *enhancement*, to compensate for defocusing effects

† A HP-2100A minicomputer was used with output on an oscilloscope having z-control.

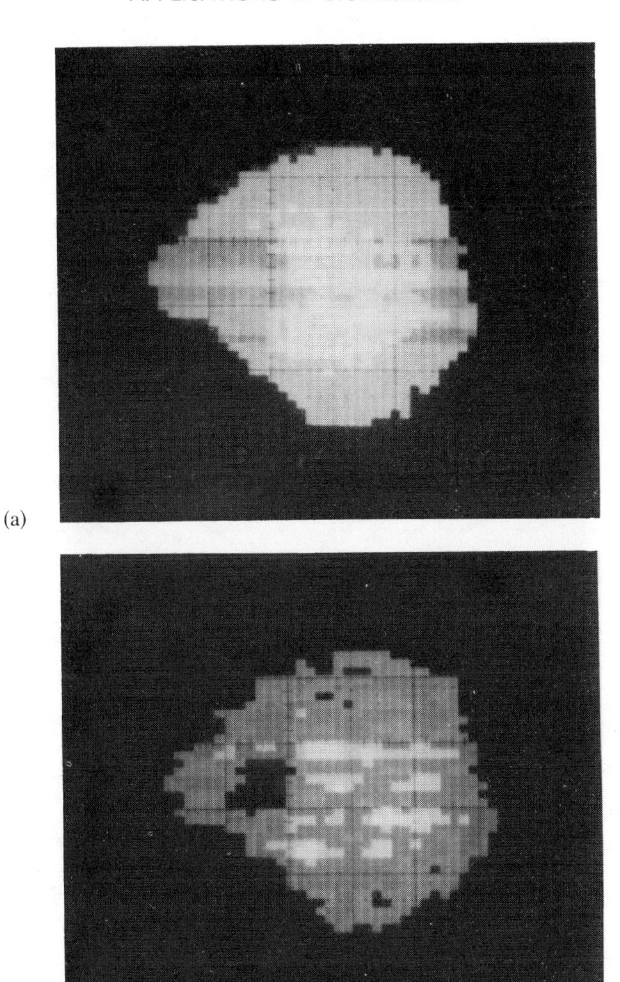

FIG. 8.55. Example of 2-D band-pass FIR digital filtering a liver scintigraphy: (a) original image, (b) filtered image (showing a cyst).

or to increase the grey level contrast, especially in the particular regions of pathological interest.[59]

An example of X-ray digital processing is shown in Fig. 8.56, developed by Fusco and Caponetti at CSATA. Figure 8.56(a) shows the original image (presented with 10 grey levels as are the following ones);† Fig. 8.58(b) shows the result of processing by means of a *smoothing* filter

† It will be observed that there is a small area in the right side of the picture (64 × 64) which is different from the other parts: this can be done for selecting that area for further processing.

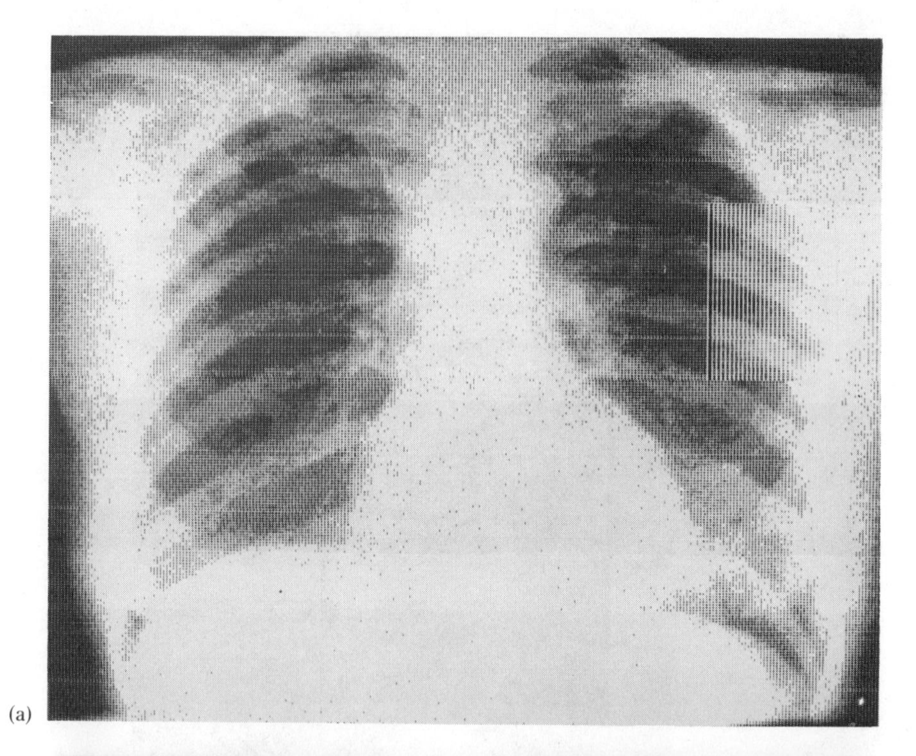
(a)

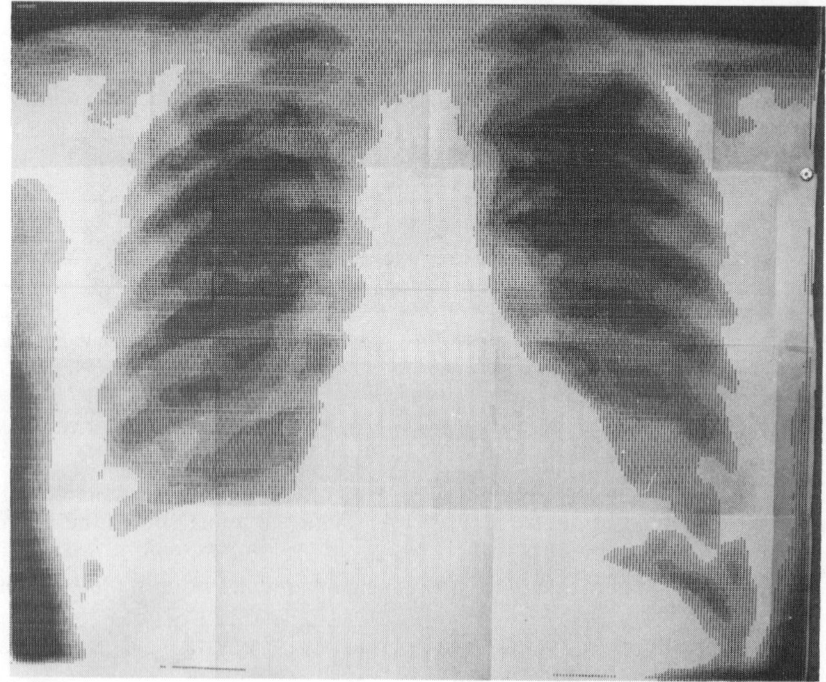
(b)

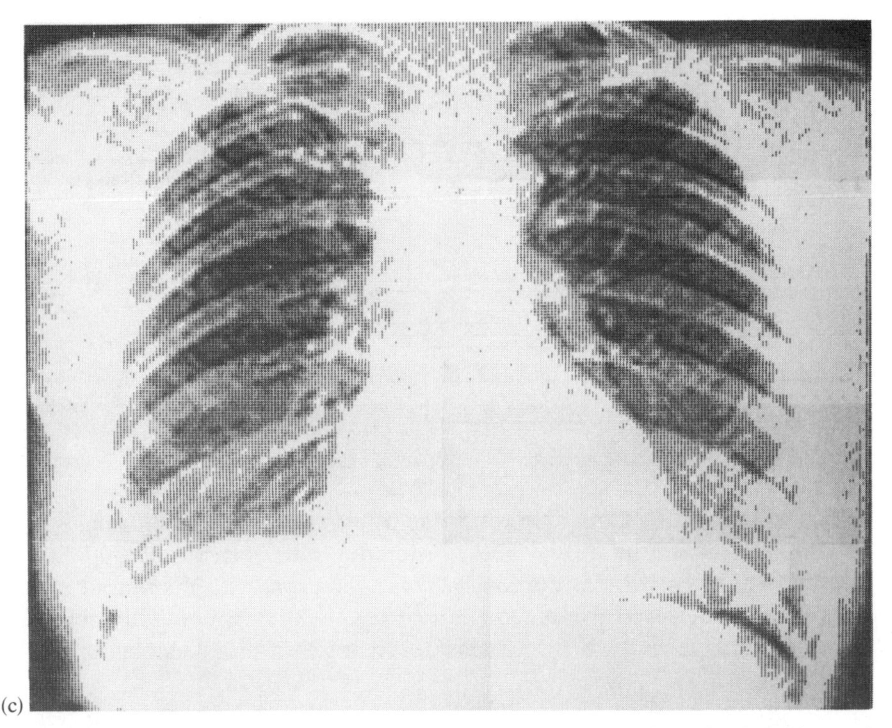
(c)

FIG. 8.56. Example of filtering a X-ray image: (a) original image; (b) first filtering; (c) second filtering.

(17×17 coefficients); Fig. 8.56(c) shows the result of processing by means of a higher order digital filter approximating the differentiation operation (17×17 coefficients)

$$y(x_1, x_2) = x(x_1, x_2) - 0.2\left(\frac{\partial^2}{\partial x_1^2} + \frac{\partial^2}{\partial x_2^2}\right)^2 x(x_1, x_2) \qquad (8.63)$$

VIIIG3 Computer tomography

A recent new advance in diagnostic analysis is represented by *computer tomography* (CT) scanning. While in standard radiography (as described with X-rays) a three-dimensional object (part of body) is represented on a plane (film), in CT the 3-D structure can be recovered.[83] Computer tomography scanning has reached the level to be currently used in many hospitals.

The basic idea of CT scanning as developed by Hounsfield at EMI in England is shown in Fig. 8.57.[81] A moving scanning system including X-ray source, collimators and detectors rotates around a part of the patient's

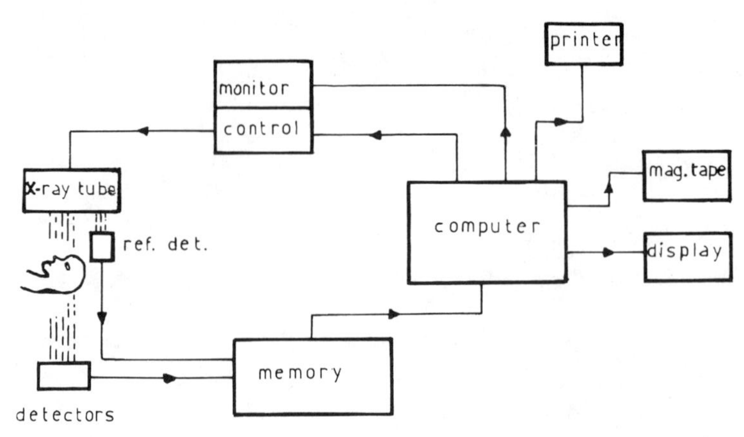

FIG. 8.57. Basic principles of CT scanning.

body, taking (e.g. 180) readings of X-ray transmissions across the analysed part (for example, the head). These different readings (typically 200–300 detector readings at each rotation pass) are then sent with the output of a reference detector to a large memory (disk) and to a computer which through suitable software programs reconstructs a cross-section of the analysed part.

The processing done by the computer is essentially a 2-D (image) reconstruction from 1-D projections (or 3-D reconstruction from 2-D projections).

According to Hounsfield[76] if we imagine the body divided into a series of small cubes, each characterized by a specific value of absorption, then the sum of the absorption values of the cubes which are contained within an X-ray beam will be equal to the *total absorption* measured for that beam. If, for example, we consider a system dividing the area to be scanned into a 160×160 matrix of such cubes, each complete scan can be considered as a series of 43 200 simultaneous equations, in which there are 25 600 variables. Since the number of equations is greater than that of variables, the system can be solved.

In general the redundancy in equations is used to reconstruct the data with some noise reduction.

The use of digital filtering and FFT techniques is crucial for the reconstruction technique of images from their projections known as *analytical reconstruction techniques.* Indeed these techniques include two main types of digital signal processing: (i) *2-D Fourier reconstruction* methods and (ii) *filtered back projection.* In the first method from the FFT of each projection it is possible to derive the complete 2-D transform of the image, from which the required image is obtained through inverse FFT transform.

In the second technique the measured projection profiles are modified—through digital filtering—before back projection is carried on in such a way that the reconstructed image now becomes a close representation of the examined object. The filtering of the profiles can be derived mathematically, in such a way as to counterbalance exactly the blurring effect generated by simple back projection: in particular while some profiles will contribute to points out of the real image other profiles will subtract from those points to obtain a close approximation of the true image.†

References

1. Alberti, U. (1976). Analisi ed elaborazione del segnale fonico: caratterizzazione in frequenza delle unita' fondamentali. Thesis, Facolta' di Ingegneria, University of Florence.
2. Archambeau, C. B., Bredford, J. C., Broome, P. W., Dean, W. C., Fleinn, E. A. and Saxe, R. L. (1965). Data processing techniques for the detection and interpretation of teleseismic signals *I.E.E.E. Proceedings* **53**, **12**, 1860–84.
3. Arkadev, A. G. and Braverman, E. M. (1970). "Teaching Computers to Recognize Patterns". Academic Press, London and New York.
4. Beauchamp, K. G. (1975). "Walsh Functions and Their Applications". Academic Press, London and New York.
5. Bedini, L., Denoth, F. and Navona, C. (1977). Un metodo per la misurazione automatica di alcuni intervalli sistolici. Symposium A.E.I., Trieste, Italy.
6. Benelli, G., Bianciardi, C., Cappellini, V. and Del Re, E. (1977). High efficiency digital communications using data compression and error correcting coding. Proceedings: EUROCON Conference, Venice, Italy.
7. Bellanger, M., Daguet, J. and Lepagnol, G. (1974). Interpolation, extrapolation and reduction of computation speed in digital filters. *I.E.E.E Trans. Acoustics, Speech and Signal Processing* **ASSP-22**, **4**.
8. Billingsby, F. C. (1970). Applications of digital image processing. *Applied Optics*, **9**.
9. Bolondi, G., Rocca, F. and Zanoletti, S. (1976). Automated contouring of faulted subsurfaces. *Geophysics* **41**, **6**, 1377–93.
10. Braccini, C. and Gambardella, G. (1975). Effects of spectral smoothing on pictorial images. 1975 Proceedings: Florence Conference on Digital Signal Processing, 204–12.
11. Brofferio, S., Cafforio, C., Rocca, S. and Ruffino, U. (1975). A dynamic programming approach to video coding using movement compensation. 1975 Proceedings: Florence Conference on Digital Signal Processing, 158–69.
12. CCITT Draft Recommendation G712.
13. Cain, G. D. (1972). Hilbert transform description of linear filtering. *Electronics Letters* **8**, 380–2.
14. Cain, G. D. and Abed, A. H. (1975). Mildly suboptimal digital filters using a host windowing approach. *Electronics Letters* **11**, **20**, 493–5.

† Interesting tests were carried out along these lines by Rocca at Politecnico di Milano.

15. Cain, G. D., Cappellini, V. and Del Re, E. (1977). A digital band-pass analysis approach for biomedical signal processing and telemetry. Proceedings: EUROCON Conference, Venice, Italy.
16. Cambridge Consultants. A–D converters with 8-bit, 30 MHz conversion rate.
17. Cambridge Electronic Design Lim. (1975). Digital synthesizer model 301, programming guide.
18. Cantoni, V., De Lotto, I., Favino, A. and Valenziano, F. (1977). An image restoration technique: application to nuclear medicine images. Conference on Digital Image Processing, Munich, Germany.
19. Cappellini, V. and D'Amico, T. (1967). Alcuni filtri numerici ottenuti dall'elaborazione dell'integrale di convoluzione e loro applicazione all'analisi spettrale. *Alta Frequenza* **36**, 835–40.
20. Cappellini, V. (1968). A class of digital filters with application to data compression. 1968 Proceedings: I.E.E.E. International Conference on Communications, Philadelphia, Pa., 788–93.
21. Cappellini, V. (1968). On-board processor for digital filtering and data compression. International Conference on Aerospace Computers and Spacecrafts, Paris.
22. Cappellini, V. (1968). Design and construction of digital filters and their applications to signal processing. Proceedings: International Conference on Automation and Instrumentation, Mondadori, Milan, Italy, 137–56.
23. Cappellini, V., Lotti, F. and Pantani, L. (1968). Radar-signal matched filtering through digital processing. *Alta Frequenza* **37**, **11**, 1103–4.
24. Cappellini, V. (1969). Some digital telemetry systems for geophysical satellites. Proceedings: Convegno Internazionale Tecnico Scientifico sullo Spazio, Rome, Italy, 129–40.
25. Cappellini, V. (1969). Design of some digital filters with application to spectral estimation and data compression. 1970 Proceedings: P.I.B. Symposium on Computer Processing in Communications, Polytechnic Institute of Brooklyn, Brooklyn, New York, **19**, 313–31.
26. Cappellini, V. (1969). Application of onboard digital filtering to spectral estimation and data compression. Proceedings: A.I.A.A. Aerospace Computer Systems Conference, Los Angeles, Ca.
27. Cappellini, V. and Emiliani, P. L. (1970). A special-purpose on-line processor for band-pass analysis. *I.E.E.E. Trans. Audio Electroacoustics* **AU-18**, 188–94.
28. Cappellini, V., Cabbanini, A., Dolara, A., Dubini, S. and Simoncini, G. (1972). Applicazione del filtraggio numerico all'analisi di fonocardiogrammi. Proceedings: Convegno Mostra di Bioingegneria, Milan, Italy.
29. Cappellini, V., Dubini, S. and Pasquini, G. (1972). Application of some data compression techniques to ECG processing. Report IROE, Florence, Italy.
30. Cappellini, V. and Gabbanini, A. (1973). A new digital interpolator of linear type. *I.E.E.E. Trans. Instrumentation* **IN-21**, **3**, 293–4.
31. Cappellini, V. (1973). Application of some electronic systems to biomedical data processing. Proceedings: 1st International Conference on Electronic Circuits, Prague, Czechoslovakia.
32. Cappellini, V., Lotti, F., Ori, C., Pasquini, C. and Pieralli, F. (1974). Application of some data compression methods to ESRO satellite telemetry data. *Alta Frequenza* **43**, **9**, 673–81.

33. Cappellini, V., Lotti, F. and Pieralli, F. (1974). Application of some data compression methods to ECG processing. Biotelemetry 2nd International Symposium, Davos, Switzerland, 88–90.
34. Cappellini, V. and Emiliani, P. L. (1974). Design of some digital filters with application to signal and image processing. Proceedings: Summer School on Circuit Theory, Prague, Czechoslovakia, 161–76.
35. Cappellini, V. (1975). 2-D digital filtering and image data compression. ATLAS Computer Laboratory Symposium No. 7, Chilton, England.
36. Cappellini, V. (1975). Two-dimensional digital signal processing. I.E.E. Colloquium on Review of Digital Signal Processing, London.
37. Cappellini, V., Chini, A. and Lotti, F. (1976). Some data compression methods for processing the images received from earth resource satellites. Proceedings: 16 Convegno sullo Spazio, Rome, Italy, 33–4.
38. Cappellini, V. and Del Re, E. (1976). Sample rate reduction for spectral estimation performed through digital filtering. *Alta Frequenza* **45**, **10**, 624–6.
39. Cappellini, V., Del Re, E., Evangelisti, A. and Pastorelli, M. (1976). Application of digital filtering and data compression to ECG processing. 11th International Conference on Medical and Biological Engineering. Ottawa, Canada, 32–3.
40. Cappellini, V. and Fondelli, M. (1977). Applicazione del filtraggio numerico e della compressione dei dati alla elaborazione di immagini per risorse terrestri. Proceedings: 24 Congresso per l'Elettronica, Rome, Italy, 91–99.
41. Cappellini, V. (1977). Some digital techniques for biomedical signal and image processing. Proceedings: Symposium on Medicine and Techniques, Zagreb, Jugoslavia.
42. Cappellini, V., Del Re, E., Bensi, A., Sferlazzo, R., Emiliani, P. L., Innocenti, A. and Martinelli, P. (1977). High efficiency frequency analysis and data compression of biomedical signals through digital filtering. 30th ACEMB, Los Angeles, Ca, 58.
43. Carey, M. J., Tattersall, G. D., Goodman, D. J., and Potter, A. R. (1976). Filtering for code conversion in digital telephone exchanges. Proceedings: International Conference ASSP, Philadelphia, Pa, 280–3.
44. Carey, M. J. and Constantinides, A. G. (1977). A simulation of a hybrid multiplex system. Digital processing of signals in communications. IERE Conference, University of Loughborough.
45. Carl, J. W. and Hall, C. F. (1972). The application of filtered transform to the general classification problem. *I.E.E.E. Trans. Comp.* **C-21**, 785–790.
46. Casini, A., Castellini, G., Emiliani, P. L., Locchi, S. and Lotti, F. (1975). Two-dimensional digital filters for nuclear medicine scintigraphies. 1975 Proceedings: Florence Conference on Digital Signal Processing, Italy, 187–95.
47. Cawpland, M. C. J. and Coll, D. C. (1969). A quantitative comparison of several adaptive equalizer algorithms. Proceedings: Symposium on Computer Processing in Communications, Polytechnic Institute of Brooklyn, New York, 683–94.
48. Champeney, D. C. (1973). "Fourier Transforms and Their Physical Applications". Academic Press, London and New York.
49. Constantinides, A. G. and Colyer, I. (1976). Digital phase-splitting network design for digital f.d.m. applications. *I.E.E. Proceedings* **123**, **12**, 1313–15.
50. Cook, C. E. and Bernfeld (1967). "Radar Signals". Academic Press, New York and London.

51. D'Amore, A., Delcaro, L., Mottola, R. and Sicuranza, G. (1977). Un sistema di elaborazione delle immagini con minicalcolatore. *Alta Frequenza* **46**, **1**, 6–11.
52. Darby, D. K. and Davies, E. B. (1967). The analysis and design of two-dimensional filters for two-dimensional data. *Geophysical Prospecting* **15**, 383–406.
53. Davies, C. (1976). A high speed multiplying ADC. *Electron*, 59.
54. Delcaro, L. and Sicuranza, G. (1976). Ottimizzazione vincolata di filtri digitali ricorrenti bidimensionali. *Tecnica Italiana* **41**, **4**, 197–204.
55. De Vita, G. (1977). Alcune tecniche di elaborazione numerica e presentazione di immagini in medicina nucleare. Thesis, Facolta' di Ingegneria, University of Florence.
56. Dilley, J. E., Naughton, M. F., Morling, R. C. S., Cain, G. D. and Abed, A. H. (1976). A microcomputer-controlled adaptive CCD transversal filter. CCD 76 Conference, Edinburgh University.
57. Duda, R. O. and Hart, P. E. (1973). "Pattern Classification and Scene Analysis". Wiley, New York.
58. Echard, J. D. and Boorstyn, R. R. (1972). Digital filtering for radar signal processing applications. *I.E.E.E. Trans. Audio Electroacoustics* **AU-20**, **1**, 42–52.
59. Ekstrom, M. P. (1974). Digital image processing at Lawrence Livermore Laboratory, Part I: Diagnostic radiography applications. *Computer* **7**, **5**, 72–8.
60. Elefesen, R. (1973). Urban land-use mapping by machine processing of ERTS-1 multispectral data: a San Francisco bay area example. LARS Inf. Note 101573, Purdue University.
61. Feleppa, E. J. (1972). Holography and medicine. *I.E.E.E. Trans. Biomed. Engineering* **BME-19**, **5**, 194–205.
62. Flanagan, J. L. (1972). "Speech Analysis, Synthesis and Perception", 2nd edn. Springer-Verlag, New York.
63. Freeman, H. (1961). On the encoding of arbitrary geometric configurations. *I.R.E. Trans. Electronic Computers* **EC-10**, **2**, 260–8.
64. Freeney, S. L., Kieburtz, R. B., Mina, K. V. and Tewksbury, S. K. (1971). Design of digital filters for an all digital frequency division multiplex-time division multiplex translator. *I.E.E.E. Trans. Circuit Theory* **CT-18**, 702–11.
65. Freeney, S. L., Kieburtz, R. B., Mina, K. V. and Tewkbury, S. K. (1971). System analysis of a TDM-FDM translator/digital A-type channel bank. *I.E.E.E. Trans. Comm. Techn.* **COM-19**, 1050–9.
66. Galpin, R. K. P. (1975). Digital processing in modems for data transmission. I.E.E. Colloquium on Review of Digital Signal Processing, London.
67. Garibotto, G. (1977). On 2-D recursive digital filter design with complex frequency specifications. Proceedings: International Conference on Communications, ICC '77, Chicago, 202–6.
68. Geddes, J. S. and Warner, H. R. (1971). A PVC detection program. *Computers and Biomedical Research*, **4**, 493.
69. Gold, B. (1962). Description of a computer program for pitch detection. Proceedings: International Conference Acoustics, Copenhagen.
70. Gold, B. (1975). The design and construction of a high speed sequential general purpose digital processor and its application to a variety of speech processing algorithms. Proceedings: Florence Conference on Digital Signal Processing.
71. Goodenough, D. J. *et al.* (1974). Optical spatial filtering of radiographic images with binary filters. *Radiology*.

72. Graham, N. (1976). Spatial frequency channels in human vision: detecting edges without edge detectors. *In* "Visual Coding and Adaptability" (C. Harris, ed.). L. Erlbaum Associates, New York.
73. Gupta, J. N. and Wintz, P. (1975). A boundary finding algorithm and its applications. *I.E.E.E. Trans. Circuits Systems* **CAS-22**, **4**, 351–62.
74. Harmuth, H. F. (1972). "Transmission of Information by Orthogonal Functions", 2nd edn. Springer-Verlag, Berlin.
75. Harmuth, H. F. (1974). Two-dimensional spatial hardware filters for acoustic imaging. Proceedings: Applications of Walsh Functions, Washington, D.C.
76. Hounsfield, G. N. (1973). Computerized transverse axial scanning tomography. Part I: description of the system. *Br. J. Radiol.* **46**, 1016–22.
77. Huang, T. S. (ed.). (1976). "Picture Processing and Digital Filtering". Springer-Verlag, New York.
78. Hunt, B. R. (1975). Digital image processing. *I.E.E.E. Proceedings* **63**, **4**, 693–708.
79. Jaffee, R. M. (1962). Digilock telemetry system for the Air Force Special Weapons Center's Blue Scout Jr., *I.R.E. Trans Space Electr. Telemetry* **SET-8**, 44–50.
80. Lebedev, D. S. and Mirkin, L. I. (1975). Digital nonlinear smoothing of images. Proceedings: Florence Conference on Digital Signal Processing.
81. Lens Van Rijn, R. A. (1976). General principles of CT scanning. *J. Belge Radiologie* **59**, **3**, 201–11.
82. Madams, P. H. C. and Witten, I. H. (1975). A speaking computer output peripheral. Proceedings: Florence Conference on Digital Signal Processing.
83. Mersereau, R. M. and Oppenheim, A. V. (1974). Digital reconstruction of multidimensional signals from their projections. *I.E.E.E. Proceedings* **62**, **10**, 1319–38.
84. Mostafavi, H. and Sakrison, D. J. (1976). Structure and properties of a single channel in the human visual system. *Vision Res.* **16**, 957–68.
85. Oppenheim, A. V. (1969). Speech Analysis-synthesis system based on homomorphic filtering. *J. Acoust. Soc. Am.* **45**, 459–62.
86. Oppenheim, A. V. and Schafer, R. W. (1975). "Digital Signal Processing". Prentice-Hall, Englewood Cliffs, New Jersey.
87. Peterson, W. W. and Weldon, E. J. Jr. (1972). "Error-Correcting Codes". M.I.T. Press, Cambridge, Boston.
88. Rabiner, L. R. and Gold, B. (1975). "Theory and Application of Digital Signal Processing". Prentice-Hall, Englewood Cliffs, New Jersey.
89. Rockland Systems Corp., West Nyack, New York. Variable-order digital filters.
90. Rosenfeld, A. (1969). "Picture Processing by Computers". Academic Press, New York and London.
91. Schwartz, M. (1970). "Information Transmission, Modulation and Noise". McGraw-Hill, New York.
92. Special issue of I.E.E.E. Proceedings (1972). Digital picture processing.
93. Special issue of I.E.E.E. Proceedings (1974). Modern radar technology.
94. Special issue of I.E.E.E. Proceedings (1975). Digital signal processing.
95. Special issue of I.E.E.E. Proceedings (1977). Optical computing.
96. Zverev, A. I. (1968). Digital MTI radar filters. *I.E.E.E. Trans. Audio Electroacoustics* **AU-16**, **3**, 422–32.

Appendix 1

Digital Harmonic Analysis: Fundamental Aspects of the DFT

A1.1 Introduction

The use of digital computers makes an analysis method based on a software Discrete Fourier Transform (DFT) implementation possible. As will now be shown, the DFT is accurate and meaningful only when the time waveforms being analysed are both band-limited and periodic (i.e. possessing finite harmonic spectra).

A1.2 Interpretation of the DFT

According to the sampling theorem, a true waveform of infinite duration can only be represented by samples if it is band-limited. With such waveform designated by $f(t)$, an infinite sequence of equally spaced samples $f(nT)$ then fully defines it:

$$f(t) = \sum_{n=-\infty}^{\infty} f(nT) \cdot \frac{\sin[(\pi/T)(t-nT)]}{(\pi/T)(t-nT)} \tag{A1.1}$$

T in the above being the sampling interval.

When $f(t)$ is truncated forming another waveform $g(t)$ that would now resemble a real observable process, the Nyquist sequence $f(nT)$ (which is now assumed zero outside an interval $0 \leq n < N$) can fully define $g(t)$ only if the latter is continually tapered down to keep the limited bandwidth requirement:

$$g(t) = \sum_{n=0}^{N-1} f(nT) \cdot \frac{\sin[(\pi/T)(t-nT)]}{(\pi/T)(t-nT)} \tag{A1.2}$$

A well-known relation exists between the value of the transform $G(\omega)$ of $g(t)$ and the complex Fourier Series coefficients of $g_p(t)$, the periodic extension of $g(t)$. For, remembering that the continuous transform is given by:

$$F(\omega)=\int_{-\infty}^{\infty} f(t)\cdot \exp(-j\omega t)\,\mathrm{d}t \tag{A1.3}$$

then for $g(t)$ of duration τ, i.e. $g(t)=0$, $t\geq\tau$ and $t<0$

$$G(\omega)=\int_{0}^{\tau} g(t)\cdot \exp(-j\omega t)\,\mathrm{d}t \tag{A1.4}$$

Also, $g(t)$ can be expressed as a Fourier Series in the interval $(0, \tau)$:

$$g(t)=\lim_{M\to\infty}\sum_{k=-M}^{M} g_k\cdot \exp(jk\omega_0 t) \tag{A1.5}$$

where

$$g_k=\frac{1}{\tau}\int_{0}^{\tau} g(t)\cdot \exp(-jk\omega_0 t)\,\mathrm{d}t \tag{A1.6}$$

and

$$\omega_0=2\pi/\tau$$

Comparing eqs (A1.4) and (A1.6) above, it can be seen that:

$$g_k=\frac{G(k\omega_0)}{\tau} \tag{A1.7}$$

Therefore, samples $G(k\omega_0)$ of the transform $G(\omega)$ determine the Fourier coefficients g_k and hence, by eq. (A1.5), the function $g(t)$ itself. Further, by eq. (A1.4), $g(t)$ uniquely determines $G(\omega)$. Therefore, the samples $G(k\omega_0)$ are sufficient to determine the transform $G(\omega)$. This result is sometimes referred to as the *sampling theorem in the frequency domain.* It can be mathematically expressed as:

$$F(\omega)=\sum_{k=-\infty}^{\infty} F(k\omega_0)\cdot\frac{\sin[(\pi/\omega_0)(\omega-k\omega_0)]}{(\pi/\omega_0)(\omega-k\omega_0)} \tag{A1.8}$$

The number of frequency samples $F(k\omega_0)$ is here seen to be infinite. For the particular case being discussed, however, $g(t)$ was assumed band-limited, as in Fig. A1.1(b), and the number of frequency samples is thus finite, as in Fig. A1.1(c).

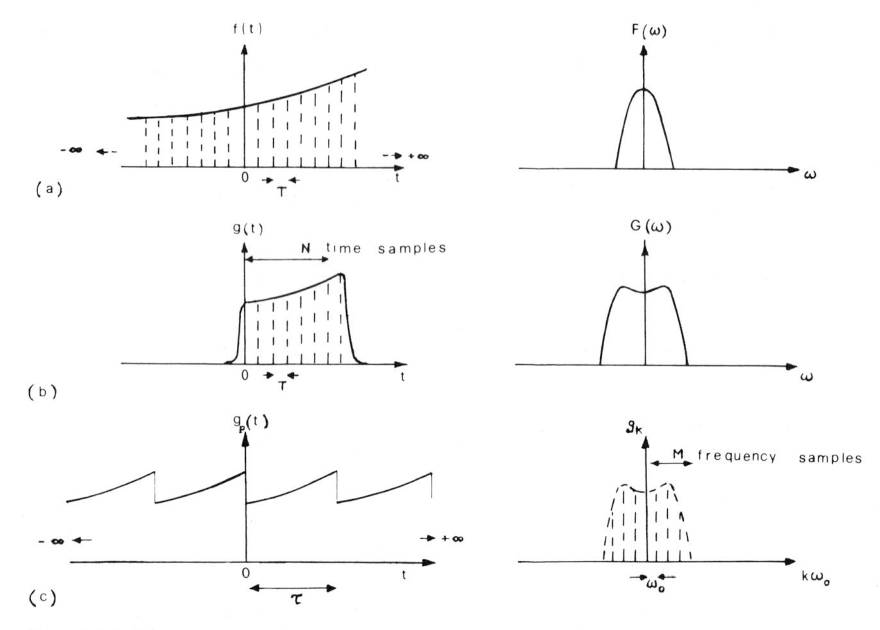

FIG. A1.1. (a) Band-limited time waveform and continuous transform. (b) Truncated waveform, smoothed at the ends to make it band-limited, and continuous transform. (c) Truncated waveform repeated. N is the number of Nyquist samples defining $g(t)$. M is the number of harmonic components g_k in the Fourier series spectrum of $g_p(t)$. g_k determines $G(k\omega_0)$, hence highest frequency in $G(\omega)$ is highest harmonic g_{kM}. Hence sampling interval T is such that: $1/T \geq f_0(2M+1)$ or $1/T \geq (2M+1)/\tau$, i.e. $1/T \geq (1/T)(2M+1)/N$; therefore $N > 2M$ to satisfy the sampling theorem.

The DFT of a sequence of Nyquist samples of length N is another sequence of the same length:

$$F(r\Omega) = \sum_{n=0}^{N-1} f(nT) \cdot \exp(-jr\Omega nT) \qquad 0 \leq r < N \qquad \text{(A1.9)}$$

where the $f(nT)$ that vanish outside the time interval $0 \leq t < NT$ represent a truncated waveform $g(t)$ identical to that already defined by eq. (A1.2). Then, from eq. (A1.7)

$$\frac{G(k\omega_0)}{\tau} = g_k$$

and from (A1.6)

$$g_k = \frac{1}{NT}\int_0^{NT} g_p(t) \cdot \exp(-jk\omega_0 t)\,\mathrm{d}t$$

$$= \frac{1}{NT}\int_0^{NT} g(t) \cdot \exp(-jk\omega_0 t)\,\mathrm{d}t$$

and using (A1.2)

$$g_k = \frac{1}{NT}\int_0^{NT}\left\{\sum_{n=0}^{N-1} f(nT)\cdot\frac{\sin[(\pi/T)(t-nT)]}{(\pi/T)(t-nT)}\right\}\exp(-jk\omega_0 t)\,\mathrm{d}t$$

$$= \frac{1}{N}\sum_{n=0}^{N-1} f(nT)\cdot\exp(-jk\omega_0 nT) \tag{A1.10}$$

where the integral of $\sin x/x$ is evaluated as in Annex A. $|k|$, the order of the harmonic, must be $\leq M$ due to the spectral bandwidth limitation implicitly assumed by the sampling theorem underlying the validity of Nyquist samples.

Comparing eqs (A1.9) and (A1.10), they are the same except for a factor NT, and (r, k) both unbounded. That is:

$$g_k = \frac{F(r\Omega)}{N} \qquad \text{for } k = r \tag{A1.11}$$

and therefore the DFT values are the same as the complex Fourier Series coefficients, as long as the sampling rate is high enough in the first place to account for the highest frequency present, i.e. $N \geq 2M+1$. If this is not so, some overlapping (aliasing) occurs in the frequency domain, and each DFT value is then the sum of a number of the Fourier series coefficients:

$$F(k\Omega) = N\sum_m g_{mN+k} \tag{A1.12}$$

In relation to Fig. A1.1, the DFT coefficients are the same as the g_k. But the latter constitute the actual amplitude spectrum of the repeated truncated waveform (see Annex B). Hence the DFT coefficients are an exact representation of the spectrum of a continuous infinite waveform only when such waveform can be regenerated by truncation over a suitable window and repetition. That is, the DFT of $g(t)$ of Fig. A1.1(b) fully represents the transform $F(\omega)$ of $f(t)$ of (a) only when $g_p(t)$ of (c) is the same as $f(t)$. It is thus apparent how the DFT is only suitable for periodic band-limited waveforms.

Annex A

The DFT coefficients and their relation to the Fourier Series coefficients

As in eq. (A1.6), the Fourier Series coefficients are given by:

$$g_k = \frac{1}{NT}\int_0^{NT} g(t)\cdot\exp(-jk\omega_0 t)\,\mathrm{d}t$$

and using the expansion of $g(t)$ given by eq. (A1.2):

$$g_k = \frac{1}{NT}\int_0^{NT}\left\{\sum_{n=0}^{N-1} f(nT)\cdot\frac{\sin[(\pi/T)(t-nT)]}{(\pi/T)(t-nT)}\right\}\exp(-jk\omega_0 t)\,dt$$

$$= \frac{1}{NT}\sum_{n=0}^{N-1} f(nT)\cdot\left\{\int_0^{NT}\frac{\sin[(\pi/T)(t-nT)]}{(\pi/T)(t-nT)}\exp(-jk\omega_0 t)\cdot dt\right\}$$

now, let $T\to 0$, then

$$\frac{\sin[(\pi/T)(t-nT)]}{(\pi/T)(t-nT)}\to T\delta(t-nT),$$

and from the definition of a delta function:

$$\int_{-\infty}^{\infty}\delta(t-\tau)\cdot\psi(t)\cdot dt = \psi(\tau)$$

$$g_k = \frac{1}{NT}\sum_{n=0}^{N-1} f(nT)\cdot T\cdot \left|\exp(-jk\omega_0 t)\right|_{t=nT}$$

$$= \frac{1}{N}\sum_{n=0}^{N-1} f(nT)\cdot\exp(-jk\omega_0 nT) \qquad \text{(A1.13)}$$

which is the DFT coefficient divided by N.

The restriction of $T\to 0$, however, means a function that is not band-limited (infinitely high sampling rate). The proof is thus not general. Another approach is as given by Oppenheim and Shafer:[2]

The repeated time waveform $g_p(t)$ is sampled, giving the sequence $f(nT)$ over one period. The line spectrum in the frequency domain thus repeats itself, and the Fourier coefficients are now evaluated by a summation instead of an integral because of the discrete time waveform:

$$g_k = \frac{1}{NT}\int_0^{NT} f(nT)\cdot\exp(-jk\omega_0 t)\cdot dt$$

$$= \frac{1}{N}\sum_{n=0}^{N-1} f(nT)\cdot\exp(-jk\omega_0 nT)$$

which is again the DFT coefficient, except for the factor $1/N$.

Annex B

The transform of a periodic function and its relation to the Fourier Series coefficients

The spectrum of a function $f(t)$ as given by the Fourier integral requires, among other limitations, that $f(t)$ should be absolutely integrable between infinite limits. Such a condition is not fulfilled when $f(t)$ is periodic. Therefore, the integral is immediately inapplicable for computing the discrete (line) spectra of periodic functions.

However, a spectral representation can be arrived at if such functions are assumed to be linear combinations of discrete cosinusoids as implied by the Fourier series:

$$
\begin{aligned}
F(\omega) &= \int_{-\infty}^{\infty} g_p(t) \cdot \exp(-j\omega t) \cdot \mathrm{d}t \\
&= \int_{-\infty}^{\infty} \sum_{k=-\infty}^{\infty} g_k \cdot \exp(jk\omega_0 t) \cdot \exp(-j\omega t) \cdot \mathrm{d}t \\
&= \sum_{k=-\infty}^{\infty} g_k \cdot \int_{-\infty}^{\infty} \exp[-j(\omega - k\omega_0)t] \cdot \mathrm{d}t \\
&= \sum_{k=-\infty}^{\infty} g_k \cdot 2\pi\delta[-(\omega - k\omega_0)] \\
&= \sum_{k=-\infty}^{\infty} g_k \cdot 2\pi \cdot \delta(\omega - k\omega_0)
\end{aligned}
\tag{A1.14}
$$

Therefore, the transform is in fact composed of the Fourier series coefficients multiplied by delta functions. This is to be expected since, in the general case of a continuous transform, such transform can be shown to be related to the amplitudes of the constituent frequencies by:[1]

$$F(\omega) = \pi \frac{\mathrm{d}C}{\mathrm{d}\omega}$$

i.e. the transform, or spectral density, is the differential of the amplitude distribution C. In the case of the constituent frequencies being discrete, C is not continuous but is composed of pulses of magnitude $2g_k$. The differential of such pulses is another set of pulses at the same discrete frequencies but multiplied by δ, as arrived at in eq. (A1.14) above.

References

1. Kharkevich, A. A. (1960). "Spectra and Analysis". Translated from Russian, Consultants Bureau Enterprises, New York.
2. Oppenheim, A. V. and Shafer, R. W. (1975). "Digital Signal Processing". Prentice Hall, Englewood Cliffs, New Jersey.

Appendix 2

The Fast Fourier Transform

The detailed theory of the Fast Fourier Transform (FFT) is outside the scope of this book and it will not be developed here. However one of the algorithms possible for FFT implementation (the *decimation in time*) is presented so as to give the reader an idea of the principles on which the FFT is based, the signal processing potential of the method and the possibility to understand the structure of a useful routine, the FORTRAN code of which is listed below.

Let us assume that we have to compute the DFT $\{X(k)\}$, $k = 0, \ldots, N-1$, of a sequence $\{x(n)\}$, $n = 0, \ldots, N-1$, for N even. The sequence $\{x(n)\}$ can be divided in two sub-sequences $\{g(n)\}$ and $\{h(n)\}$ defined as

$$\begin{aligned} g(n) &= x(2n) \\ h(n) &= x(2n+1) \end{aligned} \qquad n = 0, \ldots, \frac{N}{2} - 1 \tag{A2.1}$$

The DFT of the two sequences $\{g(n)\}$ and $\{h(n)\}$ can then be written as

$$G(k) = \sum_{n=0}^{N/2-1} g(n) \exp\left(-j\frac{4\pi nk}{N}\right) \tag{A2.2}$$

$$H(k) = \sum_{n=0}^{N/2-1} h(n) \exp\left(-j\frac{4\pi nk}{N}\right) \tag{A2.3}$$

whilst $\{X(k)\}$ can be expressed in terms of the even and odd terms and is seen to be equal to

$$X(k)=\sum_{n=0}^{N/2-1}\left\{g(n)\exp\left(-j\frac{4\pi nk}{N}\right)+h(n)\exp\left(-j\frac{2\pi k(2n+1)}{N}\right)\right\}$$

$$=\sum_{n=0}^{N/2-1} g(n)\exp\left(-j\frac{4\pi nk}{N}\right)$$

$$+\exp\left(-j\frac{2\pi k}{N}\right)\sum_{n=0}^{N/2-1} h(n)\exp\left(-j\frac{4\pi nk}{N}\right) \qquad \text{(A2.4)}$$

$$k=0,\ldots,N-1$$

On comparing (A2.4) with (A2.2) and (A2.3) we see that

$$X(k)=G(k)+\exp\left(-j\frac{2\pi k}{N}\right)H(k) \qquad 0\leq k\leq\frac{N}{2}-1 \qquad \text{(A2.5)}$$

and

$$X\left(k+\frac{N}{2}\right)=G(k)-\exp\left(-j\frac{2\pi k}{N}\right)H(k) \qquad 0\leq k\leq\frac{N}{2}-1 \qquad \text{(A2.6)}$$

Therefore the discrete Fourier transform of the N samples sequence $x(n)$ can be obtained by means of the DFT of two suitable $N/2$ sample sub-sequences, $g(n)$, $h(n)$, with N additional complex multiplications.

The time necessary for a DFT computation is proportional to the number of products and in the direct case this time is proportional to N^2. In the second case, i.e. using eq. (A2.5) and (A2.6) the computation time is proportional to $2(N/2)^2+N=(N^2/2)+N\simeq N^2/2$ for N large. Note that if the number $N/2$ is again even, the procedure can be repeated and hence if N is a power of 2, it can be repeated to arrive to the transform of one point which is obviously the point itself. It is quite straightforward in this case to see that the total number of complex multiplications required to perform the DFT reduces to $(N/2)\log_2 N$.

The routine included here, equivalent to the original FFT structure of Cooley *et al.*,[1] is modified to accept real numbers instead of complex numbers (the use of real numbers in fact makes the routine in general faster than using complex number multiplications). It implements the above algorithm by using also an iterative computation of the exponential weights necessary in the algorithm.

```
      SUBROUTINE FFT(X,Y,N,M,IND)
C - FFT ROUTINE
C - DUMMY VARIABLES:
C    X = REAL PART OF THE FUNCTION OR OF THE SPECTRUM (DIM. N)
C    Y = IMAGINARY PART OF THE FUNCTION OR OF THE SPECTRUM (DIM. N)
C    N = NUMBER OF POINTS
C    M = DEFINED BY THE THE RELATION N=2**M
C    IND = 1 - DFT, 2 - IDFT
C - NO NORMALIZATION IS PERFORMED
      DIMENSION X(N),Y(N)
      NC = N/2
      PI = 3.141592653589793 2
      DO 20 L=1,M
      LE = 2**(M+1-L)
      LE1 = LE/2
      U1 = 1.
      U2 = 0.
      ARG = PI/LE1
      C =   COS(ARG)
      S = (-1)**IND*SIN(ARG)
      DO 20 J=1,LE1
      DO 15 I=J,N,LE
      IP = I+LE1
      T1 = X(I)+X(IP)
      T2 = Y(I)+Y(IP)
      T3 = X(I)-X(IP)
      T4 = Y(I)-Y(IP)
      X(IP) = T3*U1-T4*U2
      Y(IP) = T4*U1+T3*U2
      X(I) = T1
   15 Y(I) = T2
      U3 = U1*C-U2*S
      U2 = U2*C+U1*S
   20 U1 = U3
      NV2 = N/2
      NM1 = N-1
      J = 1
      DO 30 I=1,NM1
      IF(I.GE.J) GO TO 25
      T1 = X(J)
      T2 = Y(J)
      X(J) = X(I)
      Y(J) = Y(I)
      X(I) = T1
      Y(I) = T2
   25 K = NV2
   26 IF(K.GE.J) GO TO 30
      J = J-K
      K = K/2
      GO TO 26
   30 J = J+K
      RETURN
      END
```

Reference

1. Cooley, J. W., Lewis, P. A. W. and Welch, P. D. (1969). The finite Fourier transform. *I.E.E.E. Trans. Audio Electroacoustics* **AU-17**, **2**, 77–86.

Appendix 3

Computer Programs for Designing Digital Filters

A3.1 Design of FIR digital filters with the window method

```
C - PROGRAM WINDOW
C - DESIGN OF FIR DIGITAL FILTERS WITH THE WINDOW METHOD
C - INPUTS:
C   NCOE = NUMBER OF COEFFICIENTS (MAX. VALUE 128)
C   JTYPE = 1 - SEE JTYP, 2 - DIFFERENTIATOR 3 - HILBERT TRANSFORMER
C   JTYP = USED IF JTYPE EQ 1, 1 - LOWPASS, 2 - HIGHPASS, 3 - BANDPASS,
C          4 - BANDSTOP
C   F1 =  FIRST CUTOFF FREQUENCY (NORMALIZED)
C   F2 = SECOND CUTOFF FREQUENCY (NORMALIZED)
C   JW = 1 - LANCZOS WINDOW, 2 - WEBER WINDOW
C   PAR = WINDOW PARAMETER
C - IN THE LOWPASS, HIGHPASS AND DIFFERENTIATOR CASES ONLY F1 IS USED
C - F1 AND F2 ARE EXPRESSED IN THE F/FS SCALE
      DIMENSION H(128)
      DATA A0,A1,A2,A3/.99938,.041186,-1.637363,.828217/
      DATA B0,B1,B2,B3/1.496611,-1.701521,.372793,.0650621/
      IL = 9
      IS = 12
      READ FREE(IL) NCOE,JTYPE,JTYP,F1,F2,JW,PAR
      GO TO(10,20,20),JTYPE
   10 JT = JTYP
      GO TO 30
   20 JT = JTYPE+3
   30 NODD = NCOE-(NCOE/2)*2
      NCOE1 = NCOE/2
      L = NCOE1+1
      PIG = 3.1415926
      AA = .5*FLOAT(1-NODD)
      BB = FLOAT(NCOE1)-AA
      ARG = PIG/BB
      ARGW = 1.5*PAR/BB
C - IMPULSE RESPONSE EVALUATION
      IF(NODD.EQ.0) GO TO 40
      IF(JT.EQ.1) H(L) = 2*F1
      IF(JT.EQ.2) H(L) = 1.-2.*F1
      IF(JT.EQ.3) H(L) = 2.*(F2-F1)
      IF(JT.EQ.4) H(L) = 1.-2.*(F2-F1)
      IF(JT.EQ.5.OR.JT.EQ.6) H(L) = 0.
   40 DO 50 I=1,NCOE1
      K = L-I
      AK = FLOAT(I)-AA
      AF1 = 2.*PIG*AK*F1
      AF2 = 2.*PIG*AK*F2
      D = PIG*AK
      IF(JT.EQ.1) H(K) = SIN(AF1)/D
      IF(JT.EQ.2) H(K) = (SIN(PIG*AK)-SIN(AF1))/D
      IF(JT.EQ.3) H(K) = (SIN(AF2)-SIN(AF1))/D
      IF(JT.EQ.4) H(K) = (SIN(PIG*AK)-SIN(AF2)+SIN(AF1))/D
      IF(JT.EQ.5) H(K) =-(SIN(AF1)-AF1*COS(AF1))/(D*D)
      IF(JT.EQ.6) H(K) =-(COS(AF1)-COS(AF2))/D
C - WINDOWING
      GO TO(60,70),JW
   60 ARG1 = ARG*AK
      H(K) = H(K)*ABS(SIN(ARG1)/ARG1)**PAR
      GO TO 50
   70 ARG2 = ARGW*AK
      IF(ARG2.LE..75) H(K) = H(K)*(A0+A1*ARG2+A2*ARG2**2+A3*ARG2**3)
      IF(ARG2.GT..75) H(K) = H(K)*(B0+B1*ARG2+B2*ARG2**2+B3*ARG2**3)

   50 CONTINUE
      DO 90 K=1,NCOE1
      KK = NCOE+1-K
      H(KK) = H(K)
   90 IF(JTYPE.NE.1) H(KK) = -H(KK)
      WRITE(IS,111) NCOE,(H(K),K=1,NCOE)
  111 FORMAT(1H1,1X,'IMPULSE RESPONSE LENGTH',I5,///,1X,'INPULSE RESPONS
     1E SAMPLES',//(8(F15.7)))
      STOP
      END
```

A3.2 Design of frequency sampling filters using linear programming

```
C - PROGRAM FREQUENCY SAMPLING
C - DESIGN OF FREQUENCY SAMPLING FILTERS USING LINEAR PROGRAMMING
C - TYPE A SAMPLING POINTS
C - INPUTS:
C    N = NUMBER OF FREQUENCY SAMPLES (MAX. VALUE 128)
C    N1 = LOWER BAND EDGE
C    N2 = UPPER BAND EDGE
C    NVAR = NUMBER OF VARIABLE FREQUENCY SAMPLES (MAX. VALUE 5)
C    JTYPE = 1 - SEE JTYP, 2 - DIFFERENTIATOR, 3 - HILBERT TRANSFORMER
C    JTYP = USED IF JTYPE EQ 1, 1 - LOWPASS, 2 - HIGHPASS, 3 - BANDPASS
C           4 - BANDSTOP
C    INTR = NUMBER OF INTERPOLATED FREQUENCY POINTS (MAX. VALUE 16)
C - IN THE LOWPASS, HIGHPASS AND DIFFERENTIATOR CASES ONLY N1 IS USED
C - REQUIRED SUBPROGRAMS : FINTA, APMM (IBM/360 SSP)
      DIMENSION TOP(3072),IHE(4117),PIV(21),T(49)
      DIMENSION H(128)
      COMMON /KIND/KIN(5,2),N
      EXTERNAL FINTA
      NMAX = 128
      NN = 3072
      PIG = 3.1415926
      IL = 9
      READ FREE(IL)     N,N1,N2,NVAR,JTYPE,JTYP,INTR
      GO TO(10,20,20),JTYPE
   10 JT = JTYP
      GO TO 30
   20 JT = JTYPE+3
   30 NODD = N-(N/2)*2
      NA = N/2+.5*NODD
      DF = PIG/FLOAT(NA)
      DFA = DF/FLOAT(INTR)
      NM = N/2+NODD
      DO 35 I=1,5
      KIN(I,1) = 0
   35 KIN(I,2) = 0
C - SET UP OF THE APPROXIMATION PROBLEM
C - COMPUTATION OF THE DESIRED FREQUENCY FUNCTION AND OF THE
C   CORRESPONDING GRID  (SEE APMM)
      AN1P = FLOAT(N1+NVAR)*DF
      AN1M = FLOAT(N1-NVAR-2)*DF
      AN2P = FLOAT(N2+NVAR)*DF
      AN2M = FLOAT(N2-NVAR-2)*DF
      I3 = 0
      OMEPA = 0.
      GO TO(40,50,60,40,40,60),JT
   40 I1 = 1
      I2 = N1
      OMEP = AN1P
      OMEF = PIG
      DO 44 I=1,NVAR
   44 KIN(I,1) = N1+I
      IF(JT.NE.4) GO TO 70
      I3 = N2
      I4 = NM
      OMEP = AN1P
      OMEF = AN2M
      DO 48 I=1,NVAR
   48 KIN(I,2) = N2-I
```

```
      GO TO 70
   50 I1 = N1
      I2 = NM
      OMEP = 0.
      OMEF = AN1M
      DO 55 I=1,NVAR
   55 KIN(I,1) = N1-I
      GO TO 70
   60 I1 = N1
      I2 = N2
      OMEP = 0.
      OMEF = AN1M
      OMEPA = AN2P
      OMEFA = PIG
      DO 65 I=1,NVAR
      KIN(I,1) = N1-I
   65 KIN(I,2) = N2+1
   70 IND = 0
      DO 80 I=I1,I2
      IND = IND+1
      H(IND) = 1.
      IF (JT.EQ.5) H(IND) = FLOAT(I-1)*DF
   80 IHE(IND) = I
      IF(I3.EQ.0) GO TO 100
      DO 90 I=I3,I4
      IND = IND+1
      H(IND) = 1
   90 IHE(IND) = I
  100 OMEP = OMEP-DFA
      IF1 = KIN(1,1)
      IF2 = KIN(1,2)
      KIN(1,2) = 0
      L = 0
  110 OMEP = OMEP+DFA
      L = L+1
      RISP = 0.
      DO 120 K=1,IND
      KIN(1,1) = IHE(K)
      CALL FINTA(PIV,OMEP,0)
  120 RISP = RISP+PIV(1)*H(K)
      TOP(L) = -RISP
      TOP(NN+1-L) = OMEP
      AAA = OMEP-OMEF
      IF(ABS(OMEP-OMEF).GT.1.E-3) GO TO 110
      IF(OMEPA.EQ.0.) GO TO 130
      OMEP = OMEPA-DFA
      OMEF = OMEFA
      OMEPA = 0.
      GO TO 110
  130 DO 140 I=1,L
  140 TOP(L+I) = TOP(NN+1-I)
      NOUT = L
      KIN(1,1) = IF1
      KIN(1,2) = IF2
C - OPTIMIZATION
      CALL APMM(FINTA,NOUT,NVAR,TOP,IHE,PIV,T,ITER,IER)
      WRITE(12,333) (PIV(I),I=1,NVAR)
  333 FORMAT(///1X,'OPTIMUM VALUES OF THE VARIABLE FREQUENCY SAMPLES',5X
     *,5F15.7)
      EMAX = 0.
      DO 210 I=1,NOUT
  210 EMAX = AMAX1(EMAX,ABS(TOP(I)))
      ERMAX = 20.*ALOG10(ABS(EMAX))
      WRITE(12,222) ERMAX
  222 FORMAT(///1X,'MINIMUM OUT OF BAND ATTENUATION',1X,F10.2,3X,'DB')
  112 FORMAT(2I4/)
      STOP
      END
```

A3.3 Subroutine FINTA to evaluate the interpolation functions of a frequency sampling digital filter

```
      SUBROUTINE FINTA(F,ARG,M1)
C - ROUTINE TO EVALUATE THE INTERPOLATION FUNCTIONS OF A FREQUENCY
C   SAMPLING DIGITAL FILTER
C - TYPE A SAMPLING GRIDS
      DIMENSION F(1)
      COMMON /KIND/KIN(5,2),N
      DIMENSION FA(2)
      PIG = 3.1415926
      M = M1+1
      RN = FLOAT(N)
      ARG1 = ARG/2.
      ARG2 = PIG/RN
      DO 10 I=1,M
      FA(1) = 0.
      FA(2) = 0.
      DO 20 L=1,2
      IF(L.EQ.2.AND.KIN(I,L).EQ.0) GO TO 20
      K = KIN(I,L)-1
      ARGA = FLOAT(K)*ARG2
      ARG3 = ARG1-ARGA
      ARG4 = ARG1+ARGA
      FA(L) = (SIN(RN*ARG3)/SIN(ARG3)+SIN(RN*ARG4)/SIN(ARG4))/RN
   20 CONTINUE
      F(I) = FA(1)+FA(2)
      IF(K.EQ.0) F(I) = F(I)/2.
   10 CONTINUE
      RETURN
      END
```

A3.4 Program for the design of linear phase FIR digital filters using the Remez exchange algorithm

```
C  PROGRAM FOR THE DESIGN OF LINEAR PHASE FINITE IMPULSE
C  RESPONSE (FIR) FILTERS USING THE REMEZ EXCHANGE ALGORITHM
C  JIM MCCLELLAN, RICE UNIVERSITY, APRIL 13, 1973
C  THREE TYPES OF FILTERS ARE INCLUDED--BANDPASS FILTERS
C  DIFFERENTIATORS, AND HILBERT TRANSFORM FILTERS
C
C  THE INPUT DATA CONSISTS OF 5 CARDS
C
C  CARD 1--FILTER LENGTH, TYPE OF FILTER.  1-MULTIPLE
C  PASSBAND/STOPBAND, 2-DIFFERENTIATOR, 3-HILBERT TRANSFORM
C  FILTER.  NUMBER OF BANDS, CARD PUNCH DESIRED, AND GRID
C  DENSITY.
C
C  CARD 2--BANDEDGES, LOWER AND UPPER EDGES FOR EACH BAND
C  WITH A MAXIMUM OF 10 BANDS.
C
C  CARD 3--DESIRED FUNCTION (OR DESIRED SLOPE IF A
C  DIFFERENTIATOR) FOR EACH BAND.
C
C  CARD 4--WEIGHT FUNCTION IN EACH BAND.  FOR A
C  DIFFERENTIATOR, THE WEIGHT FUNCTION IS INVERSELY
C  PROPORTIONAL TO F.
C
C  THE FOLLOWING INPUT DATA SPECIFIES A LENGTH 32 BANDPASS
C  FILTER WITH STOPBANDS 0 TO 0.1 AND 0.425 TO 0.5, AND
C  PASSBAND FROM 0.2 TO 0.35 WITH WEIGHTING OF 10 IN THE
C  STOPBANDS AND 1 IN THE PASSBAND.  THE IMPULSE RESPONSE
C  WILL BE PUNCHED AND THE GRID DENSITY IS 32.
C
C  SAMPLE INPUT DATA SETUP
C  32,1,3,1,32
C  0,0.1,0.2,0.35,0.425,0.5
C  0,1,0
C  10,1,10
C
C  THE FOLLOWING INPUT DATA SPECIFIES A LENGTH 32 WIDEBAND
C  DIFFERENTIATOR WITH SLOPE 1 AND WEIGHTING OF 1/F.  THE
C  IMPULSE RESPONSE WILL NOT BE PUNCHED AND THE GRID
C  DENSITY IS ASSUMED TO BE 16.
C
C  32,2,1,0,0
C  0,0.5
C  1.0
C  1.0
C
C
      COMMON PI2,AD,DEV,X,Y,GRID,DES,WT,ALPHA,IEXT,NFCNS,NGRID
      DIMENSION IEXT(66),AD(66),ALPHA(66),X(66),Y(66)
      DIMENSION H(66)
      DIMENSION DES(1045),GRID(1045),WT(1045)
      DIMENSION EDGE(20),FX(10),WTX(10),DEVIAT(10)
      DOUBLE PRECISION PI2,PI
      DOUBLE PRECISION AD,DEV,X,Y
      PI2=6.283185307179586
      PI=3.141592653589793
```

```
C
C  THE PROGRAM IS SET UP FOR A MAXIMUM LENGTH OF 128, BUT
C  THIS UPPER LIMIT CAN BE CHANGED BY REDIMENSIONING THE
C  ARRAYS IEXT, AD, ALPHA, X, Y, H TO BE NFMAX/2 + 2.
C  THE ARRAYS DES, GRID, AND WT MUST DIMENSIONED
C  16(NFMAX/2 + 2).
C
      NFMAX=128
  100 CONTINUE
      JTYPE=0
C
C  PROGRAM INPUT SECTION
C
      READ *,NFILT,JTYPE,NBANDS,JPUNCH,LGRID
      IF(NFILT.GT.NFMAX.OR.NFILT.LT.3) CALL ERROR
      IF(NBANDS.LE.0) NBANDS=1
C
C  GRID DENSITY IS ASSUMED TO BE 16 UNLESS SPECIFIED
C  OTHERWISE
C
      IF(LGRID.LE.0) LGRID=16
      JB=2*NBANDS
      READ *,(EDGE(J),J=1,JB)
      READ *,(FX(J),J=1,NBANDS)
      READ *,(WTX(J),J=1,NBANDS)
      IF(JTYPE.EQ.0) CALL ERROR
      NEG=1
      IF(JTYPE.EQ.1) NEG=0
      NODD=NFILT/2
      NODD=NFILT-2*NODD
      NFCNS=NFILT/2
      IF(NODD.EQ.1.AND.NEG.EQ.0) NFCNS=NFCNS+1
C
C  SET UP THE DENSE GRID.  THE NUMBER OF POINTS IN THE GRID
C  IS (FILTER LENGTH + 1)*GRID DENSITY/2
C
      GRID(1)=EDGE(1)
      DELF=LGRID*NFCNS
      DELF=0.5/DELF
      IF(NEG.EQ.0) GO TO 135
      IF(EDGE(1).LT.DELF) GRID(1)=DELF
  135 CONTINUE
      J=1
      L=1
      LBAND=1
  140 FUP=EDGE(L+1)
  145 TEMP=GRID(J)
```

```
C
C  CALCULATE THE DESIRED MAGNITUDE RESPONSE AND THE WEIGHT
C  FUNCTION ON THE GRID
C
      DES(J)=EFF(TEMP,FX,WTX,LBAND,JTYPE)
      WT(J)=WATE(TEMP,FX,WTX,LBAND,JTYPE)
      J=J+1
      GRID(J)=TEMP+DELF
      IF(GRID(J).GT.FUP) GO TO 150
      GO TO 145
  150 GRID(J-1)=FUP
      DES(J-1)=EFF(FUP,FX,WTX,LBAND,JTYPE)
      WT(J-1)=WATE(FUP,FX,WTX,LBAND,JTYPE)
      LBAND=LBAND+1
      L=L+2
      IF(LBAND.GT.NBANDS) GO TO 160
      GRID(J)=EDGE(L)
      GO TO 140
  160 NGRID=J-1
      IF(NEG.NE.NODD) GO TO 165
      IF(GRID(NGRID).GT.(0.5-DELF)) NGRID=NGRID-1
  165 CONTINUE
C
C  SET UP A NEW APPROXIMATION PROBLEM WHICH IS EQUIVALENT
C  TO THE ORIGINAL PROBLEM
C
      IF(NEG) 170,170,180
  170 IF(NODD.EQ.1) GO TO 200
      DO 175 J=1,NGRID
      CHANGE=DCOS(PI*GRID(J))
      DES(J)=DES(J)/CHANGE
  175 WT(J)=WT(J)*CHANGE
      GO TO 200
  180 IF(NODD.EQ.1) GO TO 190
      DO 185 J=1,NGRID
      CHANGE=DSIN(PI*GRID(J))
      DES(J)=DES(J)/CHANGE
  185 WT(J)=WT(J)*CHANGE
      GO TO 200
  190 DO 195 J=1,NGRID
      CHANGE=DSIN(PI2*GRID(J))
      DES(J)=DES(J)/CHANGE
  195 WT(J)=WT(J)*CHANGE
C
C  INITIAL GUESS FOR THE EXTREMAL FREQUENCIES--EQUALLY
C  SPACED ALONG THE GRID
C
  200 TEMP=FLOAT(NGRID-1)/FLOAT(NFCNS)
      DO 210 J=1,NFCNS
  210 IEXT(J)=(J-1)*TEMP+1
      IEXT(NFCNS+1)=NGRID
      NM1=NFCNS-1
      NZ=NFCNS+1
C
C  CALL THE REMEZ EXCHANGE ALGORITHM TO DO THE APPROXIMATION
C  PROBLEM
C
      CALL REMEZ(EDGE,NBANDS)
```

```
C
C  CALCULATE THE IMPULSE RESPONSE.
C
      IF(NEG) 300,300,320
  300 IF(NODD.EQ.0) GO TO 310
      DO 305 J=1,NM1
  305 H(J)=0.5*ALPHA(NZ-J)
      H(NFCNS)=ALPHA(1)
      GO TO 350
  310 H(1)=0.25*ALPHA(NFCNS)
      DO 315 J=2,NM1
  315 H(J)=0.25*(ALPHA(NZ-J)+ALPHA(NFCNS+2-J))
      H(NFCNS)=0.5*ALPHA(1)+0.25*ALPHA(2)
      GO TO 350
  320 IF(NODD.EQ.0) GO TO 330
      H(1)=0.25*ALPHA(NFCNS)
      H(2)=0.25*ALPHA(NM1)
      DO 325 J=3,NM1
  325 H(J)=0.25*(ALPHA(NZ-J)-ALPHA(NFCNS+3-J))
      H(NFCNS)=0.5*ALPHA(1)-0.25*ALPHA(3)
      H(NZ)=0.0
      GO TO 350
  330 H(1)=0.25*ALPHA(NFCNS)
      DO 335 J=2,NM1
  335 H(J)=0.25*(ALPHA(NZ-J)-ALPHA(NFCNS+2-J))
      H(NFCNS)=0.5*ALPHA(1)-0.25*ALPHA(2)
C
C  PROGRAM OUTPUT SECTION.
C
  350 PRINT 360
  360 FORMAT(1H1, 70(1H*)//25X,'FINITE IMPULSE RESPONSE (FIR)'/
     125X,'LINEAR PHASE DIGITAL FILTER DESIGN'/
     225X,'REMEZ EXCHANGE ALGORITHM'/)
      IF(JTYPE.EQ.1) PRINT 365
  365 FORMAT(25X,'BANDPASS FILTER'/)
      IF(JTYPE.EQ.2) PRINT 370
  370 FORMAT(25X,'DIFFERENTIATOR'/)
      IF(JTYPE.EQ.3) PRINT 375
  375 FORMAT(25X,'HILBERT TRANSFORMER'/)
      PRINT 378,NFILT
  378 FORMAT(15X,'FILTER LENGTH = ',I3/)
      PRINT 380
  380 FORMAT(15X,'***** IMPULSE RESPONSE *****')
      DO 381 J=1,NFCNS
      K=NFILT+1-J
      IF(NEG.EQ.0) PRINT 382,J,H(J),K
      IF(NEG.EQ.1) PRINT 383,J,H(J),K
  381 CONTINUE
  382 FORMAT(20X,'H(',I3,') = ',E15.8,' = H(',I4,')')
  383 FORMAT(20X,'H(',I3,') = ',E15.8,' = -H(',I4,')')
      IF(NEG.EQ.1.AND.NODD.EQ.1) PRINT 384,NZ
  384 FORMAT(20X,'H(',I3,') =  0.0')
      DO 450 K=1,NBANDS,4
      KUP=K+3
      IF(KUP.GT.NBANDS) KUP=NBANDS
      PRINT 385,(J,J=K,KUP)
  385 FORMAT(/24X,4('BAND',I3,8X))
      PRINT 390,(EDGE(2*J-1),J=K,KUP)
```

```
  390 FORMAT(2X,'LOWER BAND EDGE',5F15.9)
      PRINT 395,(EDGE(2*J),J=K,KUP)
  395 FORMAT(2X,'UPPER BAND EDGE',5F15.9)
      IF(JTYPE.NE.2) PRINT 400,(FX(J),J=K,KUP)
  400 FORMAT(2X,'DESIRED VALUE',2X,5F15.9)
      IF(JTYPE.EQ.2) PRINT 405,(FX(J),J=K,KUP)
  405 FORMAT(2X,'DESIRED SLOPE',2X,5F15.9)
      PRINT 410,(WTX(J),J=K,KUP)
  410 FORMAT(2X,'WEIGHTING',6X,5F15.9)
      DO 420 J=K,KUP
  420 DEVIAT(J)=DEV/WTX(J)
      PRINT 425,(DEVIAT(J),J=K,KUP)
  425 FORMAT(2X,'DEVIATION',6X,5F15.9)
      IF(JTYPE.NE.1) GO TO 450
      DO 430 J=K,KUP
  430 DEVIAT(J)=20.0*ALOG10(DEVIAT(J))
      PRINT 435,(DEVIAT(J),J=K,KUP)
  435 FORMAT(2X,'DEVIATION IN DB',5F15.9)
  450 CONTINUE
      PRINT 455,(GRID(IEXT(J)),J=1,NZ)
  455 FORMAT(/2X,'EXTREMAL FREQUENCIES'/(2X,5F12.7))
      PRINT 460
  460 FORMAT(/1X,70(1H*)/1H1)
      IF(JPUNCH.NE.0) PUNCH*,(H(J),J=1,NFCNS)
      IF(NFILT.NE.0) GO TO 100
      RETURN
      END


      FUNCTION EFF(TEMP,FX,WTX,LBAND,JTYPE)
C
C  FUNCTION TO CALCULATE THE DESIRED MAGNITUDE RESPONSE
C  AS A FUNCTION OF FREQUENCY
C
      DIMENSION FX(5),WTX(5)
      IF(JTYPE.EQ.2) GO TO 1
      EFF=FX(LBAND)
      RETURN
    1 EFF=FX(LBAND)*TEMP
      RETURN
      END


      FUNCTION WATE(TEMP,FX,WTX,LBAND,JTYPE)
C
C  FUNCTION TO CALCULATE THE WEIGHT FUNCTION AS A FUNCTION
C  OF FREQUENCY
C
      DIMENSION FX(5),WTX(5)
      IF(JTYPE.EQ.2) GO TO 1
      WATE=WTX(LBAND)
      RETURN
    1 IF(FX(LBAND).LT.0.0001) GO TO 2
      WATE=WTX(LBAND)/TEMP
      RETURN
    2 WATE=WTX(LBAND)
      RETURN
      END
```

```
      SUBROUTINE ERROR
      PRINT 1
    1 FORMAT(' ************ ERROR IN INPUT DATA **********')
      STOP
      END
```

```
      SUBROUTINE REMEZ(EDGE,NBANDS)
C
C  THIS SUBROUTINE IMPLEMENTS THE REMEZ EXCHANGE ALGORITHM
C  FOR THE WEIGHTED CHEBYCHEV APPROXIMATION OF A CONTINUOUS
C  FUNCTION WITH A SUM OF COSINES.  INPUTS TO THE SUBROUTINE
C  ARE A DENSE GRID WHICH REPLACES THE FREQUENCY AXIS, THE
C  DESIRED FUNCTION ON THIS GRID, THE WEIGHT FUNCTION ON THE
C  GRID, THE NUMBER OF COSINES, AND AN INITIAL GUESS OF THE
C  EXTREMAL FREQUENCIES.  THE PROGRAM MINIMIZES THE CHEBYCHEV
C  ERROR BY DETERMINING THE BEST LOCATION OF THE EXTREMAL
C  FREQUENCIES (POINTS OF MAXIMUM ERROR) AND THEN CALCULATES
C  THE COEFFICIENTS OF THE BEST APPROXIMATION.
C
      COMMON PI2,AD,DEV,X,Y,GRID,DES,WT,ALPHA,IEXT,NFCNS,NGRID
      DIMENSION EDGE(20)
      DIMENSION IEXT(66),AD(66),ALPHA(66),X(66),Y(66)
      DIMENSION DES(1045),GRID(1045),WT(1045)
      DIMENSION A(66),P(65),Q(65)
      DOUBLE PRECISION PI2,DNUM,DDEN,DTEMP,A,P,Q
      DOUBLE PRECISION AD,DEV,X,Y
C
C  THE PROGRAM ALLOWS A MAXIMUM NUMBER OF ITERATIONS OF 25
      ITRMAX=25
      DEVL=-1.0
      NZ=NFCNS+1
      NZZ=NFCNS+2
      NITER=0
  100 CONTINUE
      IEXT(NZZ)=NGRID+1
      NITER=NITER+1
      IF(NITER.GT.ITRMAX) GO TO 400
      DO 110 J=1,NZ
      DTEMP=GRID(IEXT(J))
      DTEMP=DCOS(DTEMP*PI2)
  110 X(J)=DTEMP
      JET=(NFCNS-1)/15+1
      DO 120 J=1,NZ
  120 AD(J)=D(J,NZ,JET)
      DNUM=0.0
      DDEN=0.0
      K=1
      DO 130 J=1,NZ
      L=IEXT(J)
      DTEMP=AD(J)*DES(L)
      DNUM=DNUM+DTEMP
      DTEMP=K*AD(J)/WT(L)
      DDEN=DDEN+DTEMP
```

```
  130 K=-K
      DEV=DNUM/DDEN
      NU=1
      IF(DEV.GT.0.0) NU=-1
      DEV=-NU*DEV
      K=NU
      DO 140 J=1,NZ
      L=IEXT(J)
      DTEMP=K*DEV/WT(L)
      Y(J)=DES(L)+DTEMP
  140 K=-K
      IF(DEV.GE.DEVL) GO TO 150
      CALL OUCH
      GO TO 400
  150 DEVL=DEV
      JCHNGE=0
      K1=IEXT(1)
      KNZ=IEXT(NZ)
      KLOW=0
      NUT=-NU
      J=1
C
C  SEARCH FOR THE EXTREMAL FREQUENCIES OF THE BEST
C  APPROXIMATION
C
  200 IF(J.EQ.NZZ) YNZ=COMP
      IF(J.GE.NZZ) GO TO 300
      KUP=IEXT(J+1)
      L=IEXT(J)+1
      NUT=-NUT
      IF(J.EQ.2) Y1=COMP
      COMP=DEV
      IF(L.GE.KUP) GO TO 220
      ERR=GEE(L,NZ)
      ERR=(ERR-DES(L))*WT(L)
      DTEMP=NUT*ERR-COMP
      IF(DTEMP.LE.0.0) GO TO 220
      COMP=NUT*ERR
  210 L=L+1
      IF(L.GE.KUP) GO TO 215
      ERR=GEE(L,NZ)
      ERR=(ERR-DES(L))*WT(L)
      DTEMP=NUT*ERR-COMP
      IF(DTEMP.LE.0.0) GO TO 215
      COMP=NUT*ERR
      GO TO 210
  215 IEXT(J)=L-1
      J=J+1
      KLOW=L-1
      JCHNGE=JCHNGE+1
      GO TO 200
  220 L=L-1
  225 L=L-1
      IF(L.LE.KLOW) GO TO 250
      ERR=GEE(L,NZ)
      ERR=(ERR-DES(L))*WT(L)
      DTEMP=NUT*ERR-COMP
      IF(DTEMP.GT.0.0) GO TO 230
      IF(JCHNGE.LE.0) GO TO 225
      GO TO 260
```

```
230 COMP=NUT*ERR
235 L=L-1
    IF(L.LE.KLOW) GO TO 240
    ERR=GEE(L,NZ)
    ERR=(ERR-DES(L))*WT(L)
    DTEMP=NUT*ERR-COMP
    IF(DTEMP.LE.0.0) GO TO 240
    COMP=NUT*ERR
    GO TO 235
240 KLOW=IEXT(J)
    IEXT(J)=L+1
    J=J+1
    JCHNGE=JCHNGE+1
    GO TO 200
250 L=IEXT(J)+1
    IF(JCHNGE.GT.0) GO TO 215
255 L=L+1
    IF(L.GE.KUP) GO TO 260
    ERR=GEE(L,NZ)
    ERR=(ERR-DES(L))*WT(L)
    DTEMP=NUT*ERR-COMP
    IF(DTEMP.LE.0.0) GO TO 255
    COMP=NUT*ERR
    GO TO 210
260 KLOW=IEXT(J)
    J=J+1
    GO TO 200
300 IF(J.GT.NZZ) GO TO 320
    IF(K1.GT.IEXT(1)) K1=IEXT(1)
    IF(KNZ.LT.IEXT(NZ)) KNZ=IEXT(NZ)
    NUT1=NUT
    NUT=-NU
    L=0
    KUP=K1
    COMP=YNZ*(1.00001)
    LUCK=1
310 L=L+1
    IF(L.GE.KUP) GO TO 315
    ERR=GEE(L,NZ)
    ERR=(ERR-DES(L))*WT(L)
    DTEMP=NUT*ERR-COMP
    IF(DTEMP.LE.0.0) GO TO 310
    COMP=NUT*ERR
    J=NZZ
    GO TO 210
315 LUCK=6
    GO TO 325
320 IF(LUCK.GT.9) GO TO 350
    IF(COMP.GT.Y1) Y1=COMP
    K1=IEXT(NZZ)
325 L=NGRID+1
    KLOW=KNZ
    NUT=-NUT1
    COMP=Y1*(1.00001)
330 L=L-1
    IF(L.LE.KLOW) GO TO 340
    ERR=GEE(L,NZ)
    ERR=(ERR-DES(L))*WT(L)
    DTEMP=NUT*ERR-COMP
```

```
      IF(DTEMP.LE.0.0) GO TO 330
      J=NZZ
      COMP=NUT*ERR
      LUCK=LUCK+10
      GO TO 235
  340 IF(LUCK.EQ.6) GO TO 370
      DO 345 J=1,NFCNS
  345 IEXT(NZZ-J)=IEXT(NZ-J)
      IEXT(1)=K1
      GO TO 100
  350 KN=IEXT(NZZ)
      DO 360 J=1,NFCNS
  360 IEXT(J)=IEXT(J+1)
      IEXT(NZ)=KN
      GO TO 100
  370 IF(JCHNGE.GT.0) GO TO 100
C  CALCULATION OF THE COEFFICIENTS OF THE BEST APPROXIMATION
C  USING THE INVERSE DISCRETE FOURIER TRANSFORM
C
  400 CONTINUE
      NM1=NFCNS-1
      FSH=1.0E-06
      GTEMP=GRID(1)
      X(NZZ)=-2.0
      CN=2*NFCNS-1
      DELF=1.0/CN
      L=1
      KKK=0
      IF(EDGE(1).EQ.0.0.AND.EDGE(2*NBANDS).EQ.0.5) KKK=1
      IF(NFCNS.LE.3) KKK=1
      IF(KKK.EQ.1) GO TO 405
      DTEMP=DCOS(PI2*GRID(1))
      DNUM=DCOS(PI2*GRID(NGRID))
      AA=2.0/(DTEMP-DNUM)
      BB=-(DTEMP+DNUM)/(DTEMP-DNUM)
  405 CONTINUE
      DO 430 J=1,NFCNS
      FT=(J-1)*DELF
      XT=DCOS(PI2*FT)
      IF(KKK.EQ.1) GO TO 410
      XT=(XT-BB)/AA
      FT=ARCOS(XT)/PI2
  410 XE=X(L)
      IF(XT.GT.XE) GO TO 420
      IF((XE-XT).LT.FSH) GO TO 415
      L=L+1
      GO TO 410
  415 A(J)=Y(L)
      GO TO 425
  420 IF((XT-XE).LT.FSH) GO TO 415
      GRID(1)=FT
      A(J)=GEE(1,NZ)
  425 CONTINUE
      IF(L.GT.1) L=L-1
```

```
430 CONTINUE
    GRID(1)=GTEMP
    DDEN=PI2/CN
    DO 510 J=1,NFCNS
    DTEMP=0.0
    DNUM=(J-1)*DDEN
    IF(NM1.LT.1) GO TO 505
    DO 500 K=1,NM1
500 DTEMP=DTEMP+A(K+1)*DCOS(DNUM*K)
505 DTEMP=2.0*DTEMP+A(1)
510 ALPHA(J)=DTEMP
    DO 550 J=2,NFCNS
550 ALPHA(J)=2*ALPHA(J)/CN
    ALPHA(1)=ALPHA(1)/CN
    IF(KKK.EQ.1) GO TO 545
    P(1)=2.0*ALPHA(NFCNS)*BB+ALPHA(NM1)
    P(2)=2.0*AA*ALPHA(NFCNS)
    Q(1)=ALPHA(NFCNS-2)-ALPHA(NFCNS)
    DO 540 J=2,NM1
    IF(J.LT.NM1) GO TO 515
    AA=0.5*AA
    BB=0.5*BB
515 CONTINUE
    P(J+1)=0.0
    DO 520 K=1,J
    A(K)=P(K)
520 P(K)=2.0*BB*A(K)
    P(2)=P(2)+A(1)*2.0*AA
    JM1=J-1
    DO 525 K=1,JM1
525 P(K)=P(K)+Q(K)+AA*A(K+1)
    JP1=J+1
    DO 530 K=3,JP1
530 P(K)=P(K)+AA*A(K-1)
    IF(J.EQ.NM1) GO TO 540
    DO 535 K=1,J
535 Q(K)=-A(K)
    Q(1)=Q(1)+ALPHA(NFCNS-1-J)
540 CONTINUE
    DO 543 J=1,NFCNS
543 ALPHA(J)=P(J)
545 CONTINUE
    IF(NFCNS.GT.3) RETURN
    ALPHA(NFCNS+1)=0.0
    ALPHA(NFCNS+2)=0.0
    RETURN
    END


      DOUBLE PRECISION FUNCTION D(K,N,M)
C
C  FUNCTION TO CALCULATE THE LAGRANGE INTERPOLATION
C  COEFFICIENTS FOR USE IN THE FUNCTION GEE.
C
      COMMON PI2,AD,DEV,X,Y,GRID,DES,WT,ALPHA,IEXT,NFCNS,NGRID
      DIMENSION IEXT(66),AD(66),ALPHA(66),X(66),Y(66)
      DIMENSION DES(1045),GRID(1045),WT(1045)
```

```
      DOUBLE PRECISION AD,DEV,X,Y
      DOUBLE PRECISION Q
      DOUBLE PRECISION PI2
      D=1.0
      Q=X(K)
      DO 3 L=1,M
      DO 2 J=L,N,M
      IF(J-K)1,2,1
    1 D=2.0*D*(Q-X(J))
    2 CONTINUE
    3 CONTINUE
      D=1.0/D
      RETURN
      END
```

```
      DOUBLE PRECISION FUNCTION GEE(K,N)
C
C  FUNCTION TO EVALUATE THE FREQUENCY RESPONSE USING THE
C  LAGRANGE INTERPOLATION FORMULA IN THE BARYCENTRIC FORM
C
      COMMON PI2,AD,DEV,X,Y,GRID,DES,WT,ALPHA,IEXT,NFCNS,NGRID
      DIMENSION IEXT(66),AD(66),ALPHA(66),X(66),Y(66)
      DIMENSION DES(1045),GRID(1045),WT(1045)
      DOUBLE PRECISION P,C,D,XF
      DOUBLE PRECISION PI2
      DOUBLE PRECISION AD,DEV,X,Y
      P=0.0
      XF=GRID(K)
      XF=DCOS(PI2*XF)
      D=0.0
      DO 1 J=1,N
      C=XF-X(J)
      C=AD(J)/C
      D=D+C
    1 P=P+C*Y(J)
      GEE=P/D
      RETURN
      END
```

```
      SUBROUTINE OUCH
      PRINT 1
    1 FORMAT(' ************ FAILURE TO CONVERGE **********'/
     1'OPROBABLE CAUSE IS MACHINE ROUNDING ERROR'/
     2'OTHE IMPULSE RESPONSE MAY BE CORRECT'/
     3'OCHECK WITH A FREQUENCY RESPONSE')
      RETURN
      END
```

A3.5 Program for the design of 2-D circularly symmetric digital filters by means of the McClellan transformation

```
C - PROGRAM ROTAZ
C - DESIGN OF 2-D CIRCULARLY SYMMETRIC DIGITAL FILTERS BY MEANS
C   OF THE MCCLELLAN TRANSFORMATION
C - INPUTS:
C   CARD 1
C   NDIM = NUMBER OF INDEPENDENT COEFFICIENTS OF THE 1-D FILTER
C          (TOTAL IMPULSE RESPONSE LENGHT 2*NDIM-1)
C   CARD 2 - N
C   COEFFICIENTS OF THE NONCAUSAL 1-D FILTER FROM 0 TO NDIM-1
C - REQUIRED SUBPROGRAMS - TCNP (IBM/360 SSP): TRANSFORMATION OF
C   A SERIES EXPANSION IN CHEBYSHEV POLYNOMIAL TO A POLYNOMIAL
      DIMENSION POL(33),AMAT(33,33),ROT(33,33),FAT(33),AM(33),BM(33),
     # CM(33),DM(33),BB(33),W(66),RPOT(33),RK(33),RP(33)
      REAL IP,IQ,IB
      DATA A,B,C,D/.5,.5,.5,-.5/AM(1),BM(1),CM(1),DM(1)/1.,1.,1.,1./
      DATA FAT(1)/1./
      OPEN 12,'SYSOUT',ATT="P"
      OPEN 9,'TMP'
      READ FREE(9)NDIM
      READ FREE(9)(POL(J),J=1,NDIM)
      NDM1 = NDIM-1
      DO 2 J=1,NDM1
      AM(J+1) = AM(J)*A
      BM(J+1) = BM(J)*B
      CM(J+1) = CM(J)*C
      DM(J+1)= DM(J)*D
      FATT = 1.
      IF(J.LE.1)GO TO 2
      DO 1 K=1,J
    1 FATT = FATT*K
    2 FAT(J+1) = FATT
      DO 3 J=2,NDIM
    3 POL(J) = 2.*POL(J)
C - TRANSFORMATION OF THE SERIES EXPANSION IN CHEBYSHEV POLYNOMIAL
C   TO A POLYNOMIAL
      CALL TCNP(1.,0.,BB,NDIM,POL,W)
C - TRANSFORMATION OF THE POLYNOMIAL TO A BIVARIATE POLYNOMIAL
C   ACCORDING TO MCCLELLAN TRANFORMATION
      DO 4 I=1,NDIM
      DO 4 J=1,NDIM
      AMAT(I,J) = 0.
    4 ROT(I,J) = 0.
      DO 5 I=1,NDIM
      DO 5 K=1,I
      COEF1 = FAT(I)/FAT(K)
      K1 = I-K+1
      DO 5 L=1,K1
      COEF2 = COEF1/FAT(L)
      K2 = I-K-L+2
      DO 5 M=1,K2
      N = I-K-L-M+3
      COEF = COEF2/FAT(M)/FAT(N)
      COEFF = AM(K)*BM(L)*CM(M)*DM(N)
    5 AMAT(K+M-1,L+M-1) = AMAT(K+M-1,L+M-1)+COEF*COEFF*BB(I)
C - EVALUATION OF THE COEFFICIENTS OF THE 2-D IMPULSE RESPONSE
C   (FIRST QUADRANT)
      DO 33 I=1,NDIM
      IM1 = I-1
      RPOT(1) = 1.
      IF(IM1.EQ.0)GO TO 8
      RPOT(2) = 1.
      IF(IM1.EQ.1) GO TO 8
      IP = 1
      DO 7 K=2,IM1
      DO 6 L=2,K
```

```
      IQ = RPOT(L)
      IB = IP+IQ
      IP = IQ
    6 RPOT(L) = IB
    7 RPOT(K+1) = 1.
    8 JK = I/2
      NODD = I-JK*2
      IF(IM1)9,9,10
    9 RN = 1.
      GO TO 11
   10 RN = 2.**(IM1-1)
   11 IF(NODD)12,12,14
   12 DO 13 J=1,JK
   13 RK(J) = RPOT(J)/RN
      JJK = JK
      GO TO 18
   14 JKK = JK+1
      DO 15 J=1,JKK
   15 RK(J) = RPOT(J)/RN
      IF(IM1)17,17,16
   16 RK(JKK) = RK(JKK)/2.
   17 JJK = JKK
   18 DO 33 L=1,NDIM
      LM1 = L-1
      RPOT(1) = 1.
      IF(LM1.EQ.0)GO TO 21
      RPOT(2) = 1.
      IF(LM1.EQ.1)GO TO 21
      IP = 1
      DO 20 K=2,LM1
      DO 19 M=2,K
      IQ = RPOT(M)
      IB = IP+IQ
      IP = IQ
   19 RPOT(M) = IB
   20 RPOT(K+1) = 1.
   21 JP = L/2
      NODD = L-JP*2
      IF(LM1)22,22,23
   22 RN = 1.
      GO TO 24
   23 RN = 2.**(LM1-1)
   24 IF(NODD)25,25,27
   25 DO 26 J=1,JP
   26 RP(J) = RPOT(J)/RN
      JJP = JP
      GO TO 31
   27 JPP = JP+1
      DO 28 J=1,JPP
   28 RP(J) = RPOT(J)/RN
      IF(LM1)30,30,29
   29 RP(JPP) = RP(JPP)/2.
   30 JJP = JPP
   31 DO 32 J=1,JJK
      M = IM1-2*J+3
      DO 32 K=1,JJP
      N = LM1-2*K+3
   32 ROT(M,N) = ROT(M,N)+RK(J)*RP(K)*AMAT(I,L)
   33 CONTINUE
      DO 34 I=2,NDIM
      ROT(I,1) = ROT(I,1)/2.
      ROT(1,I) = ROT(1,I)/2.
      DO 34 J=2,NDIM
   34 ROT(I,J) = ROT(I,J)/4.
      WRITE(12,102)
      DO 35 I=1,NDIM
   35 WRITE(12,100)(ROT(I,J),J=1,NDIM)
      CALL RESET
      STOP
  100 FORMAT(1H0,(8E15.7))
  101 FORMAT(1H0,5I3,4F10.5)
  102 FORMAT(1H1,'TWO-DIMENSIONAL IMPULSE RESPONSE')
      END
```

A3.6 Program for the design of Butterworth and Chebyshev IIR digital filters by means of the bilinear transformation

```
C - PROGRAM RICOR
C - DESIGN OF BUTTERWORTH AND CHEBYSHEV IIR DIGITAL FILTERS
C   BY MEANS OF THE BILINEAR TRANSFORMATION
C - INPUTS:
C   CARD1 - FREQUENCY TRANSFORMATION
C   JTYPE = 0 - NO TRANSFORMATION, 1 - LOWPASS TO LOWPASS, 2 - LOWPASS
C           TO HIGHPASS, 3 - LOWPASS TO BANDPASS,4 - LOWPASS TO BANDSTOP
C   F1 = LOWER BAND-EDGE AFTER TRANSFORMATION
C   F2 = UPPER BAND-EDGE AFTER TRANSFORMATION
C   IN THE 1 AND 2 CASES ONLY F1 IS USED
C   CARD2 - LOWPASS PROTOTYPE
C   ITYPE = 1 - BUTTERWORTH, 2 - CHEBYSHEV
C   FT = CUTOFF FREQUENCY
C   FATT = POINT OF THE FREQUENCY AXIS WHERE THE ATTENUATION MUST BE AT
C          LEAST 'ATTEN' DB
C   RIPPL = BANDPASS RIPPLE EXPRESSED AS THE PERCENTAGE OF THE
C           MAGNITUDE VALUE
C   ATTEN = ATTENUATION (IN DB) AT THE FATT FREQUENCY
C - ALL THE FREQUENCY VALUES ARE EXPRESSED IN THE F/FS SCALE
C - THE COEFFICIENTS OF THE CASCADE IMPLEMENTATION OF THE PROTOTYPE
C   FILTER ARE FIRST COMPUTED. IN THE LOWPASS TO BANDPASS OR BANDSTOP
C   TRANSFORMATION FOURTH ORDER SECTIONS ARE OBTAINED.
C - REQUIRED SUBPROGRAMS - TRASF
      COMPILER DOUBLE PRECISION
      DIMENSION POLRE(40),POLIM(40),A(5,20),B(5,20),CEB(41)
      OPEN 12,'SYSOUT',ATT='AP'
      CALL FOPEN(9,'TMP')
      IL = 9
      IS = 12
      READ FREE(IL)  JTYPE,F1,F2
      READ FREE(IL)  ITYPE,FT,FATT,RIPPL,ATTEN
      PIG = 3.141592653589
      NU = 3
      FCA = TAN(PIG*FT)
      FATA = TAN(PIG*FATT)
      FF = FATA/FCA
      IF(ITYPE.EQ.2) EPS = SQRT(1./(1.-RIPPL/100.)**2-1.)
      ATT = 10**(ATTEN/10.)
C - COMPUTATION OF THE ORDER OF THE FILTER
      N = 1
    2 GO TO(4,6),ITYPE
    4 RES = 1./(1.+FF**(2*N))
      GO TO 8
    6 CEB(1) = 1.
      CEB(2) = FF
      IF(N.LE.1) GO TO 7
      F = FF+FF
      DO 5 I=2,N
    5 CEB(I+1) = F*CEB(I)-CEB(I-1)
    7 RES = 1./(1.+(EPS*CEB(N+1))**2)
    8 IF(RES.LT.ATT) GO TO 9
      N = N+1
      GO TO 2
    9 IF(MOD(N,2).EQ.1) N = N+1
C - POLE POSITION EVALUATION
      AT = FCA
      BT = FCA
      IF(ITYPE.EQ.1) GO TO 10
      AA = ((1.+SQRT(1.+EPS**2))/EPS)**(1./FLOAT(N))
      AT = (AA-1./AA)*AT/2.
      BT = (AA+1./AA)*BT/2.
   10 NMAX = N/2
      DO 20 JX=1,NMAX
      TETAP = PIG*(2.*FLOAT(JX)-1.)/(2.*FLOAT(N))
      POLRE(JX) = -AT*COS(TETAP)
```

```
   20 POLIM(JX) = BT*SIN(TETAP)
      DO 30 JZ=1,NMAX
      P1 = POLIM(JZ)**2
      P2 = (1.-POLRE(JZ))**2
      P3 = POLRE(JZ)**2
      POLRE(JZ) = (1.-P3-P1)/(P2+P1)
   30 POLIM(JZ) = (2.*POLIM(JZ))/(P2+P1)
      DO 40 I=1,NMAX
C - EVALUATION OF THE SECOND ORDER SECTION COEFFICIENTS
      B(1,I) = 1.
      B(2,I) = -2.*POLRE(I)
   40 B(3,I) = POLRE(I)*POLRE(I)+POLIM(I)*POLIM(I)
      IF(ITYPE.EQ.1) COST = 1.
      IF(ITYPE.EQ.2) COST = (1.-RIPPL/100.)**(1./FLOAT(NMAX))
      DO 50 I=1,NMAX
      TOT = 4./(B(1,I)+B(2,I)+B(3,I))
      CC = TOT/COST
      A(1,I) = 1./CC
      A(2,I) = 2./CC
   50 A(3,I) = 1./CC
      IF(JTYPE.EQ.0) GO TO 60
C - FREQUENCY TRANSFORMATION
      CALL TRASF(A,B,5,20,NMAX,NU,FT,F1,F2,JTYPE)
   60 WRITE(IS,111) N
  111 FORMAT(1H1,1X,'FILTER ORDER N = ',I3)
      WRITE(IS,112)
  112 FORMAT(///1X,'NUMERATOR COEFFICIENTS'//)
      DO 70 I=1,NMAX
   70 WRITE(IS,114) (A(IL,I),IL=1,NU)
      WRITE(IS,113)
  113 FORMAT(///1X,'DENOMINATOR COEFFICIENTS'//)
      DO 80 I=1,NMAX
   80 WRITE(IS,114) (B(IL,I),IL=1,NU)
  114 FORMAT(5(4X,E21.14))
      STOP
      END
```

A3.7 Subroutine for the frequency transformation of IIR digital filters

```
      COMPILER DOUBLE PRECISION
      SUBROUTINE TRASF(A,B,NA,NB,NCEL,NU,FP,F1,F2,JTYPE)
C - ROUTINE FOR THE FREQUENCY TRANSFORMATION OF IIR DIGITAL FILTERS
      DIMENSION A(NA,NB),B(NA,NB),AN(5,3),FINA(3),FINB(3)
      PIG = 3.141592653589
      GO TO(10,20,30,40),JTYPE
   10 AL = SIN(PIG*(FP-F1))/SIN(PIG*(FP+F1))
      A0 = -AL
      A1 = 1.
      A2 = 0.
      B0 = A1
      B1 = A0
      B2 = 0.
      NU = 3
      GO TO 50
   20 AL = -COS(PIG*(FP+F1))/COS(PIG*(FP-F1))
      A0 = -AL
      A1 = -1.
      A2 = 0.
      B0 = -A1
      B1 = -A0
      B2 = 0.
      NU = 3
      GO TO 50
   30 AL = COS(PIG*(F2+F1))/COS(PIG*(F2-F1))
      CAPPA = TAN(PIG*FP)/TAN(PIG*(F2-F1))
      A0 = -(CAPPA-1.)/(CAPPA+1.)
      A1 = 2.*AL*CAPPA/(CAPPA+1.)
      A2 = -1.
      B0 = -A2
      B1 = -A1
      B2 = -A0
      NU = 5
      GO TO 50
   40 AL = COS(PIG*(F2+F1))/COS(PIG*(F2-F1))
      CAPPA = TAN(PIG*FP)*TAN(PIG*(F2-F1))
      A0 = (1.-CAPPA)/(1.+CAPPA)
      A1 = -2.*AL/(CAPPA+1.)
      A2 = 1.
      B0 = A2
      B1 = A1
      B2 = A0
      NU = 5
   50 AN(1,1) = B0*B0
      AN(2,1) = 2.*B0*B1
      AN(3,1) = B1*B1+2.*B0*B2
      AN(4,1) = 2.*B1*B2
      AN(5,1) = B2*B2
      AN(1,2) = A0*B0
      AN(2,2) = A0*B1+A1*B0
      AN(3,2) = A0*B2+A1*B1+A2*B0
      AN(4,2) = A1*B2+A2*B1
      AN(5,2) = A2*B2
      AN(1,3) = A0*A0
      AN(2,3) = 2.*A0*A1
      AN(3,3) = A1*A1+2.*A0*A2
      AN(4,3) = 2.*A1*A2
      AN(5,3) = A2*A2
      DO 60 L=1,NCEL
      DO 65 K=1,3
      FINA(K) = A(K,L)
   65 FINB(K) = B(K,L)
      DO 60 K=1,5
      A(K,L) = 0.
      B(K,L) = 0.
      DO 60 I=1,3
      A(K,L) = AN(K,I)*FINA(I)+A(K,L)
   60 B(K,L) = AN(K,I)*FINB(I)+B(K,L)
      DO 70 K=1,NCEL
      RN = B(1,K)
      DO 70 I=1,NU
      A(I,K) = A(I,K)/RN
   70 B(I,K) = B(I,K)/RN
      RETURN
      END
```

Appendix 4

Computer Programs for the Implementation of Digital Filters

A4.1 Function FLTR to implement the direct convolution algorithm for 1-D FIR digital filters

```
      FUNCTION FLTR(F,A,X,N)
C - FUNCTION TO IMPLEMENT THE DIRECT CONVOLUTION ALGORITHM FOR 1-D
C   FIR DIGITAL FILTERS
C - DUMMY VARIABLES:
C   F = INPUT SAMPLE
C   A = VECTOR OF COEFFICIENTS (DIM. N)
C   X = REGISTER OF THE FILTER (DIM. N+1)
C   N = LENGHT OF THE IMPULSE RESPONSE
C - THE REGISTER OF THE FILTER MUST BE CLEARED BEFORE THE FILTERING
      DIMENSION A(1),X(1)
      NN = N+1
      X(NN) = F
      Y = 0.
      DO 10 II=1,N
      X(II) = X(II+1)
      NNN = NN-II
   10 Y = Y+A(NNN)*X(II)
      FLTR = Y
      RETURN
      END
```

A4.2 Subroutine FLTRM to implement the direct convolution algorithm for 1-D FIR digital filters

```
      SUBROUTINE FLTRM(SIG,NP,A,N)
C - SUBROUTINE TO IMPLEMENT THE DIRECT CONVOLUTION ALGORITHM FOR 1-D
C   FIR DIGITAL FILTERS
C - DUMMY VARIABLES:
C   SIG = VECTOR OF INPUT SAMPLES (DIM. NP)
C   NP = NUMBER OF SAMPLES TO BE FILTERED
C   A = VECTOR OF COEFFICIENTS (DIM. N)
C   N = LENGHT OF THE IMPULSE RESPONSE
      DIMENSION SIG(1),A(1)
      NNP = NP-N+1
      DO 10 I=1,NNP
      KI = I-1
      Y = 0.
      DO 20 J=1,N
   20 Y = Y+A(J)*SIG(KI+J)
   10 SIG(I) = Y
      NNP = NNP+1
      DO 30 I=NNP,NP
   30 SIG(I) = 0.
      RETURN
      END
```

A4.3 Subroutine FLTRF to implement the 1-D select-save algorithm

```
      SUBROUTINE FLTRF(SIG,M,XRE,XIM,HRE,HIM,N,NM,L)
C - SUBROUTINE TO IMPLEMENT THE 1-D SELECT SAVE ALGORITHM
C - DUMMY VARIABLES:
C    SIG = VECTOR OF INPUT SAMPLES (DIM. M)
C    M = NUMBER OF SAMPLES TO BE FILTERED
C    XRE = SERVICE VECTOR (DIM. N)
C    XIM = SERVICE VECTOR (DIM. N)
C    HRE = VECTOR OF COEFFICIENTS (DIM. N)
C    HIM = SERVICE VECTOR (DIM. N)
C    N = LENGHT OF THE ALGORITHM SEGMENTS
C    NM = DEFINED BY THE RELATION N=2**NM
C    L = LENGHT OF THE IMPULSE RESPONSE
      DIMENSION SIG(1),XRE(1),XIM(1),HRE(1),HIM(1)
      LL = L+1
      NZ = L-1
      RN = N
C - EVALUATION OF THE FILTER FFT
      DO 10 I=LL,N
   10 HRE(I) = 0.
      DO 20 I=1,N
   20 HIM(I) = 0.
      CALL FFT(HRE,HIM,N,NM,1)
C - INPUT SIGNAL SELECTION
      K = 0
      K1 = 0
   25 DO 30 I=1,N
      K = K+1
      IF(K.GT.M) GO TO 35
      XRE(I) = SIG(K)
      GO TO 30
   35 XRE(I) = 0.
   30 CONTINUE
      K = K-NZ
      DO 40 I=1,N
      K = K+1
      IF(K.GT.M) GO TO 45
      XIM(I) = SIG(K)
      GO TO 40
   45 XIM(I) = 0.
   40 CONTINUE
C - FILTERING OF A SEGMENT
      CALL FFT(XRE,XIM,N,NM,1)
      DO 50 I=1,N
      XR = XRE(I)*HRE(I)-XIM(I)*HIM(I)
      XIM(I) = XRE(I)*HIM(I)+XIM(I)*HRE(I)
   50 XRE(I) = XR
      CALL FFT(XRE,XIM,N,NM,2)
C - RECOMBINATION OF THE FILTERED SAMPLES
      DO 55 I=L,N
      K1 = K1+1
      IF(K1.GT.M) GO TO 65
   55 SIG(K1) = XRE(I)/RN
      DO 60 I=L,N
      K1 = K1+1
      IF(K1.GT.M) GO TO 65
   60 SIG(K1) = XIM(I)/RN
      IF(K.GE.M) GO TO 65
      K = K-NZ

      GO TO 25
   65 K1 = K1+1
      IF(K1.GT.M) RETURN
      DO 70 I=K1,M
   70 SIG(I) = 0.
      RETURN
      END
```

A4.4 Subroutine FILRI to implement a digital IIR filter as a cascade of canonic second order sections

```
      SUBROUTINE FILRI(FF,YY,A,B,M,W,NCEL)
C - ROUTINE TO IMPLEMENT A DIGITAL IIR FILTER AS A CASCADE OF CANONIC
C   SECOND ORDER SECTIONS
C - DUMMY VARIABLES:
C   FF = INPUT SAMPLE
C   YY = OUTPUT SAMPLE
C   A = NUMERATOR COEFFICIENTS (DIM. 3,NCEL)
C   B = DENOMINATOR COEFFICIENTS (DIM. 3,NCEL)
C   M = ORDER OF THE SECTIONS
C   W = REGISTERS OF THE SECTIONS (DIM. 3,NCEL)
C   NCEL = NUMBER OF SECTIONS
C - IF SECTIONS OF ORDER GREATER THAN 2 MUST BE USED, IT IS NECESSARY
C   TO CHANGE THE DIMENSION STATEMENT AND THE VALUE OF M
C - THE MATRIX W MUST BE CLEARED BEFORE THE FILTERING
      DIMENSION A(3,NCEL),B(3,NCEL),W(3,NCEL)
      MM = M+1
      Y = FF
      DO 20 I=1,NCEL
      W(MM,I) = Y
      Y = 0.
      DO 10 JJ=1,M
      MMM = MM-JJ+1
      W(MM,I) = W(MM,I)-B(MMM,I)*W(JJ,I)
      Y = Y+A(MMM,I)*W(JJ,I)
   10 W(JJ,I) = W(JJ+1,I)
   20 Y = Y+A(1,I)*W(MM,I)
      YY = Y
      RETURN
      END
```

A4.5 Subroutine COND to implement the forward path of the shunt capacitor

```
      SUBROUTINE COND(A,B,C,D,E,ALF)
C - WAVE DIGITAL FILTERS
C - ROUTINE TO IMPLEMENT THE FORWARD PATH OF THE SHUNT CAPACITOR
C - DUMMY VARIABLES:
C   A = INPUT
C   B = REGISTER OF THE SECTION (D IN CON1)
C   C = INTERMEDIATE RESULT OUTPUT (E IN CON1)
C   D = INTERMEDIATE RESULT OUTPUT (C IN CON1)
C   E = OUTPUT
C   ALF - COEFFICIENT (SAME AS IN CON1)
      C = B-A
      D = C*(1-ALF)
      E = D+A
      RETURN
      END
```

A4.6 Subroutine CON1 to implement the backward path of the shunt capacitor

```
      SUBROUTINE CON1(A,B,C,D,E,ALF)
C - WAVE DIGITAL FILTERS
C - ROUTINE TO IMPLEMENT THE BACKWARD PATH OF THE SHUNT CAPACITOR
C - DUMMY VARIABLES:
C    A = OUTPUT
C    B = INPUT
C    C = INTERMEDIATE RESULT INPUT (D IN COND)
C    D = REGISTER OF THE SECTION (B IN COND)
C    E = INTERMEDIATE RESULT INPUT (C IN COND)
C    ALF = COEFFICIENT (SAME AS IN COND)
      A = B+C
      D = B-E*ALF
      RETURN
      END
```

A4.7 Subroutine ZL11 to implement the forward path of the series inductor

```
      SUBROUTINE ZL11(A,B,C)
C - WAVE DIGITAL FILTERS
C - ROUTINE  TO IMPLEMENT THE FORWARD PATH OF THE SERIES INDUCTOR
C - DUMMY VARIABLES:
C    A = INPUT (B IN ZLLL)
C    B = REGISTER OF THE SECTION (F AND E IN ZLLL)
C    C = OUTPUT
      C = A+B
      RETURN
      END
```

A4.8 Subroutine ZLLL to implement the backward path of the series inductor

```
      SUBROUTINE ZLLL(A,B,C,D,E,F,G,ALF)
C - WAVE DIGITAL FILTERS
C - ROUTINE TO IMPLEMENT THE BACKWARD PATH OF THE SERIES INDUCTOR
C - DUMMY VARIABLES:
C    A = INPUT
C    B = SAME AS THE INPUT OF ZL11
C    C,D = INTERNAL RESULTS
C    E,F = REGISTER OF THE SECTION (B IN ZL11)
C    G = OUTPUT
C    ALF = COEFFICIENT
      C = B-A
      D = C+E
      F = C-D*ALF
      G = A+F
      RETURN
      END
```

A4.9 Subroutine AFILT for the implementation of the direct convolution algorithm for 2-D FIR digital filters

```
      SUBROUTINE AFILT(ANUM,MNUM,NNUM,AINPT,MINP,NINP,MTAIL,NTAIL,IDIR,
     # JDIR,VETT,NMAX)
C - IMPLEMENTATION OF THE DIRECT CONVOLUTION ALGORITHM
C   FOR 2-D FIR DIGITAL FILTERS
C - DUMMY VARIABLES:
C   ANUM = MATRIX OF THE FILTER COEFFICIENTS
C   MNUM,NNUM = DIMENSIONS OF ANUM
C   AINPT = SERVICE MATRIX(DIM. MNUM,NMAX)
C   MINP,NINP = DIMENSIONS OF THE INPUT IMAGE (WITHOUT TAILS)
C   MTAIL,NTAIL = DIMENSIONS OF THE TAILS
C   IDIR,JDIR = PARAMETERS WHICH CONTROL THE ANGLE OF FILTERING:
C   IF IDIR<0 THE PROGRAM PERFORMS THE REVERSAL OF EACH INPUT ROW
C   IF JDIR<0 THE PROGRAM PERFORMS THE REVERSAL OF EACH INPUT COLUMN
C   NMAX= NINP+NTAIL+NNUM-1
C - THE DISK INPUT-OUTPUT DEPENDS ON THE USED COMPUTER
      DIMENSION ANUM(MNUM,NNUM),AINPT(MNUM,NMAX),VETT(NMAX)
      NMEZ = (NNUM+1)/2
      DO 1 I=1,MNUM
      DO 1 J=1,NMEZ
      SAVE = ANUM(I,J)
      ANUM(I,J) = ANUM(I,NNUM-J+1)
    1 ANUM(I,NNUM-J+1) = SAVE
      MMEZ = (MNUM+1)/2
      DO 2 I=1,MMEZ
      DO 2 J=1,NNUM
      SAVE = ANUM(I,J)
      ANUM(I,J) = ANUM(MNUM-I+1,J)
    2 ANUM(MNUM-I+1,J) = SAVE
      MM = MINP+MTAIL
      NN = NINP+NTAIL
      NN1 = NN+NNUM-1
      DO 10 L=1,MM
      LL = L
      IF(JDIR.GT.0) GO TO 3
      LL = MM-L+1
    3 CALL FSEEK(11,LL)
      READ BINARY(11)(VETT(J),J=1,NN)
      IF(IDIR.GT.0) GO TO 5
      NIM = NN/2
      DO 4 J=1,NIM
      AA = VETT(J)
      VETT(J) = VETT(NN-J+1)
    4 VETT(NN-J+1) = AA
    5 CONTINUE
      DO 6 J=1,NN
    6 AINPT(MNUM,J+NNUM-1) = VETT(J)
      DO 8 K=1,NN
      SUM =0.
      DO 7 I=1,MNUM
      DO 7 J=1,NNUM
    7 SUM = SUM+ANUM(I,J)*AINPT(I,J+K-1)
      VETT (K) = SUM
    8 CONTINUE
      WRITE(12,1000)(VETT(K),K=1,NN)
      DO 9 I=2,MNUM
      DO 9 J=1,NN1
    9 AINPT(I-1,J) = AINPT(I,J)
   10 CONTINUE

      RETURN
 1000 FORMAT(/,(8E15.7))
      END
```

A4.10 Subroutine BFILT for the implementation of the direct recursion algorithm for 2-D IIR digital filters

```
      SUBROUTINE BFILT(ADEN,MDEN,NDEN,OUT,MOUT,NOUT,AINPT,MTAIL,NTAIL,
     # IDIR,JDIR,VETT,NMAX)
C - IMPLEMENTATION OF THE DIRECT RECURSION ALGORITHM
C   FOR 2-D IIR DIGITAL FILTERS
C - DUMMY VARIABLES:
C   ADEN = MATRIX OF THE FILTER COEFFICIENTS
C   MDEN,NDEN = DIMENSIONS OF ADEN
C   OUT = SERVICE MATRIX(DIM. MDEN,NMAX)
C   MOUT,NOUT = DIMENSIONS OF THE INPUT AND OUTPUT MATRICES
C               (WITHOUT TAILS)
C   AINPT = SERVICE VECTOR(DIM. NMAX)
C   MTAIL,NTAIL = DIMENSIONS OF THE TAILS
C   IDIR,JDIR = PARAMETERS WHICH CONTROL THE ANGLE OF FILTERING:
C   IF IDIR<0 THE PROGRAM PERFORMS THE REVERSAL OF EACH ROW
C   IF JDIR<0 THE PROGRAM PERFORMS THE REVERSAL OF EACH COLUMN
C   VETT = SERVICE VECTOR(DIM.NMAX)
C   NMAX = NOUT+NTAIL+NDEN-1
C - THE DISK INPUT-OUTPUT DEPENDS ON THE USED COMPUTER
      DIMENSION ADEN(MDEN,NDEN),OUT(MDEN,NMAX),AINPT(NMAX),VETT(NMAX)
      NMEZ = (NDEN+1)/2
      DO 1 I=1,MDEN
      DO 1 J=1,NMEZ
      SAVE = ADEN(I,J)
      ADEN(I,J) = ADEN(I,NDEN-J+1)
    1 ADEN(I,NDEN-J+1) = SAVE
      MMEZ = (MDEN+1)/2
      DO 2 I=1,MMEZ
      DO 2 J=1,NDEN
      SAVE = ADEN(I,J)
      ADEN(I,J) = ADEN(MDEN-I+1,J)
    2 ADEN(MDEN-I+1,J) = SAVE
      MM = MOUT + MTAIL
      NN = NOUT + NTAIL
      NN1 = NN + NDEN -1
      DIV = ADEN(MDEN,NDEN)
      ADEN(MDEN,NDEN) = 0.
      DO 10 L=1,MM
      LL = L
      LM = L
      IF(JDIR.LT.0) LL=MM-L+1
      CALL FSEEK(11,LL)
      READ BINARY(11)(VETT(J),J=1,NN)
      DO 3 J=NDEN,NN1
    3 AINPT(J) = VETT(J-NDEN+1)
      IF(IDIR.EQ.1)GO TO 5
      N1D = NN/2
      DO 4 K=1,N1D
      AA = AINPT(K+NDEN-1)
      AINPT(K+NDEN-1) = AINPT(NN1-K+1)
    4 AINPT(NN1-K+1) = AA
    5 CONTINUE
      DO 7 K=1,NN
      SUM = 0.
      DO 6 I=1,MDEN
      DO 6 J=1,NDEN
    6 SUM = SUM+ADEN(I,J)*OUT(I,J+K-1)
    7 OUT(MDEN,K+NDEN-1) = (AINPT(NDEN+K-1)-SUM)/DIV
      DO 8 J=1,NN

    8 VETT(J) = OUT(MDEN,J+NDEN-1)
      CALL FSEEK(11,LM)
      WRITE BINARY(11)(VETT(J),J=1,NN)
      WRITE(12,1000)(VETT(J),J=1,NN)
      DO 9 I=2,MDEN
      DO 9 J=1,NN1
    9 OUT(I-1,J) = OUT(I,J)
   10 CONTINUE
      RETURN
 1000 FORMAT(/,(8E15.7))
      END
```

Appendix 5

The DFT of Special Waveforms: Application to Power Systems Protection

A5.1 Introduction

Power systems contain large numbers of devices such as transformers, machines such as generators plus thousands of kilometres of transmission lines. Even with routine maintenance, faults that result in abnormal system condition do occur. Protective devices are therefore installed to isolate faulty sections of the system and limit damage.

The reliability and operating times of such devices are of critical importance, and while hitherto the performance of conventional electromechanical and static relays has been sufficient,[1] it is felt that the new limits and more stringent requirements demanded by the increasingly complex power systems can only be met by the use of digital minicomputers and perhaps microprocessors. Such use is seen to be attractive if it is to be remembered that minicomputers are already becoming common sights in control centres and substations for monitoring steady-state performance.

It would thus seem worthwhile to try and exploit or further develop existing minicomputer installations in substations to handle protection and relaying under faulty conditions. This would require a digital method for the analysis of power systems waveforms, since it is on such analysis (in an analogue form) that some existing relaying principles are based.

Choice of parameters

Power system waveforms are known to possess harmonic spectra, with the random appearance of non-harmonic components that are here ignored. If such waveforms are thus fed into a low-pass filter, the band-limited output of this is suitable for analysis by the DFT. The cutoff frequency of the filter can be chosen so as to attenuate all the high frequencies that prove to be of no value in relaying.

Remembering that one fundamental period is required (so as to yield the original waveform when periodically extended), the number of samples over such period must be high enough to account for the highest frequency present. Limiting the latter to the 10th harmonic means that 20 samples at least must be known over the period of the fundamental. In practice, it is more advisable to choose 32 samples, such a number being the smallest power of 2 that is greater than 20. Any further processing based on an FFT algorithm is thus made possible.

For a fundamental of 50 c/s, then, a sampling rate of 1600 samples/s gives 32 samples over 20 ms, with a sampling interval of 625 μs. This allows all harmonics up to the 15th in the waveform, the 16th being doubled or folded over in the spectrum. See Figs A5.1 and A5.2. Hence, the cutoff frequency of the low-pass filter is to be around 750–800 c/s.

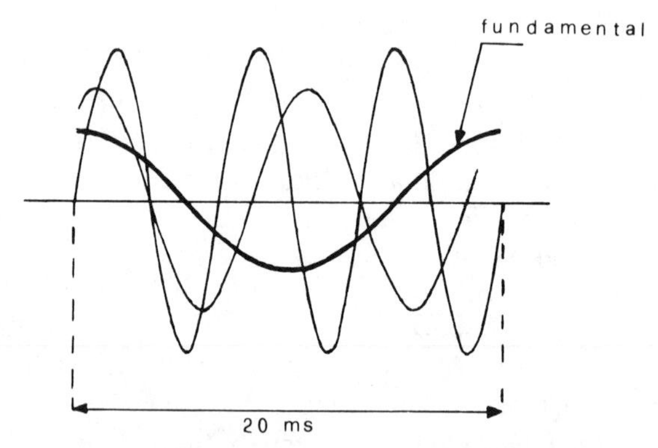

FIG. A5.1. Time waveforms.

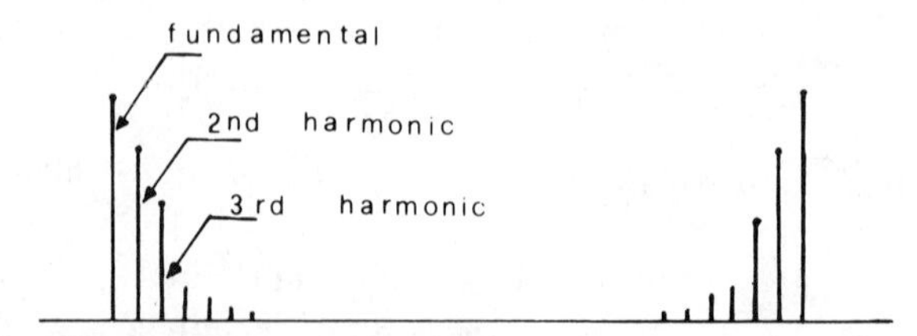

FIG. A5.2. Spectra of time waveforms in Fig. A5.1, as obtained by the DFT.

A5.2 The full-cycle scheme

Distance relays use as a criterion for operation the fundamental impedance of the transmission line. This is easily determined from the knowledge of the fundamental components of both the voltage and current waveforms. It would therefore be sufficient to calculate only one DFT coefficient: that corresponding to the fundamental frequency.

A sample-and-hold circuitry running at 1600 samples/s feeds a sample every 625 μs into the right-most location of a shift register of length 32. As this is done, the left-most sample is discarded. The processor then calculates the fundamental component. However, this may prove to be impossible to carry out every sampling interval, as some intervals may be too short. In the case of the PDP-15 minicomputer, for example, with multiplication times of as long as 25 μs, the time required for the calculation of one DFT coefficient is just below 5 ms, corresponding to quarter of a cycle or 8 samples.

Incoming samples, before they can be used for updating the shift register, must thus be stored until the processing cycle being performed is completed. For this, a smaller shift register R_L is needed, Fig. A5.3. This updates from the sample-and-hold every sampling interval (T seconds), but only updates the main register R_N every quarter fundamental cycle or so (LT seconds).

The master program examines the fundamental impedance as determined by the processor and sends a trip signal to the circuit breaker when such impedance is found to lie outside some pre-set permissible limits.

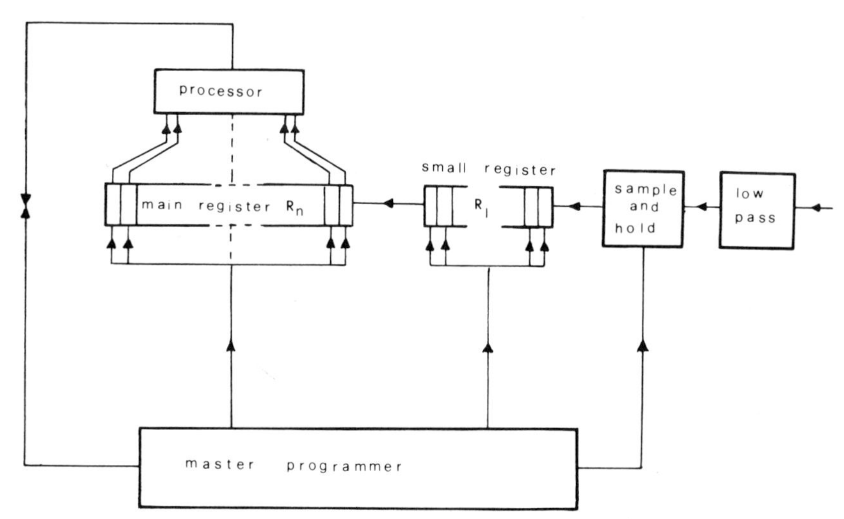

FIG. A5.3. Block diagram of the system.

The half- and quarter-cycle methods

Protection times with the full-cycle method described above can run well above the 20 ms mark in the worst case when post-fault data must cover the whole of a fundamental period before fault detection takes place. This is when all the contents of the main register must be post-fault before the digital analysis of such contents can indicate a substantial change in the line impedance.

A digital relay with fault detection time of more than 20 ms does not compete with analogue and electromechanical relays. The aspect of the problem concerning data acquisition is thus reviewed. The study of some of the geometric properties of the sampled cosinusoidal oscillation shows that a unique definition of such an oscillation need not extend over the whole of one cycle. In fact, a half-cycle window is sufficient in that the full-cycle can be then reconstructed by repetition, see Fig. A5.4(b). With the phase of the cosinusoid being a multiple of $\pi/2$, even a window length of quarter a cycle will suffice, Fig. A5.4(c).

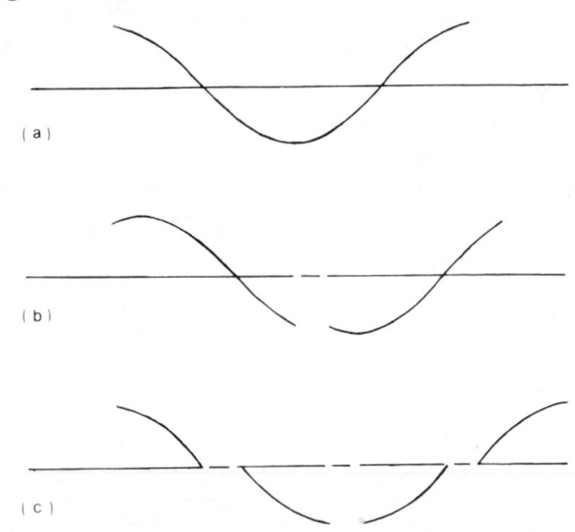

FIG. A5.4. (a) Full-cycle; (b) half-cycle, negatively repeated; (c) quarter-cycle, mirror imaging and repetition.

Using such techniques for completing the full cosinusoidal cycle from the knowledge of a half or a quarter of a cycle makes it possible to fill the whole of the main register with post-fault data only quarter or half a cycle after fault incidence. Data acquisition times thus improve, and, as it turns out, considerable simplifications in the mathematics of the discrete transform also result because of the induced symmetry in the waveform.

It must be appreciated, though, that while such "predictive" methods for completing the data required for analysis are valid for the fundamental on its own, such validity is questionable in the presence of any harmonics. Figure A5.5 below, for example, shows a cycle reconstructed from a

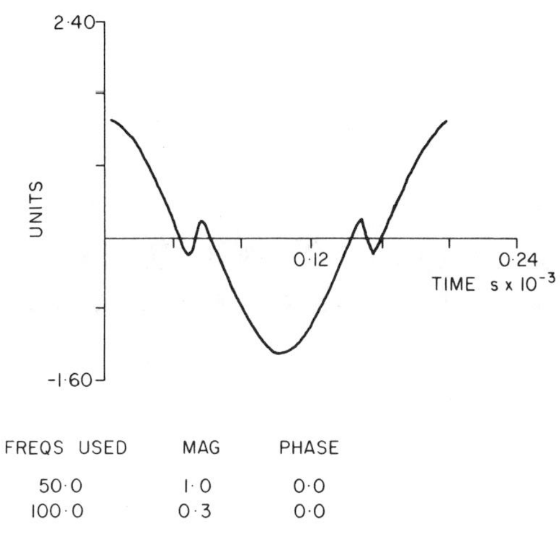

FIG. A5.5. Time waveform.

quarter-cycle. The second harmonic, as can be seen, causes some discontinuity in the completed waveform. Such discontinuity manifests itself in the spectrum as high frequency components. Nevertheless, the fundamental and second harmonic components remain a fairly accurate representation of the original (quarter-cycle) waveform.

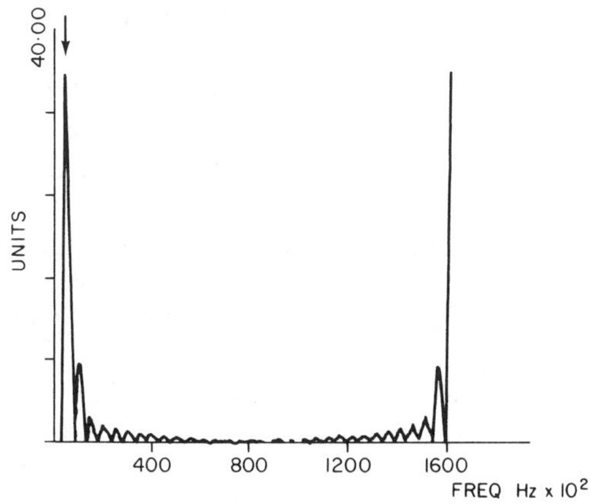

FIG. A5.6. Magnitude transform.

Reference

1. The Electricity Council (1975). "Power System Protection". MacDonald and Jane's, London.

Appendix 6

Digital Processing of 2-D Data Received From Earth's Resource Satellites and Aircraft Photos

Recently efforts have been directed towards the evaluation of earth resources (e.g. agriculture, water, energy) and also towards the evaluation of such aspects of human life as pollution, environmental and living conditions.

Several methods of observation either using aircrafts or indeed satellites are available based on normal photographic systems (visible) or multi-spectral systems (MSS) covering infra-red, visible and ultraviolet spectral regions.

The LANDSAT satellites 1 and 2 were specifically designed to collect data related to earth resources. Multi-spectral systems are used aboard: the radiation reflected at the earth in the ultraviolet, visible and infra-red is collected and the 2-D data are transmitted as 1-D sequences to the ground receiving stations. Here the maps are processed to extract useful information related to some specific aspect of earth resource under examination.

Digital signal processing and in particular 2-D digital filtering can be applied with advantage to process these satellite maps, as is the case with aerial photographs. Different operations are useful: low-pass filtering for reducing the amplitude of high space frequency components, high-pass filtering for enhancing certain attributes of the map (e.g. lines) or extracting contours. Of course correction filtering can be performed by the same 2-D filter.

Figure A6.1 shows an example of different high-pass 2-D digital filterings on an aerial photograph, to extract boundaries and regions: 2-D FIR digital filters with Cappellini window were used.

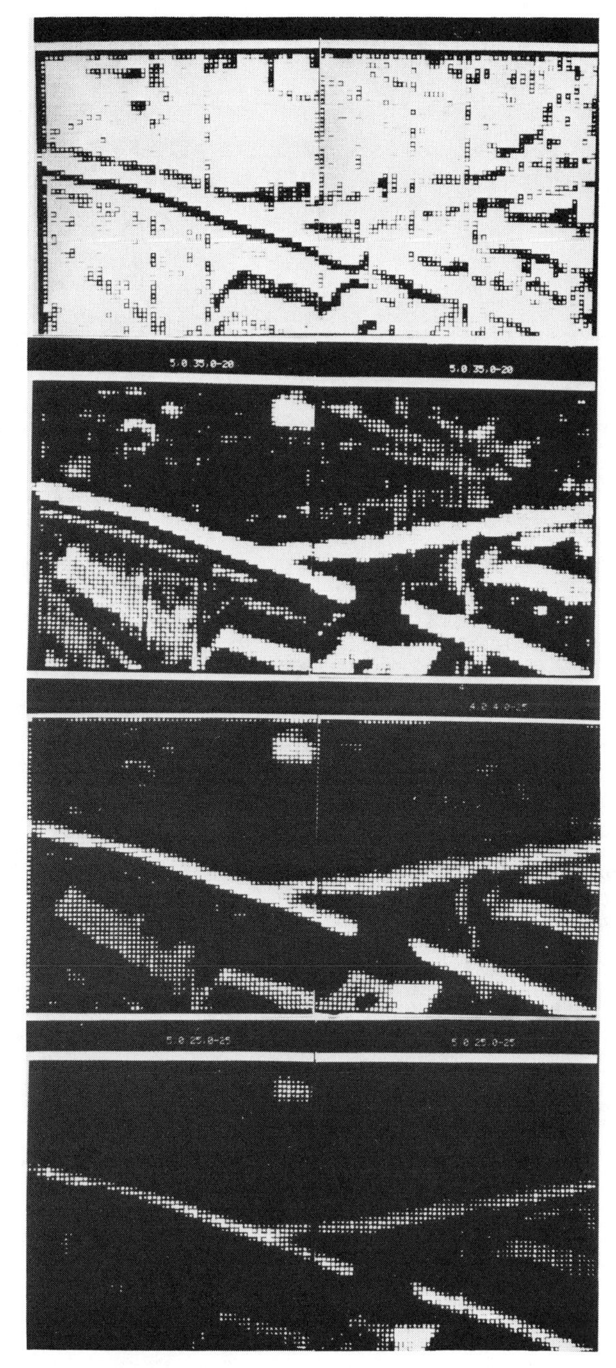

FIG. A6.1. Some examples of 2-D FIR high-pass filtering on aerial photographs of a particular region. Edge enhancement is evident from this processing.

Figure A6.2 shows an example of digital processing for data compression (which can be done, as described in Chapter 8, very suitably after digital filtering) on a LANDSAT 1 image (given by TELESPAZIO): the same algorithm of data compression described in Section VIIID was used here.

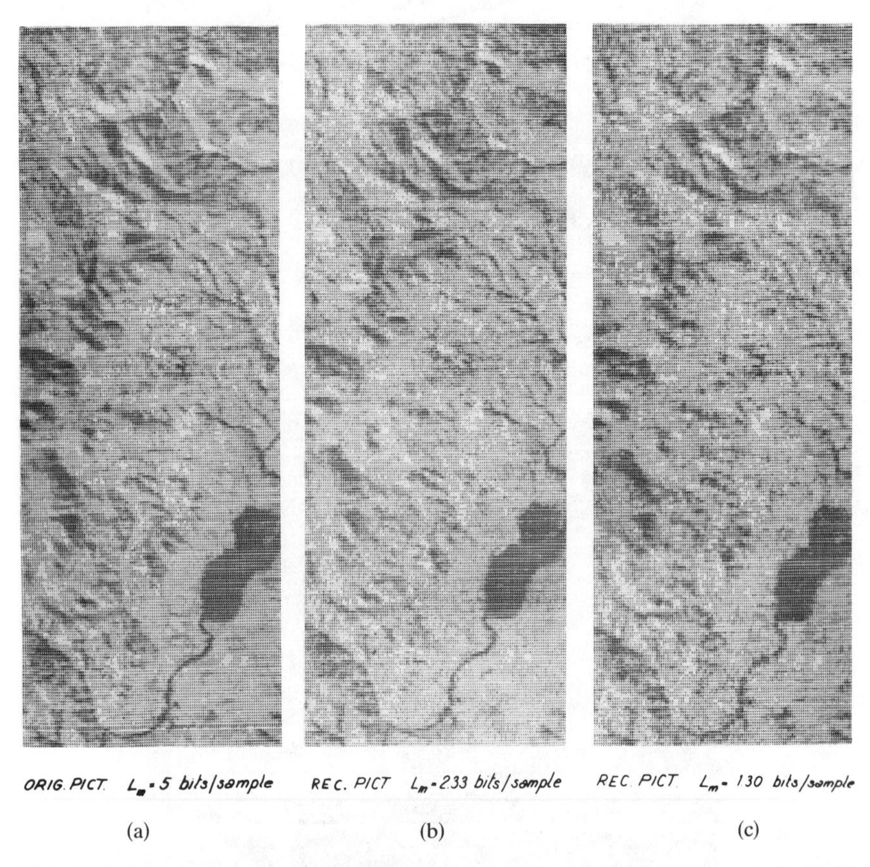

(a) (b) (c)

FIG. A6.2. Data compression on LANDSAT 1 image: (a) original picture presented with 10 grey levels (5 bits/sample are required); (b) reconstructed image after 2-D FWT (2.33 bits/sample are required); (c) other reconstructed image after 2-D FWT (1.30 bits/sample are required).

Index

R

S